Resistive, Capacitive, Inductive, and Magnetic Sensor Technologies

Series in Sensors

Series Editors: Barry Jones and Haiying Huang

Other recent books in the series:

Semiconductor X-Ray Detectors
B. G. Lowe and R. A. Sareen

Portable Biosensing of Food Toxicants and Environmental Pollutants
Edited by Dimitrios P. Nikolelis, Theodoros Varzakas, Arzum Erdem,
and Georgia-Paraskevi Nikoleli

Optochemical Nanosensors
Edited by Andrea Cusano, Francisco J. Arregui, Michele Giordano,
and Antonello Cutolo

Electrical Impedance: Principles, Measurement, and Applications
Luca Callegaro

Biosensors and Molecular Technologies for Cancer Diagnostics
Keith E. Herold and Avraham Rasooly

Compound Semiconductor Radiation Detectors
Alan Owens

Metal Oxide Nanostructures as Gas Sensing Devices
G. Eranna

Nanosensors: Physical, Chemical, and Biological
Vinod Kumar Khanna

Handbook of Magnetic Measurements
S. Tumanski

Structural Sensing, Health Monitoring, and Performance Evaluation
D. Huston

Chromatic Monitoring of Complex Conditions
Edited by G. R. Jones, A. G. Deakin, and J. W. Spencer

Principles of Electrical Measurement
S. Tumanski

Novel Sensors and Sensing
Roger G. Jackson

Hall Effect Devices, Second Edition
R. S. Popovic

Sensors and Their Applications XII
Edited by S. J. Prosser and E. Lewis

Resistive, Capacitive, Inductive, and Magnetic Sensor Technologies

Winncy Y. Du
San Jose State University

CRC Press
Taylor & Francis Group
Boca Raton London New York

CRC Press is an imprint of the
Taylor & Francis Group, an **informa** business

CRC Press
Taylor & Francis Group
6000 Broken Sound Parkway NW, Suite 300
Boca Raton, FL 33487-2742

First issued in paperback 2019

© 2015 by Taylor & Francis Group, LLC
CRC Press is an imprint of Taylor & Francis Group, an Informa business

No claim to original U.S. Government works

ISBN-13: 978-1-4398-1244-0 (hbk)
ISBN-13: 978-0-367-86465-1 (pbk)

Library of Congress Cataloging-in-Publication Data

Du, Winncy.
   Resistive, capacitive, inductive, and magnetic sensor technologies / author, Winncy Du.
       pages cm. -- (Series in sensors)
   Includes bibliographical references and index.
   ISBN 978-1-4398-1244-0 (hardback)
   1. Detectors. 2. Transducers. I. Title.

TK7872.D48D83 2014
681'.2--dc23   2013045996

Visit the Taylor & Francis Web site at
http://www.taylorandfrancis.com

and the CRC Press Web site at
http://www.crcpress.com

To my late parents: Xianmin Du (杜贤民) and Shujie Chao (曹淑杰)
who not only gave me life, but also taught me how to live;
To my dear brothers: Chengzhi Du (杜承志) and Yingzhi Du (杜英志)
who never ask me for anything, but always give me everything;
To my beloved husband: Scott Yelich
who enriches my life and multiplies my efforts.

Contents

Preface...xvii
Acknowledgments...xix
Author xxi

Chapter 1 RCIM Sensor Characteristics and Terminology 1

    1.1 RCIM Sensor History... 1
    1.2 RCIM Sensor Definition...2
    1.3 RCIM Sensor Characteristics and Terminology 2
        1.3.1 Transfer Function 2
        1.3.2 Sensitivity...2
        1.3.3 Offset...5
        1.3.4 Full Span and Full-Span Output 5
        1.3.5 Accuracy..7
        1.3.6 Hysteresis 8
        1.3.7 Nonlinearity 9
        1.3.8 Noise and Signal-to-Noise Ratio 9
        1.3.9 Resolution 11
        1.3.10 Precision and Repeatability Error............................. 12
        1.3.11 Calibration and Calibration Error 14
        1.3.12 Response Time and Bandwidth................................. 16
        1.3.13 Sensor Lifespan 18
        1.3.14 Other Sensor Characteristics.....................................20
    Exercises.. 21
    References 24

Chapter 2 Resistive Sensors 25

    2.1 Introduction 25
    2.2 Potentiometric Sensors 25
        2.2.1 Sensing Principle...25
        2.2.2 Configuration and Circuitry 26
        2.2.3 Potentiometric Sensor Design 29
            2.2.3.1 Linear Potentiometers.................................29
            2.2.3.2 Rotary Potentiometers 30
        2.2.4 Potentiometric Sensor Applications 30
            2.2.4.1 Potentiometric Pressure Sensors.................30
            2.2.4.2 Potentiometric Airflow Sensor 31
            2.2.4.3 Potentiometric Gas Sensor.......................... 31
            2.2.4.4 Potentiometric Biosensor.............................32

2.3   Resistive Temperature Sensors 33
       2.3.1   Thermoresistive Effects 33
               2.3.1.1   Thermoresistive Effect for Metals 33
               2.3.1.2   Thermoresistive Effect for
                         Semiconductors 37
       2.3.2   Wiedemann–Franz Law for Metals 39
       2.3.3   Resistance Temperature Devices (RTDs) 40
               2.3.3.1   RTD Characteristics 40
               2.3.3.2   RTD Measurement 41
               2.3.3.3   RTD Design 42
               2.3.3.4   RTD Applications 43
       2.3.4   Thermistors 44
               2.3.4.1   Thermistor Characteristics 44
               2.3.4.2   Thermistor Design 46
               2.3.4.3   Thermistor Applications 48
2.4   Photoresistive Sensors 49
       2.4.1   Photoresistive Effect 49
       2.4.2   Photoresistor Characteristics 54
       2.4.3   Photoresistive Sensor Design 56
       2.4.4   Photoresistive Sensor Applications 57
2.5   Piezoresistive Sensors 58
       2.5.1   Piezoresistive Effect in Metals and Alloys 58
       2.5.2   Piezoresistive Effect in Semiconductors 60
       2.5.3   Characteristics of Piezoresistive Sensors 63
       2.5.4   Piezoresistive Sensor Design 67
               2.5.4.1   Types and Structures of Strain Gauges 67
               2.5.4.2   Strain Gauge Materials 70
               2.5.4.3   Supporting Structure and Bonding
                         Methods of Strain Gauges 71
       2.5.5   Piezoresistive Sensor Applications 74
               2.5.5.1   Piezoresistive Accelerometers 74
               2.5.5.2   Piezoresistive Pressure Sensor 75
               2.5.5.3   Piezoresistive Flow Rate Sensor 75
               2.5.5.4   Piezoresistive Blood Pressure Sensor 76
               2.5.5.5   Piezoresistive Force Sensor 76
               2.5.5.6   Piezoresistive Imaging Sensor 76
2.6   Chemoresistive Sensors 77
       2.6.1   Chemoresistive Effect 77
       2.6.2   Characteristics of Chemoresistive Sensors 79
               2.6.2.1   Characteristics of Mixed Metal Oxide
                         Semiconductor Sensors 79
               2.6.2.2   Characteristics of Polymer or Organic
                         Material Sensors 80
       2.6.3   Chemoresistive Sensor Design 81
       2.6.4   Chemoresistive Sensor Applications 84
               2.6.4.1   Hygristor 84

          2.6.4.2   Groundwater Monitoring System 84

          2.6.4.3   Electronic Nose 84

2.7   Bioresistance/Bioimpedance Sensors 85

     2.7.1   Sensing Principles 85

          2.7.1.1   Types of Bioresistance/Bioimpedance

                    Sensors 85

          2.7.1.2   Modeling of Bioresistance/

                    Bioimpedance Sensors 86

     2.7.2   Sensor Materials and Characteristics 87

     2.7.3   Designs and Applications of Bioresistance/

          Bioimpedance Sensors 88

          2.7.3.1   Body Composition Monitor 90

          2.7.3.2   Resistance Measurement at

                    Acupuncture Points 90

          2.7.3.3   Ovulation Predictor 91

          2.7.3.4   Venous Blood Volume Measurement 91

Exercises 91

References 96

Chapter 3   Capacitive Sensors 99

3.1   Introduction 99

3.2   Capacitors and Capacitance 100

3.3   Physical Laws and Effects Governing Capacitive

     Sensors 106

     3.3.1   Coulomb's Law 106

     3.3.2   Gauss's Law for Electric Field 107

     3.3.3   Piezoelectric Effect 110

     3.3.4   Effect of Excitation Frequencies 111

3.4   Parallel-Plate (Flat-Plate) Capacitive Sensors 112

     3.4.1   Spacing-Variation-Based Sensors 112

          3.4.1.1   Sensing Principle and Characteristics 112

          3.4.1.2   Sensor Design 113

          3.4.1.3   Sensor Applications 117

     3.4.2   Area-Variation-Based Sensors 120

          3.4.2.1   Sensing Principle and Characteristics 120

          3.4.2.2   Sensor Design 120

          3.4.2.3   Sensor Applications 122

     3.4.3   Dielectric-Constant-Variation-Based Sensors 125

          3.4.3.1   Sensing Principle and Characteristics 125

          3.4.3.2   Sensor Design 127

          3.4.3.3   Sensor Applications 127

     3.4.4   Electrode-Property-Variation-Based Sensors 132

          3.4.4.1   Sensing Principle and Characteristics 132

          3.4.4.2   Electrode Materials and Design 133

          3.4.4.3   Sensor Applications 133

3.5   Cylindrical Capacitive Sensors 135
        3.5.1   Electrode-Movement-Based Sensors 136
                3.5.1.1   Sensing Principle and Characteristics 136
                3.5.1.2   Sensor Design and Applications 136
        3.5.2   Dielectric-Media-Movement-Based Sensors 140
                3.5.2.1   Sensing Principle and Characteristics 140
                3.5.2.2   Design and Applications 141
        3.5.3   Dielectric-Constant-Variation-Based Sensors 141
                3.5.3.1   Temperature Sensor 141
                3.5.3.2   Oil Analyzer 142
                3.5.3.3   Water Quality Detector 142
3.6   Spherical Capacitive Sensors 143
        3.6.1   Geophysical Fluid Flow Cell 143
        3.6.2   Coating Thickness Detector 144
        3.6.3   Ultra-Precision Spherical Probe 144
3.7   Capacitive Sensor Arrays 144
        3.7.1   Spherically Folded Pressure Sensor Array 145
        3.7.2   Capacitive Finger Print Detector 146
        3.7.3   Capacitive Touchscreen 146
Exercises 147
References 151

Chapter 4   Inductive Sensors 153

4.1   Introduction 153
4.2   Inductors, Inductance, and Magnetic Field 154
        4.2.1   Inductors 154
        4.2.2   Inductance and Magnetic Field 155
4.3   Physical Laws and Effects Governing Inductive Sensors 159
        4.3.1   Lorentz Force 159
        4.3.2   Faraday's Law of Electromagnetic Induction 161
        4.3.3   Biot–Savart Law 162
        4.3.4   Ampere's Law 162
        4.3.5   Magnetomotive Force 163
        4.3.6   Eddy Current 164
        4.3.7   Skin Effect 165
        4.3.8   Proximity Effect 166
        4.3.9   Electric and Magnetic Field Analogies 168
4.4   Characteristics and Materials of Inductive Sensors 168
        4.4.1   Terminologies 168
        4.4.2   Coil Materials 169
        4.4.3   Core Materials 169
                4.4.3.1 B–H Characteristics of Magnetic Core
                           Materials 170
                4.4.3.2   Magnetic Hysteresis Loop 171

4.4.4   Housing, Cable, and Target Materials.................... 172
4.4.5   Power Losses in Inductive Sensors 173
4.4.6   Quality Factor Q.. 175
4.4.7   Number of Turns N... 175
4.4.8   Frequency Response of Inductive Sensors 175
4.4.9   Stability of Inductive Sensors................................. 176
         4.4.9.1   Thermal Stability.................................... 176
         4.4.9.2   Long-Term Stability............................... 176
4.5   Types and Operating Principles of Inductive
Sensors... 176
    4.5.1   Types of Inductive Sensors..................................... 176
    4.5.2   Operating Principles.. 178
4.6   Inductive Air Coil Sensors 179
    4.6.1   Types of Air Coils 179
    4.6.2   Features of Air Coil Sensors 180
    4.6.3   Design Considerations of Air Coil Sensors............. 181
         4.6.3.1   Coil Diameter 181
         4.6.3.2   Target Thickness and Temperature
                    Stability of Target Material 181
         4.6.3.3   EMI Interference 181
         4.6.3.4   Other Considerations 182
    4.6.4   Applications of Air Coil Sensors 183
4.7   Inductive Sensors with Ferromagnetic Cores........................ 184
    4.7.1   Characteristics of Inductive Sensors with
             Magnetic Cores 184
    4.7.2   Sensor Design... 185
         4.7.2.1   Ferromagnetic Core Design..................... 185
         4.7.2.2   Probe Design.. 186
    4.7.3   Applications of Inductive Sensors with
             Ferromagnetic Cores 189
         4.7.3.1   Inductive Proximity Sensor 190
         4.7.3.2   Inductive Displacement Sensor................ 190
         4.7.3.3   Eddy-Current Force Sensor 191
         4.7.3.4   Thread Detector...................................... 191
4.8   Transformer-Type Inductive Sensors 192
    4.8.1   Introduction 192
    4.8.2   Sensing Principles of LVDTs, Fluxgate Sensors,
             RVDTs, Synchros, and Resolvers............................ 195
         4.8.2.1   LVDTs/RVDTs 195
         4.8.2.2   Fluxgate Sensors.................................... 197
         4.8.2.3   Synchros and Resolvers 198
    4.8.3   Features of Transformer-Type Sensors.................... 201
         4.8.3.1   Features of LVDTs and RVDTs................ 201
         4.8.3.2   Features of Fluxgate Sensors.................. 202
         4.8.3.3   Features of Synchros and Resolvers 203

4.8.4 Design and Applications of LVDTs, RVDTs,
Fluxgate Sensors, Synchros, and Resolvers 204

4.8.4.1 Design and Applications of LVDTs/
RVDTs/PVDTs 204

4.8.4.2 Design and Applications of Fluxgate
Sensors 206

4.8.4.3 Design and Applications of Synchros
and Resolvers 209

4.9 Oscillator and Signal Processing Circuits of Inductive
Sensors 211

4.9.1 Colpitts Oscillator 212

4.9.2 Balanced Wheatstone Bridge Circuit 212

4.9.3 Phase Circuit 214

Exercises 214

References 220

Chapter 5   Magnetic Sensors 223

5.1 Introduction 223

5.2 Hall Sensors 224

5.2.1 Hall Effect 224

5.2.1.1 Hall Effect in Metals 226

5.2.1.2 Hall Effect in Semiconductors 227

5.2.2 Operating Principle of Hall Sensors 228

5.2.3 Characteristics of Hall Sensors 228

5.2.3.1 Transfer Function 229

5.2.3.2 Sensitivity (or Gain) 229

5.2.3.3 Ohmic Offset 230

5.2.3.4 Nonlinearity 230

5.2.3.5 Input and Output Resistance and Their
Temperature Coefficient 230

5.2.3.6 Noise 230

5.2.4 Types and Design of Hall Sensors 231

5.2.4.1 Vertical Configurations 232

5.2.4.2 Cylindrical Configurations 233

5.2.4.3 Multiaxis Configurations 234

5.2.5 Applications of Hall Sensors 234

5.2.5.1 Hall Position Sensor 234

5.2.5.2 Hall Current Sensor 235

5.2.5.3 Door Security System 236

5.2.5.4 Flow Rate Meter 237

5.2.5.5 Motor Control 237

5.3 Magnetoresistive Sensors 239

5.3.1 Magnetoresistance Effects 239

5.3.1.1 Ordinary Magnetoresistance (OMR)
Effect 240

5.3.1.2   Anisotropic Magnetoresistance (AMR)
Effect..241

5.3.1.3   Giant Magnetoresistance (GMR)
Effect..243

5.3.1.4   Tunneling Magnetoresistance (TMR)
Effect..244

5.3.1.5   Ballistic Magnetoresistance (BMR)
Effect..245

5.3.1.6   Colossal Magnetoresistance (CMR)
Effect..246

5.3.2   AMR Sensors and the Barber-Pole Structure 246

5.3.3   AMR Sensor Materials and Circuit
Configurations 249

5.3.4   GMR Sensors and Their Multilayer Structures 250

5.3.5   MR Sensor Design 252

5.3.6   MR Sensor Applications 253

5.3.6.1   Magnetic Recording System....................254

5.3.6.2   MR Biochips...254

5.3.6.3   Currency Counter 254

5.4   Magnetostrictive/Magnetoelastic Sensors............................256

5.4.1   Magnetostrictive Effects 256

5.4.1.1   Joule and Villari Effects 256

5.4.1.2   Wiedemann and Matteuci Effects 257

5.4.2   Operating Principles of Magnetostrictive
Sensors 258

5.4.3   Materials and Characteristics of
Magnetostrictive Sensors 258

5.4.4   Design and Applications of Magnetostrictive
Sensors 259

5.4.4.1   Magnetostrictive Position Sensor 260

5.4.4.2   Magnetostrictive Level Transmitter..........260

5.4.4.3   Magnetostrictive Force/Stress Sensors.....262

5.4.4.4   Magnetostrictive Torque Sensors..............263

5.4.4.5   Magnetic Field Sensor 265

5.4.4.6   Magnetostrictive Fiber-Optic
Magnetometers 265

5.5   Nuclear Magnetic Resonance (NMR)/Magnetic
Resonance Imaging (MRI) Sensors 266

5.5.1   Nuclear Magnetic Resonance..................................266

5.5.2   Magnetic Resonance Imaging..................................267

5.5.3   Operating Principles of Magnetic Resonance
Sensors 268

5.5.4   Design and Applications of Magnetic Resonance
Sensors 268

5.5.4.1   Resonant Magnetic Field Sensors.............268

5.5.4.2   MRI Devices..270

5.6   Barkhausen Sensors..270
        5.6.1   Barkhausen Effect..270
        5.6.2   Operating Principle of Barkhausen Sensors 271
        5.6.3   Design and Applications of Barkhausen Sensors.....272
                 5.6.3.1   Impact Toughness Tester 272
                 5.6.3.2   Barkhausen Stress Sensor........................272
5.7   Wiegand Sensors 273
        5.7.1   Wiegand Effect...273
        5.7.2   Operating Principle of Wiegand Sensors 273
        5.7.3   Design and Applications of Wiegand Sensors 274
5.8   Magneto-Optical Sensors 276
        5.8.1   Magneto-Optical Effects..276
                 5.8.1.1   Faraday Effect..276
                 5.8.1.2   Voigt and Cotton–Mouton Effects............277
                 5.8.1.3   Malus's Law..278
                 5.8.1.4   Magneto-Optical Kerr Effect 279
        5.8.2   Operating Principles of Magneto-Optical (MO)
                 Sensors 281
        5.8.3   Materials and Characteristics of MO Sensors.........284
        5.8.4   Design and Applications of Magneto-Optical
                 Sensors 285
                 5.8.4.1   Faraday Current Sensor 286
                 5.8.4.2   MO Disk Reader 286
                 5.8.4.3   MOKE Sensor for Magnetization Study....286
                 5.8.4.4   Fiber-Terfenol-D Hybrid Sensor 287
5.9   Superconducting Quantum Interference Devices
        (SQUIDs) 287
        5.9.1   Superconductor Quantum Effects 287
                 5.9.1.1   Meissner Effect..287
                 5.9.1.2   Josephson Effect 288
        5.9.2   Operating Principle of SQUIDs................................290
                 5.9.2.1   DC SQUID 290
                 5.9.2.2   RF SQUID 291
        5.9.3   Materials and Characteristics of SQUID
                 Sensors 292
        5.9.4   SQUID Noise 293
        5.9.5   Design and Applications of SQUIDs 294
                 5.9.5.1   SQUID for Biomagnetism 295
                 5.9.5.2   SQUID Battery Monitors 295
Exercises...296
References 299

Chapter 6   RCIM Sensor Circuitry 303

6.1   Introduction 303
6.2   RCIM Sensor Signal Characteristics.....................................303

6.3   Sensor Noise Sources and Forming Mechanisms 305
        6.3.1   Noise Sources 306
        6.3.2   Noise Transmission Mechanisms 306
                6.3.2.1   Conductive Coupling 306
                6.3.2.2   Capacitive Coupling............................. 306
                6.3.2.3   Inductive Coupling 307
                6.3.2.4   Electromagnetic Coupling 308
        6.3.3   Types of Noise 309
                6.3.3.1   Crosstalk 310
                6.3.3.2   Popcorn Noise................................... 310
                6.3.3.3   Thermal (Johnson or Nyquist) Noise 310
                6.3.3.4   1/f (Flicker or Pink) Noise 312
                6.3.3.5   Shot Noise 313
6.4   Grounding and Shielding Techniques 313
        6.4.1   Grounding and Ground Loops 314
        6.4.2   Shielding.. 314
                6.4.2.1   Capacitive Shielding 314
                6.4.2.2   Inductive Shielding 315
                6.4.2.3   Electromagnetic Shielding...................... 316
6.5   DC Bridges for Resistance Measurements 316
        6.5.1   Wheatstone Bridge and Its Balance Condition 317
        6.5.2   Sensitivity of a Wheatstone Bridge 319
        6.5.3   Wheatstone Bridge-Driven Means 321
                6.5.3.1   Wheatstone Bridge Driven by a
                          Constant Voltage................................ 321
                6.5.3.2   Wheatstone Bridge Driven by a
                          Constant Current................................ 322
        6.5.4   Kelvin Bridge 325
        6.5.5   Megaohm Bridge 326
6.6   AC Bridges for Capacitance and Inductance
        Measurements 327
        6.6.1   AC Bridge and Its Balance Condition 328
        6.6.2   Comparison Bridge 330
        6.6.3   Schering Bridge... 332
        6.6.4   Maxwell Bridge.. 333
        6.6.5   Hay Bridge... 333
        6.6.6   Owen Bridge.. 335
        6.6.7   Wien Bridge.. 335
        6.6.8   Bridge Selection Considerations 336
6.7   RCIM Sensor Output Circuits 337
        6.7.1   Voltage Output Circuits..................................... 337
        6.7.2   Current Output Circuits..................................... 339
        6.7.3   Charge Output Circuits 340
        6.7.4   Resistance Output Circuits.................................. 341
                6.7.4.1   Resistance-to-Voltage Conversion 341
                6.7.4.2   Resistance-to-Current Conversion........... 342

6.7.4.3   Resistance-to-Time Conversion (RC
          Decay) 343
6.7.4.4   Resistance-to-Frequency Conversion:
          RC Oscillator 344
6.7.5   Capacitance Output Circuits 345
6.7.5.1   Capacitance-to-Voltage Conversion 345
6.7.5.2   Capacitance-to-Time Conversion: RC
          Decay 346
6.7.5.3   Capacitance-to-Time-to-Voltage
          Conversion 346
6.7.5.4   Capacitance-to-Frequency Conversion:
          RC Oscillator 346
6.7.6   Inductance Output Circuits 347
6.8   Sensor Compensation Circuits 349
6.8.1   Temperature Compensation 349
6.8.2   Nonlinearity Compensation 350
6.8.3   Offset Error Compensation 350
6.9   Sensor Signal Conditioning, Passive and Active Filters 351
6.9.1   Filtering 351
6.9.2   Characteristics of a Filter 352
6.9.3   Passive Filters 354
6.9.4   Active Filters 357
6.9.5   A Design Example of Sensor Signal
        Conditioning Circuits 359
6.9.5.1   Amplification Circuit 359
6.9.5.2   High-Pass Filter 359
6.9.5.3   Low-Pass Filter 360
6.9.5.4   Notch/Band-Reject Filter 361
Exercises 364
References 371

Index 375

Preface

Resistive, capacitive, inductive, and magnetic (RCIM) sensors comprise more than 70% of the sensor market today. For many years, there has been a strong need for a comprehensive book on RCIM sensors that combines the most important physical principles, designs, and practical applications of the RCIM sensors. This book was designed to fill that need. It is based on sensor information that the author has collected over the years, her research work, and lecture materials she has developed while teaching sensor technology-related courses with the Department of Mechanical Engineering at San Jose State University.

This book is a complete and comprehensive overview of RCIM sensing technologies. It contains six chapters beginning with RCIM sensor characteristics and terminology (Chapter 1), followed by resistive (Chapter 2), capacitive (Chapter 3), inductive (Chapter 4), and magnetic (Chapter 5) sensors. Sensor signal characteristics, noise types, bridge and compensation circuits, passive/active filters and signal conditioning are also covered (Chapter 6). The unique features of this book include: (1) *Completeness*: It covers all the dominating principles for RCIM sensors. For instance, on electromagnetic sensors alone (including inductive and magnetic sensors), more than 15 different physical laws, phenomena, and effects are presented. No other single book or review paper has such a complete coverage of inductive/magnetic sensing principles. (2) *Conciseness*: Many sensing principles involve abstruse theories and complex mathematical equations, some of which could fill an entire book. Here, each principle and its associated mathematical model and theory are described in a way that makes it easy for readers to follow and understand. In addition, many useful illustrations are provided to help readers visualize the material. (3) *Comprehensiveness*: Unlike some sensor books that are either too theoretical or too practical, this book provides a good balance. The mathematical equations have been chosen based on which equations have the most influence on the sensor's performance, whereas the practical examples have been selected based on which examples best represent a typical scenario or cover a new area of application. (4) *Practicality*: The calculation examples and exercise problems provided are realistic and represent real-world sensor data and performance. Each parameter value and every curve presented in the examples or exercise problems are from experiments performed by the author, published research papers, or manufacturers' testing data. This allows readers to learn theory along with the true levels of sensor performance.

The information provided in this book will not only help readers to understand RCIM sensors, but also to understand many other types of sensors quickly and efficiently since there are many overlaps in terms of principles, material characteristics, noise types, design considerations, sensor circuitry and signal conditioning. The book is interdisciplinary in nature and will be helpful in subjects or courses such as sensor technologies and principles, applied physics, semiconductor materials and applications, engineering design, mechatronics, robotics, automatic control using sensor feedback, nondestructive inspection (NDI) technologies, analog signal processing, sensor circuitry, and instrumentation and measurements. This book is intended for advanced undergraduate and graduate-level engineering students. It would also be a useful reference for professional engineers at all levels and scientists involved in sensor research and development.

Winncy Y. Du
San Jose, California, USA

Acknowledgments

I would like to thank the students who pilot-tested this book in my sensor classes for their valuable feedback and working on the exercise problems; and the companies and publishers who allowed me to use their data, drawings, and materials. I am especially indebted to my husband, Scott Yelich, for his tireless proofreading of the book draft, his meticulous and thoughtful suggestions, and his support and encouragement during this challenging project. Last, but not the least, I would like to acknowledge the excellent services and support provided by the editorial and production staff at Taylor & Francis Group, LLC.

Winncy Y. Du
San Jose, California, USA

Author

Dr. Winncy Y. Du is a professor with the Department of Mechanical Engineering at *San Jose State University* (SJSU) where she is director of the Robotics, Sensor, and Machine Intelligence Laboratory. She is a visiting professor with the Department of Mechanical Engineering at MIT. Her PhD in mechanical engineering and MS degree in electrical and computer engineering were awarded by the *Georgia Institute of Technology*. She earned MS and BS degrees in mechanical engineering from *West Virginia University* and Jilin University (China), respectively. She also holds *Professional Engineer (PE), Engineer-in-Training (EIT)*, and *Computer Integrated Manufacturing Systems (CIMS)* licenses/certificates. Prior to joining SJSU, she was an assistant professor at *Georgia Southern University*. She also worked as a journal editor, engineer, and researcher with the *Science and Technology Research and Information Center, Ministry of Mechanical and Electronics Industry of China*. Her research and teaching areas are sensors, robotics, mechatronics, automation, and control, with an emphasis on health care and biomedical applications. She teaches a wide range of undergraduate and graduate courses, has published more than 40 peer-reviewed journal/conference papers, and contributed to the books *Smart Sensors and Sensing Technology* (Springer-Verlag, 2008) and *Modern Sensors, Transducers and Sensor Networks* (International Frequency Sensor Association, 2012). She is a principal investigator (PI) or co-PI of more than 10 research grants, and the leader of more than 20 industry sponsored research projects. In 2010, she was awarded the rank of Fellow with American Society of Mechanical Engineers in recognition of her exceptional engineering achievements and contribution to the engineering profession. She is also an active member of IEEE. Her professional honors include the International ASME Diversity & Outreach Award (2004), Richard A. Fitz Outstanding Faculty Advisor Award (2005), ASME District D Student Section Advisor Award (2007, 2011, and 2013), Charles W. Davidson College of Engineering Faculty Award for Excellence in Scholarship at SJSU (2012), and Newnan Brothers Award for Faculty Excellence (2014).

1 RCIM Sensor Characteristics and Terminology

Resistive, capacitive, inductive, and magnetic (RCIM) sensors are the most widely used sensors, comprising more than 70% of the sensor market today. Their easily understood principles, robust features, simple electronic circuits, low cost, and mature design and manufacturing techniques have made them the first choice in many sensor designs and applications. RCIM sensors can measure pressure, humidity, acceleration, flow rate, radiation, electrical current, electric and magnetic fields, or chemical and biological substances, and are used in a variety of applications, including homeland security, environmental protection, home automation, health care, aviation, as well as food, chemical, and medical industries. This chapter discusses the main characteristics and terminologies of RCIM sensors.

1.1 RCIM SENSOR HISTORY

RCIM sensors have assisted mankind in analyzing, controlling, measuring, and monitoring thousands of functions for over centuries. Examples include the compass developed by the Chinese about 4000 years ago, the carbon track potentiometer invented by Thomas Edison in 1872, an induction coil sensor made by Arthur Chattock in 1887, and a nonlinear rheostat patented by Mary Hallock-Greenewalt in 1920. As the automobile industry grew, many RCIM sensors were developed to measure air flow rate, throttle position, coolant/air temperature, oxygen volume, wheel speed, and so forth. With increasing aviation safety standards and higher reliability requirements, hundreds of RCIM sensors have been adapted for use in aircrafts. Today, RCIM sensors are also increasingly used in the medical and biological fields. For instance, by measuring the electrical properties of the human body, such as the electrical resistance of saliva [1] or skin impedance [2], a doctor can diagnose diseases, monitor female fertility or ovulation cycle [3], and measure the transepidermal water loss of the skin [4].

The simplest RCIM sensors have only one element. For example, some resistive pressure sensors are nothing more than a variable resistor; some inductive motion sensors are just a coil. More sophisticated RCIM sensors may contain circuits and components for temperature/humidity compensations; power regulation or stabilization; biasing and offset reduction; self-calibration; communication and interface; application-specific signal processing, conversion, and discrimination; digital displays; programmable functions; or even other sensors. For example, *Analog Device*'s ADXL-50 accelerometer contains oscillator-, reference-, self-testing, and output-amplifier circuits; the most advanced Hall sensors incorporate digital signal processing and programmable functions; some capacitance ultrasonic flow sensors and complementary metal-oxide semiconductors (CMOS) biocell sensors are the integration of ultrasonic, radio frequency (RF), CMOS, and fiber-optics sensors.

The progress in new materials, microelectromechanical systems (MEMS) and micromachining, nano, and VLSI (very-large-scale integration) patterning technologies have advanced RCIM sensors to a new level—miniaturization, high sensitivity, high reliability and durability, low power consumption, rapid response time, better capabilities, and improved performance-to-cost ratios. Many RCIM sensors can be batch-fabricated simultaneously (e.g., by optically repeating the patterns on a wafer) and packaged with thousands of electronic circuits into a single chip to reduce the cost and increase the consistency of manufacturing.

1.2   RCIM SENSOR DEFINITION

The word *sensor* came from the Latin *sentire* meaning "to perceive." *Webster's Collegiate Dictionary* defines a sensor as "a device that responds to a physical (or chemical) stimulus (such as heat, light, sound, pressure, magnetism, or motion) and transmits a resulting impulse (as used for measurement or operating a control)." An RCIM sensor is a sensor that results in a change in electrical resistance, capacitance, inductance, electrical field, or magnetic field when it is exposed to a physical, chemical, or biological stimulus. Its output could be a voltage, current, electrical charge, time, or frequency to reflect these changes.

1.3   RCIM SENSOR CHARACTERISTICS AND TERMINOLOGY

1.3.1   TRANSFER FUNCTION

The transfer function of an RCIM sensor describes the relationship between its input and its output, and is often expressed by an equation, a graph, or a table. For example, an *Infineon Technologies'* KP125 capacitive pressure sensor has a transfer function of [5]

$$V_{out} = (aP_{in} + b)V_{dd} \tag{1.1}$$

This transfer function contains important information about the sensor: the sensitivity a, the offset b, and the relationship between the input pressure P_{in} and its output voltage V_{out} under a constant power supply V_{dd}. This input–output relationship can also be described graphically (Figure 1.1) or as a table (Table 1.1).

A sensor's transfer function is also used for the sensor's calibration or performance prediction. Ideally, a sensor should have a linear transfer function, that is, the sensor's output is linearly proportional to its input. In reality, most sensors display different degrees of nonlinearity; thus, linearization of its transfer function at an operating point is involved in sensor design, modeling, and control.

1.3.2   SENSITIVITY

Sensitivity, the slope of a sensor's transfer function, is defined as the ratio of a small change in the sensor's output to a small change in its input. In some sensors, the

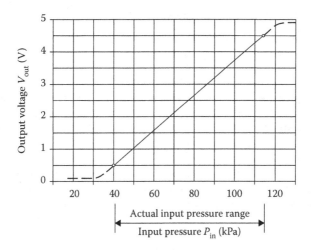

FIGURE 1.1 The input–output characteristic curve of a KP125 capacitive pressure sensor. (From *Infineon Technologies*, Milpitas, California, USA. With permission.)

TABLE 1.1
Parameters of a KP125 Pressure Sensor

Input Pressure (kPa)		Output Voltage (V)	
$P_{in,min}$	40	$V_{out,min}$	0.5
$P_{in,max}$	115	$V_{out,max}$	4.5

sensitivity is expressed as the input parameter change required to produce a standardized output change. In others, sensitivity is described by an output change for a given input change under the same excitation condition or a constant voltage power supply.

EXAMPLE 1.1

What is the sensitivity of the KP125 pressure sensor shown in Figure 1.1 and Table 1.1?

SOLUTION

Since sensitivity is the slope of the input–output characteristic curve (the sensor's transfer function), thus

$$a = \frac{V_{out,max} - V_{out,min}}{P_{in,max} - P_{in,min}} = \frac{4.5\ \text{V} - 0.5\ \text{V}}{115 \times 10^3\ \text{Pa} - 40 \times 10^3\ \text{Pa}} = 0.0533\ \text{V} \cdot \text{kPa}^{-1}$$

EXAMPLE 1.2

Two capacitive sensors have sensitivities of 1 V/0.1 mm and 1 V/0.05 mm, respectively. Which sensor is more sensitive?

SOLUTION

In order to produce a 1 V output change, the first sensor must experience a 0.1 mm gap change between the capacitor's two plates, while the second sensor only needs a 0.05 mm gap change. Thus, the second sensor is more sensitive.

EXAMPLE 1.3

Two resistive blood pressure sensors are available: one has a sensitivity of 10 mV/V/mmHg; the other has a 7.5 mV/V/mmHg sensitivity. Which sensor should be selected for an application that requires a higher sensitivity?

SOLUTION

The first sensor produces 10 mV output for each volt of excitation potential and each mmHg of the input pressure, which is higher than 7.5 mV output from the second sensor. Thus, the first sensor should be selected.

Sensitivity error or *sensitivity drift* is the actual sensitivity deviation from the ideal sensitivity, usually expressed as a percentage:

$$\text{Sensitivity error or drift} = \frac{\text{actual sensitivity} - \text{ideal sensitivity}}{\text{ideal sensitivity}} \times 100\% \qquad (1.2)$$

For many sensors, a change in sensitivity is often caused by temperature fluctuation. A curve that shows how sensitivity changes as the temperature varies is often found in a sensor's datasheet. If such a sensor is used at different temperatures, a calibration at each of these temperatures must be performed. Figure 1.2 shows the sensitivity (in percentage) versus temperature under a constant voltage power supply of an HMC1021 magnetoresistive sensor. The sensor has a minimum sensitivity of 0.75 mV/V/Oe at 125°C, a maximum sensitivity of 1.30 mV/V/Oe at −40°C, and a typical sensitivity of 1.06 mV/V/Oe at 25°C [6].

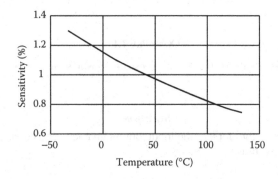

FIGURE 1.2 HMC1021 magnetoresistive sensor's sensitivity versus temperature. (Reproduced by the author using an HMC1021 sensor.)

EXAMPLE 1.4

A resistive blood pressure sensor has an actual sensitivity of 9.5 mV/V/mmHg. Its ideal value is 10 mV/V/mmHg. What is its sensitivity error or drift?

SOLUTION

The sensor's sensitivity error in percentage is

$$\frac{9.5 \text{ mV/V/mmHg} - 10 \text{ mV/V/mmHg}}{10 \text{ mV/V/mmHg}} \times 100\% = -5\%$$

1.3.3 OFFSET

When the measured property is zero but a sensor's output value is not zero, then the sensor has an *offset*. Offset is also called *zero* or *null offset*, *zero or null drift*, *offset error*, *bias*, *DC offset*, or *DC component*. An offset can be described by either a sensor's output value at its zero input (Figure 1.3a) or the difference between the sensor's actual output value and a specified ideal output value (Figure 1.3b). In Figure 1.3a, the sensor's offset is indicated by the value *b* measured from the zero input point (origin) along the vertical (output) axis. In Figure 1.3b, the sensor's offset *b* is indicated by the difference between the sensor's actual output and its ideal output on the vertical (output) axis.

Offset occurs because of calibration errors, sensor degradation over time, or environmental changes. Among these, temperature change is the primary factor causing the drift. Usually, the transfer function curve at room temperature (e.g., 25°C) is used as the reference or ideal curve, while curves at other temperatures are considered as actual curves. Offset error can be easily removed from a sensor's outputs by subtracting the constant *b* from the sensor's actual outputs or through a calibration process.

1.3.4 FULL SPAN AND FULL-SPAN OUTPUT

Full span or *full scale* (FS), or *span* for short, is a term used to describe a sensor's measurement range or limitation. It represents the broadest range, from minimum to maximum, of an input parameter that can be measured by a sensor without causing

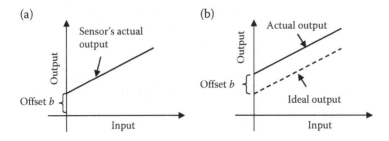

FIGURE 1.3 Sensor's offset.

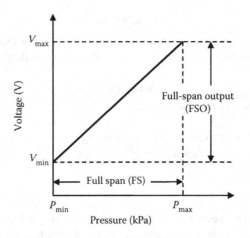

FIGURE 1.4 Full span (FS) and full-span output (FSO).

unacceptable inaccuracies. *Full-span output* (FSO) is a term used to describe a sensor's *dynamic range* or *output range*—the difference between a sensor's outputs measured with the minimum input stimulus and the maximum input stimulus. Figure 1.4, taking a piezoresistive pressure sensor as an example, illustrates FS and FSO in an ideal transfer function. FS is the range between the minimum pressure P_{min} and the maximum pressure P_{max} that can be measured by the sensor; FSO is the difference between the output voltages V_{min} and V_{max} corresponding to P_{min} and P_{max}, respectively.

For an actual transfer function, FSO should include all deviations from the ideal transfer function as shown in Figure 1.5.

The FS of a sensor can be unipolar (either positive or negative values of the measurand) or bipolar. Examples of sensors with unipolar ranges are pressure sensors (e.g., 0 ~ 200 MPa) or force sensors for compressive forces (e.g., −30 kN ~ 0 N).

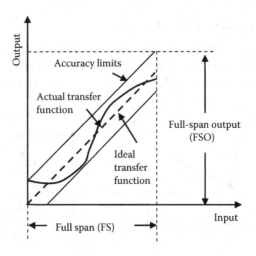

FIGURE 1.5 FS and FSO for an actual transfer function.

Bipolar ranges may be symmetric or asymmetric. For instance, an acceleration sensor has a symmetrical range (e.g., $\pm25g$, where $g = 9.81$ m·s^{-2} is the gravitational acceleration), while a shock sensor has an asymmetrical range of $-200g$ to $+10,000g$).

Some sensor datasheets also provide *overload* in addition to FS and FSO ranges. The overload can be specified as a value (e.g., 140 MPa) or as a percent of the span (e.g., 140%). If a sensor is overloaded, it will no longer perform within the specified tolerances or accuracy. An overload operation is the justification for a sensor manufacturer to reject a warranty claim.

1.3.5 ACCURACY

Accuracy is defined as the maximum deviation of a value measured by the sensor from its true value. Accuracy can be represented in one of the following forms:

Absolute accuracy: in terms of (1) measured parameters (e.g., pressure or acceleration), or (2) output parameters (e.g., voltage and resistance)

Relative accuracy: in terms of (1) a percentage of the maximum measurement error versus the true value:

$$\frac{\text{Max}\{\text{measured value} - \text{true value}\}}{\text{true value}} \times 100\% \tag{1.3}$$

or (2) a percentage of the maximum measurement error versus the full span:

$$\frac{\text{Max}\{\text{measured value} - \text{true value}\} \times 100\%}{\text{full span}} \tag{1.4}$$

For example, the accuracy of a piezoresistive pressure sensor may be specified as ±0.5 kPa (in terms of measured parameter), ±0.05 Ω (in terms of the output resistance value), or $\pm0.5\%$ (relative accuracy to its true value).

Many factors can affect the accuracy of a sensor, including temperature fluctuation, linearity, hysteresis, repeatability, stability, zero offset, A/D (analog-to-digital) conversion error, and display resolution. The overall accuracy of a sensor can be expressed by the following equation:

$$\text{Overall accuracy} = \pm\sqrt{e_1^2 + e_2^2 + e_3^2 + \cdots} \tag{1.5}$$

where $e_1, e_2, e_3, \ldots$ represents each contributing component of accuracy.

EXAMPLE 1.5

A Type-T thermocouple has a measurement error of $\pm1.1°$F. Its transmitter, A/D conversion, display resolution, and wire temperature effect is $\pm0.8°$F. What is the sensor's overall accuracy?

SOLUTION

$$\text{Overall accuracy} = \pm\sqrt{(1.1°F)^2 + (0.8°F)^2} = \pm1.36°F$$

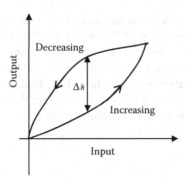

FIGURE 1.6   Hysteresis error.

1.3.6 HYSTERESIS

A sensor may produce different outputs when measuring the same quantity depending on the "direction" in which the value has been approached. The maximum difference in terms of the measured quantity is defined as *hysteresis error*, denoted as δ_h, or the maximum deviation in value between the increasing and decreasing cycle measured at the same input point (see Figure 1.6).

The *percentage hysteresis error* is determined by taking the maximum deviation and dividing it by FS:

$$\delta_h = \frac{\text{Max}(\Delta_h)}{\text{FS}} \times 100\% \tag{1.6}$$

Hysteresis errors are typically caused by friction, sensor structure, sensor material properties, or temperature and humidity changes.

EXAMPLE 1.6

A potentiometric displacement sensor produces a maximum of 20 mV difference when measuring the distance of an object passing a certain point: one measurement is taken when the object moves from the left to the right to the point, and the other is taken when the object moves from the right to the left to the point. Find the hysteresis error if the sensitivity of the sensor is 10 mV/mm.

SOLUTION

Since the hysteresis error is defined as the maximum deviation in terms of the measured quantity, which is displacement in this case, we need to convert the 20 mV output into the displacement input in millimeters:

$$\text{Hysteresis error (mm)} = \frac{\text{maximum output voltage difference}}{\text{sensitivity}} = \frac{20 \text{ mV}}{10 \text{ mV} \cdot \text{mm}^{-1}} = 2 \text{ mm}$$

1.3.7 NONLINEARITY

Nonlinearity describes the "straightness" of a sensor's transfer function. A nonlinearity can be expressed by the maximum deviation of a real transfer function from its best-fit straight line, Δ_l, or in percentage of Δ_l over the full span:

$$\text{Nonlinearity } (\%) = \frac{\text{Max}\{\Delta_l\}}{\text{FS}} \times 100\% \tag{1.7}$$

Nonlinearity should not be confused with accuracy. The latter indicates how close a measured value is to its true value, but not how straight the transfer function is.

EXAMPLE 1.7

A capacitive displacement sensor's linearity error at its worst point is 0.01 mm. If the full-scale range of its transfer function is 1 mm, what is its nonlinearity expressed as a percentage?

SOLUTION

From Equation 1.7

$$\text{Nonlinearity } (\%) = \frac{\text{Max}\{\Delta_l\}}{\text{FS}} \times 100\% = \frac{0.01\,\text{mm}}{1\,\text{mm}} \times 100\% = 1\%$$

1.3.8 NOISE AND SIGNAL-TO-NOISE RATIO

All sensors generate noise in addition to their output signals. Sensor noise can be defined as any deviation or fluctuation from an expected value. To describe these fluctuations, a simple average is meaningless since the average of the random variations is zero. Instead, the root mean square (RMS) of the deviations from V_{average} over a time interval T is used:

$$\sqrt{V_n^2} = \sqrt{\frac{1}{T} \int_0^T (V - V_{\text{average}})^2 \, dt} \tag{1.8}$$

where V_n^2 is the mean-square noise in voltage. $\sqrt{V_n^2}$ is the RMS noise in voltage.

The extent to which noise becomes significant in a sensing or a measurement process depends on the relative amplitude of the signal of interest to the unwanted noise value, called the *signal-to-noise ratio* (*SNR* or *S/N ratio*). If the noise value is small compared to the signal level, then the SNR is large, and the noise becomes unimportant. In sensor design, one always tries to maximize the SNR.

The SNR can be calculated by

$$\text{SNR} = \frac{\bar{P}_s}{\bar{P}_n} = \frac{V_s^2}{V_n^2} \tag{1.9}$$

or in decibels (dB)

$$\text{SNR} = 10\log_{10}\left(\frac{V_s^2}{V_n^2}\right) = 20\log_{10}\left(\frac{V_s}{V_n}\right) \tag{1.10}$$

where $\bar{P}_s$ and $\bar{P}_n$ are the average power of the signal and noise, respectively, and V_s and V_n are the RMS voltages of the signal and noise, respectively.

Another quantitative measurement of noise is *noise factor*, F_n, defined as

$$F_n = \frac{\text{SNR at input}}{\text{SNR at output}} \tag{1.11}$$

The noise factor F_n specifies how much additional noise the sensor contributes to besides the noise already received from the source. Ideally, $F_n = 1$.

EXAMPLE 1.8

The expected output voltage of a sensor is 100 mV. However, a noise voltage of 18 μV is superimposed on its output signal. Find the SNR of the sensor.

SOLUTION

$$\text{SNR} = \frac{(100 \times 10^{-3} \text{ V})^2}{(18 \times 10^{-6} \text{ V})^2} = 3.09 \times 10^7$$

or

$$\text{SNR} = 20\log_{10}\left(\frac{100 \times 10^{-3}}{18 \times 10^{-6}}\right) = 74.90 \text{ dB}$$

EXAMPLE 1.9

An amplifier in a sensor circuit has a signal voltage level of 3 μV and a noise voltage level of 1 μV at input. If the gain of the amplifier is 20 and a 5 μV of noise is added by the amplifier at the output, determine the noise factor of the amplifier.

SOLUTION

SNR at input:

$$\text{SNR} = \frac{(3 \times 10^{-6} \text{ V})^2}{(1 \times 10^{-6} \text{ V})^2} = 9$$

SNR at output:

$$SNR = \frac{(3 \times 20 \times 10^{-6} \text{ V})^2}{\left[(1 \times 20 + 5) \times 10^{-6} \text{ V}\right]^2} = 5.76$$

Note: The amplifier also amplifies the input noise. The noise factor therefore is

$$F_n = \frac{SNR \text{ at input}}{SNR \text{ at output}} = \frac{9}{5.76} = 1.56$$

Different sensors produce different types of noise. Chapter 6 will discuss the five main types of sensor noise and the methods to reduce them.

EXAMPLE 1.10

A photodetector has an RMS signal current of 20 μA, a shot noise current of 2.27 pA, and a thermal noise current of 0.42 pA. Find the current SNR of the photodetector.

Solution

The total current noise in RMS can be found from $i_n = \sqrt{i_{shot}^2 + i_{thermal}^2}$. Thus

$$SNR = \frac{(i_s)^2}{\left(\sqrt{i_{shot}^2 + i_{therm}^2}\right)^2} = \frac{(20 \times 10^{-6})^2}{(2.27 \times 10^{-12})^2 + (0.42 \times 10^{-12})^2} = 7.65 \times 10^{13}$$

1.3.9 Resolution

The resolution of a sensor specifies the smallest change of input parameter that can be detected and reflected in the sensor's output. In some sensors, when an input parameter continuously changes, the sensor's output may not change smoothly but in small steps. This typically happens in potentiometric sensors and infrared occupancy detectors with grid masks.

Resolution can be described either in absolute terms or relative terms (in percentage of FS). For example, a *Honeywell's* HMC100 magnetic sensor has a resolution of 27 μgauss (microgauss) [6] in terms of an absolute value, meaning that a minimum change of 27 μgauss can be detected by the sensor; an *Analog Devices'* ADT7301 digital temperature sensor has a resolution of 0.03°C/LSB (least significant bit) [7], meaning that one bit can distinguish 0.03°C temperature change; and a *SENSIRION's* SHT1 humidity sensor has a resolution of 0.05% relative humidity (RH) [8], meaning that the sensor's resolution is 0.05% of its span.

The resolution of a sensor must be higher than the accuracy required in the measurement. For instance, if a measurement requires an accuracy within 0.02 μm, then the resolution of the sensor must be better than 0.02 μm.

Factors that affect resolution vary from sensor to sensor. For most capacitive sensors, the primary affecting factor is electrical noise. Taking a capacitive displacement sensor as an example, even though the distance between the sensor and the target is constant, the voltage output will still fluctuate slightly due to the "white" noise of the system. Assume no signal conditioning, one cannot detect a shift in the voltage output that is less than the peak-to-peak voltage value of noise. Because of this, most resolutions of capacitive sensors are evaluated by the peak-to-peak value of noise divided by sensitivity:

$$\text{Resolution} = \text{peak-to-peak value of noise/sensitivity} \qquad (1.12)$$

EXAMPLE 1.11

A capacitive displacement sensor's peak-to-peak value of noise is 0.01 V. If the sensor's sensitivity is 10 V·mm^{-1}, what is its resolution?

SOLUTION

Resolution = peak-to-peak value of noise/sensitivity = 0.01 V/(10 V·mm^{-1}) = 1 µm

1.3.10 PRECISION AND REPEATABILITY ERROR

Precision refers to the degree of *repeatability* or *reproducibility* of a sensor. That is, if the exactly same value is measured a number of times, an ideal sensor would produce exactly the same output every time. But sensors actually output a range of values distributed in some manner relative to the actual correct value. Precision can be expressed mathematically as

$$\text{Precision} = 1 - \left| \frac{x_n - \bar{x}}{\bar{x}} \right| \qquad (1.13)$$

where x_n is the value of the nth measurement and $\bar{x}$ is the average of the set of n measurements.

EXAMPLE 1.12

An exact pressure of 150 mmHg was applied to a static pressure sensor. Even though the applied pressure never changed, the output values from the sensor over 10 measurements varied (see Table 1.2). Calculate the average output of the sensor and precision of the sixth measurement.

SOLUTION

The average output of the sensor, $\bar{x}$, is equal to the sum of the 10 measurements divided by 10:

$$\bar{x} = \frac{1}{10} \sum_{i=1}^{10} x_i = \frac{1}{10}(149 + 151 + 152 + 149 + 150 + 153$$
$$+ 152 + 149 + 147 + 151) \text{ mmHg}$$
$$= 150.3 \text{ mmHg}$$

TABLE 1.2

Ten Outputs from a Static Pressure Sensor with a 150 mmHg Input

Reading Number	Sensor Output Value (mmHg)
1	149
2	151
3	152
4	149
5	150
6	153
7	152
8	149
9	147
10	151

The precision of the sixth measurement is

$$\text{Precision} = 1 - \left| \frac{x_6 - \bar{x}}{\bar{x}} \right| = 1 - \left| \frac{(153 - 150.3)\,\text{mmHg}}{150.3\,\text{mmHg}} \right| = 1 - 0.02 = 0.98$$

Repeatability error describes the inability of a sensor to recognize the same value under identical conditions. Figure 1.7 shows the two runs of the same measurement performed by sensor A and sensor B. Apparently, sensor B has a less repeatability error than sensor A. Usually, repeatability error is expressed in a percentage of full scale

$$\delta_r = \frac{\text{Max}(\Delta_r)}{\text{FS}} \times 100\% \tag{1.14}$$

For example, a *Tekscan*'s A201 force sensor typically has a repeatability error of ±2.5% FS [9]. Repeatability error may be caused by thermal noise, buildup charges, material plasticity, and friction.

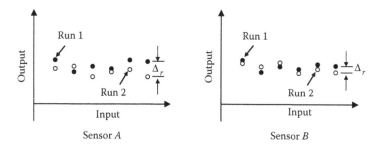

FIGURE 1.7 Repeatability error of sensors A and B.

1.3.11 CALIBRATION AND CALIBRATION ERROR

Many sensors require calibration prior to their use. For some sensors, calibration is to determine the coefficients in their transfer functions (offset, sensitivity, or both); for others, calibration means to convert the sensor's electrical output (e.g., voltage) to a measured value (e.g., temperature). Although detailed calibration procedures vary from sensor to sensor, the general calibration procedure involves taking a number of measurements over the full operating range and comparing the measured values with actual values (or plotting the sensor's input–output curve against a standard/reference curve), and then adjusting the parameters to match the actual values (or the standard curve). The actual values (or the standard curve) may be (1) obtained using a more accurate sensor as the calibration reference or (2) provided by the sensor's manufacturer. Smart or advanced sensors may have self-calibration and adaptive features to adjust themselves to different operating conditions.

Following examples demonstrate the one-point and two-point calibration procedures of a *National Semiconductor*'s LM335 linear temperature sensor [10]. Its transfer function specified by the manufacturer is

$$V_{out} = (T_n + 273.15\ \text{K})(10\ \text{mV/K}) \tag{1.15}$$

where V_{out} is its output voltage proportional to the absolute temperature to be measured over its useful range ($-40°C$ to $+100°C$) and T_n (in °C) is the nominal temperature.

EXAMPLE 1.13

One-point calibration procedures for an LM335 temperature sensor

One-point calibration is to determine the offset (intercept) of a sensor's transfer function, b, with the assumption that the slope of the transfer function, a, is known (e.g., $a = 1$).

Step 1: Let the temperature being measured by the sensor, T_s, be equal to

$$T_s = aT_n + b = T_n + b \tag{1.16}$$

where T_n is the nominal value of the temperature being measured, a ($=1$) is the known slope, and b is the offset before calibration.

Step 2: Expose the LM335 sensor to a known and stable temperature, T_a (e.g., $T_a = 35°C$). Read and record the temperature measured by the sensor, T_s (e.g., $T_s = 34.6°C$).

Step 3: Determine the new offset value b' using the formula

$$b' = b + T_a - T_s = b + 35°C - 34.6°C = b + 0.4°C$$

Step 4: Replace b with b' in Equation 1.16:

$$T_s = aT_n + b + 0.4°C = T_n + b + 0.4°C$$

EXAMPLE 1.14

Two-point calibration procedures for an LM335 temperature sensor

Two-point calibration is to determine both the slope a and the offset b of a sensor's transfer function.

Step 1: Let the temperature being measured by the sensor, T_s, be equal to

$$T_s = aT_n + b \tag{1.17}$$

where T_n is the nominal value of the temperature being measured, and a and b are the slope and the offset, respectively, before calibration.

Step 2: Expose the sensor to an accurately known and stable temperature, T_{a1} (e.g., $T_{a1} = 35°C$). Read and record the sensor output T_{s1} (e.g., $T_{s1} = 34.6°C$).

Step 3: Expose the sensor to another accurately known and stable temperature, T_{a2} (e.g., $T_{a2} = 75°C$). Read and record the sensor output T_{s2} (e.g., $T_{s2} = 75.3°C$).

Step 4: Determine the new slope value a' and the new offset value b' using the formula

$$a' = m\,a$$

$$b' = n + m\,b$$

where the intermediate values m and n are

$$m = \frac{T_{a1} - T_{a2}}{T_{s1} - T_{s2}} \qquad n = T_{a1} - mT_{s1}$$

Thus

$$a' = m\,a = \frac{T_{a1} - T_{a2}}{T_{s1} - T_{s2}}\,a = \frac{35 - 75}{34.6 - 75.3}\,a = 0.98a$$

$$b' = n + m\,b = (T_{a1} - mT_{s1}) + mb = T_{a1} + \frac{T_{a1} - T_{a2}}{T_{s1} - T_{s2}}(b - T_{s1})$$

$$= 35 + \frac{35 - 75}{34.6 - 75.3}(b - 34.6) = 1.00 + 0.98b$$

Step 5: Replace a with a' and b with b':

$$T_s = (0.98a)T_n + (1.00 + 0.98b) = 0.98(aT_n + b) + 1.00$$

The calibration accuracy in both one-point and two-point methods depends on the accuracy of the known temperatures. As a rule of thumb, if a sensor's measurement error is within the maximum tolerable error, no calibration is necessary; otherwise, calibration is required. For a lower accuracy requirement, the one-point calibration is sufficient; for the higher accuracy requirement, the two-point calibration should be used. If a greater accuracy is desired, several two-point calibrations, or section-by-section calibrations (each has a narrow range) over the full span may be used.

Calibration error is the amount of inaccuracy permitted when a sensor is calibrated. This error is often of a systematic nature, meaning that it can occur in all real transfer functions. It may be constant or it may vary over the measurement range, depending on the type of error in the calibration.

1.3.12   RESPONSE TIME AND BANDWIDTH

Strictly speaking, sensors do not immediately respond to an input stimulus. Rather, they have a finite *response time* or *rise time* t_r to an instantaneous change in stimulus. The response time can be defined as the "time required for a sensor's output to change from its previous state to a final settled value within a tolerance band of the correct new value" [11]. Usually, t_r is determined as the time required for a sensor signal to change from a specified low value to a specified high value of its step response. Typically, these values are 10% and 90% of its final value for overdamped, 5% and 95% of its final value for critically damped, and 0% and 100% of its final value for underdamped step response. A *FlexiForce's* A201 force sensor has a response time less than 5 µs [9], which is typical for most sensors. The *time constant* τ represents the time it takes for the sensor's step response to reach $1 - 1/e \approx 0.632$ or 63.2% of its final value. Figure 1.8 indicates a sensor's response/rise time t_r and its time constant τ under a positive step stimulus in an overdamped case.

Overshoot is a sensor's maximum output over its final steady output in response to a step input (see Figure 1.9). In this case, t_r is the time required for a sensor's output to rise from 0% to 100% of its steady-state value; *peak time t_p* is the time required for the sensor to reach the first peak of its step response; *settling time t_s* is the time required for the sensor's output to reach and stay within ±2% of its final value. Steady-state error is the error between the desired and actual output value.

The overshoot in percentage (called *percentage overshoot*) can be obtained by

$$\text{P.O. (percentage overshoot)} = \frac{\text{maximum peak value} - \text{final value}}{\text{final value}} \times 100\% \quad (1.18)$$

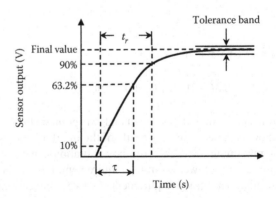

FIGURE 1.8   Sensor's overdamped step response.

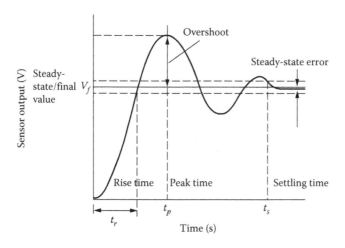

FIGURE 1.9   Sensor's underdamped step response.

EXAMPLE 1.15

Figure 1.10 is a step response of a renal vascular resistive (RVR) sensor after sudden release from the renal artery pressure [12]. Find its rise time, peak time, settling time, and percentage overshoot.

SOLUTIONS

From Figure 1.10, it is found that
Rise time (0% ~ 100%): $t_r = 36$ s
Peak time: $t_p = 43$ s
Settling time: $t_s = 82$ s

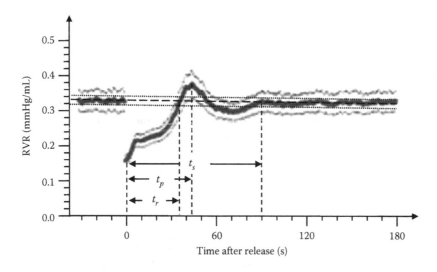

FIGURE 1.10   Step response of a renal vascular resistance (RVR) to a renal artery pressure after sudden release. (From Just, A. et al., *J. Physiol.* 538, 167, 2002. With permission.)

Percentage overshoot:

$$\frac{0.38 - 0.335}{0.335} \times 100\% = 13.43\%$$

The inverse of response time is the *cutoff frequency, f_c,* indicating the lowest or highest frequency of input signal that a sensor can take and respond properly. Mathematically, the cutoff frequency is defined as the frequency at which the magnitude of the response curve falls to 0.707 (or -3 dB $= 20 \log 0.707$) of its steady-state value. The bandwidth of a sensor is the frequency range of the input signal that the sensor can process—usually $0 \sim f_c$, or the range between the two cutoff frequencies ($f_{c1} \sim f_{c2}$).

EXAMPLE 1.16

The frequency response of an MC95 magnetic field sensor to a magnetic field is recorded in Figure 1.11a. Find its cutoff frequencies and the bandwidth.

SOLUTION

The frequencies that cross 0.707 are $f_{c1} = 23$ Hz and $f_{c2} = 3700$ Hz (see Figure 1.11b). Thus, the bandwidth is 23 Hz $\leq f \leq 3700$ Hz.

EXAMPLE 1.17

A CMOS photocurrent detector measures the total charge collected at the detector terminal. Its overdamped response to a short-pulse optical input is shown in Figure 1.12a [13]. Find its rise time.

SOLUTION

For an overdamped response, 10–90% of its final values should be used to determine T_r. Refer to Figure 1.12b: at 10% point, $T_1 = 10 \times 10^{-12}$ s; at 90% point, $T_2 = 125 \times 10^{-12}$ s. Thus, the rise time is $T_r = 125 \times 10^{-12}$ s $- 10 \times 10^{-12}$ s $= 115 \times 10^{-12}$ s or 115 ps.

1.3.13 SENSOR LIFESPAN

All sensors have a finite life, indicated by operating/service life, cycling life, continuous rating, intermittent rating, storage life, or expiration date. Several factors may affect a sensor's life: its type, design, material, frequency and duration of use, concentration levels to be measured, manufacturing process, maintenance efforts, application, storage, and environmental conditions (e.g., temperature, humidity) [14]. Sensors that consume internal materials during the sensing process (e.g., certain glucose or oxygen sensors) can be used only once or just a few times. Some gas sensors (e.g., *Biosystems'* CO and H_2S sensors), although nonconsumptive, only have a 2–4 year life limit due to the factors such as evaporation (drying out), leakage, and catalyst contamination. Oxygen sensors used in automobile engines that were built prior to 1995 have a lifespan of 50,000 miles, while newer designs (1996 and later) only need to be replaced every 100,000 miles. A sulfur dioxide sensor exposed to 5 ppm (parts per million) concentration continuously may last 10,000 hours or over a year;

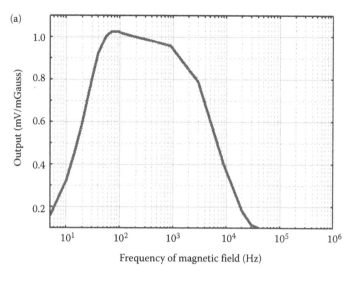

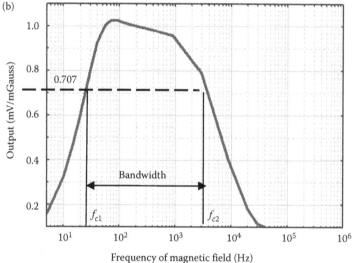

FIGURE 1.11 (a) A MC95 sensor's response to a magnetic field; (b) the sensor's cutoff frequencies and bandwidth.

if the concentration is lower than 5 ppm, a longer life will result. Most mechanical sensors have a long lifespan under normal operating conditions. For example, *FlexiForce*'s load sensors have a life of one million load cycles at loads of under 50 lb (cycling life). Many temperature sensors have a service life of over 10 years.

Aging also affects some sensors' accuracy and cause sensors to slowly lose sensitivity over time. For instance, certain types of temperature sensors decrease their sensitivities by 0.1°C each year. Most chemical and biosensors' accuracy depends on their ages. In addition, rough handling of a sensor could also shorten its useful

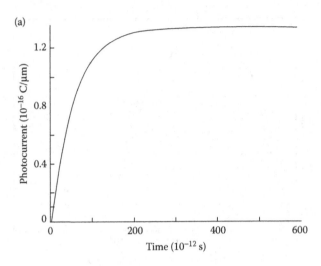

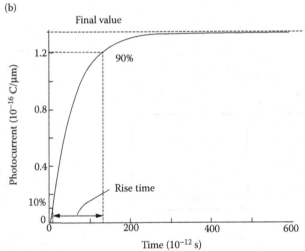

FIGURE 1.12 (a) A photocurrent sensor's response to a short-pulse optical input; (b) determination of its rise time.

life. A sensor that is repeatedly installed and removed will have a shorter life than a sensor that is installed and left in place. The best way to extend sensors' lives is to store them properly, regularly test and verify their accuracy, and recalibrate them whenever necessary. More advanced sensors today are equipped with sensor life monitoring systems to remind the users when these sensors need to be replaced. Sensors should be replaced when they can no longer be calibrated or zeroed easily.

1.3.14 OTHER SENSOR CHARACTERISTICS

Many sensors have their unique characteristics and terminologies to describe their performance. Examples include thermistors' *dissipation constant*, the Hall current sensors' *output current sink*, the magnetic sensors' *noise density*, and the photo-electric sensors' *receiver excess gain*. Interested readers should learn these specific characteristics when using these sensors.

EXERCISES

1. All of the following information about a sensor can be obtained from the sensor's transfer function *except*
   A. Sensitivity
   B. Offset
   C. Input and output relationship
   D. Accuracy
2. Which of the following is a unit that could be used for sensitivity?
   A. mV/V/Oe
   B. %
   C. °C
   D. mm·s^{-1}
3. Which of the following terms is not synonymous with *offset*?
   A. Bias
   B. AC component
   C. Null drift
   D. Offset error
4. Two sensors are available to measure a temperature using the same circuit. Sensor *A* produces a voltage output of 4.5 mV, while Sensor *B*'s output is 3.8 mV. Which sensor has higher sensitivity to temperature?
5. Which of the following full-span ranges for an accelerometer is bipolar?
   A. $-10g$ to $-50g$
   B. $\pm25g$
   C. $0g$ to $21g$
   D. $-50g$ to $0g$
6. Hysteresis errors are typically caused by the following factors *except*
   A. Temperature and humidity variation
   B. Features of the measured parameter
   C. Sensor structure
   D. Sensor material properties
7. A sensor's accuracy can be expressed using all of the following terms *except*
   A. Measured parameter
   B. Output parameter
   C. Ratio of maximum measurement error versus measured value
   D. Percentage of maximum measurement error versus the full span
8. In sensor design, the objective is to
   A. Minimize the S/N ratio
   B. Minimize the noise factor F_n
   C. Make the S/N ratio close to 1
   D. Make the noise factor F_n close to 1

9. If a sensor system has several noise sources (e_1, e_2, e_3, ...), then the total noise will be
   A. $e_{\text{Total}} = e_1 + e_2 + e_3 + \cdots$
   B. $e_{\text{Total}} = |e_1| + |e_2| + |e_3| + \cdots$
   C. $e_{\text{Total}} = e_1^2 + e_2^2 + e_3^2 + \cdots$
   D. $e_{\text{Total}} = \sqrt{e_1^2 + e_2^2 + e_3^2 + \cdots}$

10. Resolution of a sensor specifies the
    A. Maximum deviation of a value measured by the sensor from its true value
    B. Smallest change of input parameter that can be reflected in the sensor's output signal
    C. Degree of reproducibility of a sensor
    D. Ability of a sensor to recognize the same value under identical conditions

11. Mathematically, *precision* can be expressed as
    A. $\dfrac{\Delta}{\text{FS}} \times 100\%$
    B. $\left| \dfrac{x_n - \bar{x}}{\bar{x}} \right|$
    C. $1 - \left| \dfrac{x_n - \bar{x}}{\bar{x}} \right|$
    D. $\left(1 - \left| \dfrac{x_n - \bar{x}}{\bar{x}} \right| \right) \times 100\%$

12. Calibration of a sensor may involve all of the followings *except*
    A. Determining or resetting the parameters in its transfer function
    B. Adjusting S/N ratio
    C. Converting the sensor's output to its measured value
    D. Comparing the measured values with actual values

13. A photoresistor has a unit step response shown in Figure 1.13. Find its settling time at 90% of its final value and its rise time (10–90%).

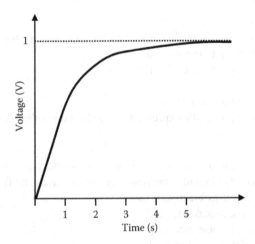

FIGURE 1.13   Unit step response of a photoresistor.

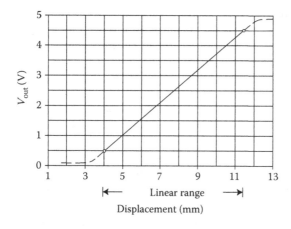

FIGURE 1.14 Input–output curve of a potentiometer.

14. A potentiometric displacement sensor has an input–output curve shown in Figure 1.14. Determine its sensitivity.

Exercises 15 through 19 are related to the three-run measurement results of an angular velocity sensor as shown in Table 1.3.

15. Find the relative accuracy of the sensor when measuring the motor speed at 720 rpm if the true value is 9 V.

TABLE 1.3

Three-Run Test[a] Data of an Angular Velocity Sensor

Input (Motor Speed, rpm)	Output (V) – Run 1		Output (V) – Run 2		Output (V) – Run 3	
	Up	Down	Up	Down	Up	Down
0	0.05	0.48	0.06	0.99	0.38	0.49
80	1.10	1.50	1.25	1.68	1.38	1.64
160	1.34	2.26	1.76	2.59	2.46	2.35
240	3.43	3.47	2.87	3.53	3.14	3.69
320	3.77	5.17	4.53	3.52	4.25	3.94
400	5.45	5.27	4.92	5.91	4.79	5.67
480	5.87	5.92	6.18	6.41	5.95	6.54
560	7.30	6.58	6.63	7.11	7.49	7.33
640	8.34	8.16	7.98	7.77	8.35	8.46
720	9.05	9.20	8.87	9.41	9.24	9.88
800	10.25	10.03	10.02	11.01	9.73	10.41

[a] A *three-run* test is a broadly accepted method to test a sensor's performance and accuracy over its entire operating range, starting with the input at the low extreme and increasing it (in a uniform step) until the high extreme is reached (Up), and then returning via the same input points to the low extreme again (Down) to complete a cycle. Repeat this procedure for three cycles.

16. Find the precision of the sensor when measuring the motor speed at 400 rpm (using all six data at 400 rpm and given $x_n = 5.3$ V).
17. Draw the sensor's hysteresis curve using Run 2 data (Up and Down), and find its hysteresis error δ_h.
18. Determine the repeatability error among Run 1 Up and Run 2 Up measurements.
19. Find the sensor's transfer function using Run 3 Down data (assume a linear transfer function, that is, $y = ax + b$; use Excel to find a straight line with a and b values that best fits the sensor data).
20. The nominal transfer function of an MPX4250A piezoresistive pressure sensor provided by the manufacturer is $V_{out} = V_s(0.004P_{in} - 0.04)$, where V_s is the supply voltage (in V), P_{in} is the input pressure (in kPa), and V_{out} is the sensor's output (in V). (1) If $V_s = 5.1$ V, find the sensor's nominal sensitivity and nominal offset. (2) If the supply voltage applied to the sensor fluctuates from 4.85 V DC to 5.35 V DC, that is, $V_s = 5.1 \pm 0.25$ V, find the maximum and minimum absolute output error caused by the unstable power supply when measuring a 100 kPa pressure.

REFERENCES

1. Nakada, M., Stress evaluation apparatus, United States Patent 7,155,269, 2006.
2. Archer, W.I., Kohli, R., Roberts, J.M., and Spencer, T.S., Skin impedance measurement, in *Method for Cutaneous Investigation*, Marcel Dekker, Inc., New York, chap. 7, 1990.
3. Fernando, R. and Betz, G., Ovulation prediction by monitoring salivary electrical resistance with the CUE fertility monitor, *J. Obstet. Gynecol.*, 22, 282, 1988.
4. Cui, Y., Xiao, P., Ciortea, L.I., De Jesus, M.E.P., Berg, E.P., and Imhof, R.E., Mathematical modeling for the condenser method of trans-epidermal water loss measurements, *J. Nondestruct. Test. Eval.*, 22, 229, 2007.
5. Absolute pressure sensor: KP125, Datasheet, Infineon Technologies, Milpitas, CA, USA, 2007.
6. 1- and 2-axis magnetic sensors, Datasheet 900248, Honeywell, Morristown, New Jersey, USA.
7. Preliminary technical data for ADT7301, Analog Devices, Inc., Norwood, Massachusetts, USA, 2004.
8. Datasheet SHT1x (SHT10, SHT11, SHT15), Sensirion AG, Switzerland, 2009. Available at: www.sensirion.com.
9. FlexiForce® A201 standard force and load sensors, Tekscan Inc., South Boston, Massachusetts, 2009. Available at: www.tekscan.com.
10. AD590 and LM335 sensor calibration, Application Note #17, ILX Lightwave Corporation, Bozeman, Montana, 2006.
11. Carr J.J. and Brown, J.M., *Introduction to Biomedical Equipment Technology*, 3rd ed., 1998. ISBN: 0-13-849431-2.
12. Just, A. et al., Role of angiotensin II in dynamic renal blood flow autoregulation of the conscious dog, *J. Physiol.*, 538, 167, 2002.
13. Bhatnagar, A., Latif, S., Debaes, C., and Miller, D.A.B., Pump-probe measurements of CMOS detector rise time in the Blue, *J. Lightwave Technol.*, 22, 2213, 2004.
14. McDermott, H.J. and Ness, S.A., *Air Monitoring for Toxic Exposures*, 2nd ed., John Wiley & Sons, Inc., Hoboken, New Jersey, 2004.

2 Resistive Sensors

2.1 INTRODUCTION

Electrical resistance is the easiest electrical property to be measured precisely over a wide range at a moderate cost. Resistive sensors have many desirable features: reliability, adjustable resolution, simple construction, and ease of maintenance, which have made them the preferred choice in sensor design and application. Based on their sensing principles, resistive sensors are classified as

- *Potentiometric sensors*: resistance change due to linear or angular position change
- *Resistive temperature sensors*: resistance change caused by temperature variation (*thermoresistive effect*)
- *Photoresistive sensors*: decrease in resistance when light strikes a photoconductive material (*photoresistive effect*)
- *Piezoresistive sensors*: resistance change when a force is applied to a piezoresistive conductor (*piezoresistive effect*)
- *Magnetoresistive sensors*: resistance change in the presence of an external magnetic field (*magnetoresistive effect*)
- *Chemoresistive sensors*: conductivity change in a material or solution due to chemical reactions that alter the number of electrons or concentration of ions
- *Bioresistive sensors*: bioresistance change in proteins or cells induced by structural variations and biological interactions

The following sections will discuss each of these sensor types, except magnetoresistive sensors (to be covered in Chapter 5). The primary emphases are sensing principles, material characteristics, design, and applications.

2.2 POTENTIOMETRIC SENSORS

2.2.1 Sensing Principle

Potentiometric sensors are designed based on Equation 2.1—a conductor's resistance R (in ohms, Ω) is a function of the resistivity of the conductor material ρ (in ohm-meter, $\Omega \cdot m$), its length l (in meter, m), and its cross-sectional area A (in meter square, m^2):

$$R = \rho \frac{l}{A} \qquad (2.1)$$

Although any change in l, A, and ρ will cause a change in resistance, potentiometric sensors (also called *potentiometers* or *pots* for short) are often designed by varying the length l only for the sake of simplicity and for saving cost. Some chemoresistive sensors are designed based on materials' resistivity ρ change caused by chemical reactions.

The resistance of a potentiometer can be evaluated using *Ohm's Law* by applying an electric current I (in amperes, A) and measuring the voltage V (in volts, V) across the potentiometer:

$$R = \frac{V}{I} \tag{2.2}$$

2.2.2 CONFIGURATION AND CIRCUITRY

Potentiometric sensors are available in two configurations: *linear* and *rotary*, as shown in Figure 2.1a and b, respectively. In both configurations, resistance change is the result of position variation (x or θ) of a movable contact (wiper) on a fixed resistor, resulting in an output voltage change.

The circuit symbols and typical circuits of potentiometric sensors are shown in Figure 2.2a, through c.

The potentiometer R_2 in Figure 2.2b functions as a voltage divider. The voltage across R_2 is the measured output:

$$V_{out} = \frac{R_2}{(R_1 + R_2)} V_S \tag{2.3}$$

If a load R_L is placed across R_2, as shown in Figure 2.2c, the amount of current "diverted" from R_2 will depend on the magnitude of R_L relative to R_2. The output voltage across R_2 (which is also the load voltage) is then

$$V_{out} = \frac{R_2 R_L}{R_1 R_L + R_2 R_L + R_1 R_2} V_S \tag{2.4}$$

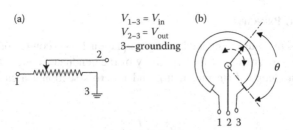

FIGURE 2.1   Linear (a) and rotary (b) potentiometer configurations.

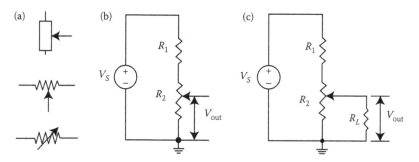

FIGURE 2.2 Potentiometer circuits: (a) circuit symbols; (b) a voltage divider circuit; (c) a voltage divider circuit with a load.

If $R_L \gg R_2$, ideal measurement conditions exist where the power extracted by the load, such as a meter, oscilloscope, or data acquisition unit, is negligible, then

$$\frac{V_{out}}{V_S} = \frac{R_2}{R_1 + R_2}$$

EXAMPLE 2.1

Given $V_S = 6$ V, $R_1 = 0.5$ kΩ, $R_2 = 1$ kΩ, $R_L = 10$ kΩ. Find V_L.

SOLUTION

Apply Equation 2.4:

$$V_L = V_{out} = \frac{R_2 R_L}{R_1 R_L + R_2 R_L + R_1 R_2} V_S$$

$$= \frac{(1 \times 10^3 \, \Omega)(10 \times 10^3 \, \Omega)(6 \, V)}{(0.5 \times 10^3 \, \Omega)(10 \times 10^3 \, \Omega) + (1 \times 10^3 \, \Omega)(10 \times 10^3 \, \Omega) + (0.5 \times 10^3 \, \Omega)(1 \times 10^3 \Omega)}$$

$$= 3.87 \, V$$

The voltage divider circuit in Figure 2.2b can be used by many resistive sensors, that is, R_2 can be a photodetector, a thermistor, a magnetoresistive, or a piezoresistive sensor. One drawback of this circuit is that the relationship between V_{out} and R_2 is nonlinear, since $V_{out} = V_S/(R_1/R_2 + 1)$. This means that V_{out} is not proportional to R_2. In practice, an operational amplifier (op-amp) circuit in Figure 2.3 is commonly used to provide a linear relationship between V_{out} and R_2.

EXAMPLE 2.2

Derive V_{out} as a function of R_2 in Figure 2.3 (assume an ideal op-amp). V_S and R_1 are known.

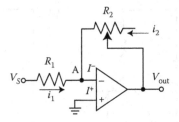

FIGURE 2.3 An op-amp measurement circuit providing a linear relationship between V_{out} and R_2.

SOLUTION

An ideal op-amp has infinite input impedance, thus current in both inverting input (−), I_-, and noninverting input (+), I_+, should be zero, that is, $I_- = I_+ = 0$. Also, an ideal op-amp has the same voltage potential at its two inputs, that is, $V_- = V_+$. Apply Kirchhoff's current law at node A:

$$i_1 + i_2 = I_- = 0 \Rightarrow \frac{V_S - V_A}{R_1} + \frac{V_{out} - V_A}{R_2} = 0$$

and $V_A = V_- = V_+ = 0$ (since V_+ is grounded), resulting in

$$V_{out} = -\frac{R_2}{R_1} V_S \tag{2.5}$$

Equation 2.5 provides the linear relationship between V_{out} and R_2. The ratio R_2/R_1 is the gain of the amplifier. The negative sign indicates that it is an inverting amplifier. R_1 is often chosen equal to the sensor's resistance range (i.e., $R_1 = R_{2,Max}$). This op-amp circuit is also applicable to many other resistive sensors.

EXAMPLE 2.3

Figure 2.4 shows a potentiometric sensor for water level measurement. If the resistance changes linearly from 0 to 2 kΩ over the entire water level range, (1) develop

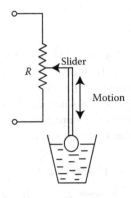

FIGURE 2.4 A potentiometric water level sensor.

a measurement circuit to provide a linear, 0 to 10 V output when the water level changes from 0 to 80 mm; (2) if the sensor's output is 7.5 V, what is the water level?

SOLUTIONS

1. The same op-amp circuit as shown in Figure 2.3 can be used for this measurement with $R_1 = 2$ kΩ (the maximum resistance of the sensor), $V_S = 10$ V, and R_2 being the potentiometer.
   Apply Equation 2.5 and ignore the sign (for easy explanation):

$$V_{out} = \frac{R_2}{R_1} V_S = \frac{R_2}{2000}(10) = 0.005\, R_2$$

As R_2 varies from 0 to 2 kΩ, the output voltage V_{out} will change linearly from 0 to 10 V.

2. When V_{out} is 7.5 V, R_2 is

$$R_2 = \frac{V_{out}}{0.005} = \frac{7.5}{0.005} = 1.5 \text{ k}\Omega$$

Since the water level x is proportional to the resistance change R_2, therefore

$$\frac{x - X_{min}}{X_{max} - X_{min}} = \frac{R_2 - R_{min}}{R_{max} - R_{min}} \Rightarrow \frac{x - 0}{80 - 0} = \frac{R_2 - 0}{2000 - 0} \Rightarrow$$
$$x = 0.04\, R_2 = 0.04 \times 1500 = 60 \text{ mm}$$

2.2.3 POTENTIOMETRIC SENSOR DESIGN

2.2.3.1 Linear Potentiometers

Major components of a linear potentiometer include three terminals (power input, ground, and sensor signal output), a fixed resistor, a wiper, a sliding track, and the housing. The terminals are often gold plated to prevent corrosion or tarnishing. The resistor is usually made of high-quality conductive materials (e.g., copper) with a lower temperature coefficient for stability and long life. The wiper uses high-quality metal or alloy (e.g., a platinum alloy) for long life and low noise. Some wipers even have a multifinger shape to prevent intermittence when used in high shock or vibration environments. The wiper is connected to the moving object being measured through threads, chamfers, or spring returns. Some potentiometric sensors have other components, such as guide rails (to enhance stability during wiper sliding) and wave washers for antibacklash control. The housing holds the components in place and protects the sensor from harsh environments.

The major considerations in a linear potentiometric sensor design include

- The length of the stroke to be measured
- Power rating and resistance value
- The space limitations

- The quality of the conductive element
- The means of connecting the sensor to the moving device being measured

2.2.3.2 Rotary Potentiometers

There are two different designs of rotary potentiometers: *single-turn* and *multiturn*. A single-turn potentiometer rotates less than one full revolution (i.e., <360°) to reach its full resistance range. It is often used as an angular position sensor. A multiturn potentiometer rotates more than one revolution (e.g., 5, 10, 20, or 25 turns) to reach its full resistance range, and it is often used where a higher resolution adjustment or multiple revolutions are required. Multiturn potentiometers are more expensive, but very stable with high precision. Similar to linear potentiometers, both single-turn and multiturn potentiometers have three terminals, a fixed resistor, a rotary wiper, a shaft, and the housing. The wire-wound resistor in a rotary pot often uses a material that has a lower temperature coefficient to improve the stability. A conductive plastic is sometimes molded over the wire-wound resistor to reduce inductance and protect the resistor.

The major considerations in a single-turn or multiturn rotary potentiometer design are

- Angle or rotation to be measured
- Power rating and resistance value
- Resolution. Choose a single-turn design if resolutions above 10% are acceptable (less expensive). Choose a multiturn design if higher resolution is required (e.g., 1%—finer control, better linearity and stability)

2.2.4 POTENTIOMETRIC SENSOR APPLICATIONS

Potentiometric sensors are broadly used to measure position, displacement, level, motion, pressure, airflow, and many other physical parameters. They are also integrated into other sensors to monitor chemicals, gases, or biocells. The major advantages of potentiometers are simplicity, low cost, adaptability to many applications, and high output signal level (thus, eliminating the need for signal amplification and conditioning required by many other sensors). Their disadvantages include high hysteresis due to sliding friction, sensitivity to vibration, and finite lifetime associated with wiping elements. Some application examples of potentiometric sensors are described as follows.

2.2.4.1 Potentiometric Pressure Sensors

Figure 2.5a shows a rotary potentiometric pressure sensor developed by *SFIM SAGEM*, France. When the pressure of an input liquid or gas expands the diaphragm, the wiper connected to the diaphragm will sweep across the potentiometer. The movement of the wiper indicates the magnitude of the pressure. Figure 2.5b is a linear potentiometric pressure sensor. When the wiper arm moves up or down due to a pressure change, the resistance varies, causing the output voltage to change. A similar design can also be used to measure water level, motion, or displacement.

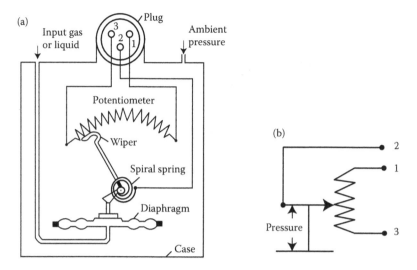

FIGURE 2.5   (a) A rotary potentiometric pressure sensor; (Courtesy of *SFIM SAGEM,* France.) (b) a linear potentiometric pressure sensor.

2.2.4.2   Potentiometric Airflow Sensor

Figure 2.6 shows a potentiometric airflow meter used in Toyota vehicles. It converts the air flow volume to a vane opening angle that is measured by a potentiometer. The sensor's output voltage is then sent to the vehicle's electronic control unit (ECU) to determine the volume of air that is getting into the engine.

2.2.4.3   Potentiometric Gas Sensor

A typical potentiometric gas sensor is shown in Figure 2.7. It consists of a measuring (working) electrode, a gas selective membrane, and a reference electrode—together

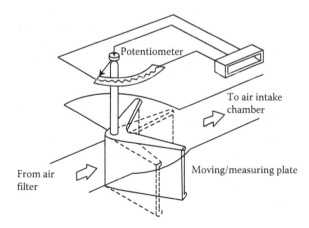

FIGURE 2.6   A potentiometric airflow sensor used in Toyota vehicles. (Courtesy of *Toyota Corporation*, Japan.)

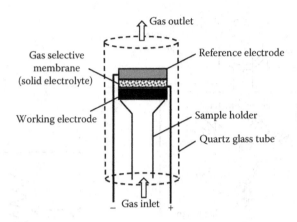

FIGURE 2.7   A potentiometric gas sensor.

TABLE 2.1

Typical Gas Detection Membranes

Membrane	Typical Gas to Be Detected
Glass membrane	CO_2, SO_2, NH_3
Ag_2S membrane	HCN, H_2S
Crystalline LaF_3 membrane	HF

Source: From Bunce, N., *CHEM7234/CHEM 720 Fundamentals of Electrochemistry Lecture Notes*, University of Guelph, Ontario, Canada, Spring 2003. With permission.

---

forming a sandwich with the membrane in the middle. The gas selective membrane binds the gas of interest. The bounded gas then reacts with the analyte on each side of the membrane, causing a change in conductivity of the membrane. This change is indicated by an output voltage change between the two electrodes. By convention, the measuring electrode is considered as the cathode in potentiometric sensors. Table 2.1 shows several typical membranes for gas detection [1].

2.2.4.4   Potentiometric Biosensor

Figure 2.8a is a potentiometric biosensor. It consists of a measuring electrode (Ag-AgCl electrode) and a reference electrode. Both electrodes are placed on a person's skin but at different locations to measure electrical signals generated by the flexion and extension of muscles. Each electrode acts as a transducer that converts ion flow in the body into electron flow (current) in the conductive electrode (see Figure 2.8b). This transduction takes place at the electrode–electrolyte interface where an oxidation or reduction reaction occurs (determining the direction of current

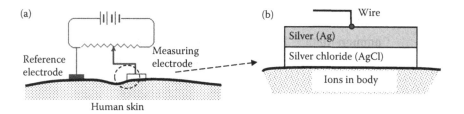

FIGURE 2.8   A potentiometric biosensor (a) and its Ag/AgCl electrodes (b).

flow). The biopotential between the two electrodes is proportional to the size of the muscle and the amount of flexion or extension.

2.3   RESISTIVE TEMPERATURE SENSORS

2.3.1   THERMORESISTIVE EFFECTS

2.3.1.1   Thermoresistive Effect for Metals

The electrical resistance of a metal conductor increases as temperature increases. This is because the electrical conductivity of a metal relies on the movement of electrons through its crystal lattice. Due to thermal excitation, the vibration of electrons increases, which slows the electrons' movement, thus causing the resistance to increase. The relationship between resistance R and temperature T can be expressed by a polynomial equation:

$$R_T = R_0[1 + A(T - T_0) + B(T - T_0)^2 + C(T - T_0)^3 + \cdots] \qquad (2.6)$$

Its simplified version is

$$R_T = R_0[1 + A(T - T_0)] \qquad (2.7)$$

where R_0 is resistance at the reference temperature T_0 (usually either 0°C, 20°C, or 25°C); $A, B, C, \ldots$ are material-dependent temperature coefficients (in $\Omega \cdot \Omega^{-1} \cdot °C^{-1}$). Metals have *positive temperature coefficients* (PTC), because their resistance increases as the temperature increases. All resistance temperature devices (RTDs), made of metals, are PTC sensors. The temperature coefficient for all pure metals is of the order of 0.003–0.007 $\Omega \cdot \Omega^{-1} \cdot °C^{-1}$. Temperature coefficients A for common metals are listed in Table 2.2 [2].

The temperature coefficient of an alloy is often very different from that of the constituent metals. Small traces of impurities can greatly change the temperature coefficients. For example, an alloy of 84% Cu, 12% Mn, and 4% Ni has almost zero response to temperature. Thus, it is used to manufacture precision resistors. Figure 2.9 shows a typical resistance–temperature curve of an RTD.

TABLE 2.2
Common Metals' Temperature Coefficients at 20°C

Metal	Temperature Coefficients A at 20°C ($\Omega \cdot \Omega^{-1} \cdot °C^{-1}$)
Gold	0.003715
Silver	0.003819
Copper	0.004041
Aluminum	0.004308
Tungsten	0.004403
Iron	0.005671
Nickel	0.005866

Source: From Temperature coefficient of resistance, Creative Commons, Stanford, California, USA, 2008. With permission.

EXAMPLE 2.4

A copper wire has a resistance of 5 Ω at 20°C. Calculate its resistance if the temperature is increased to 65°C.

SOLUTION

From Table 2.2, $A = 0.004041\ \Omega \cdot \Omega^{-1} \cdot °C^{-1}$ at 20°C. Use Equation 2.7:

$$R_T = R_0[1 + A(T - T_0)] \Rightarrow R_{65} = (5\ \Omega)[1 + 0.004041\ \Omega \cdot \Omega^{-1} \cdot °C^{-1}(65°C - 20°C)]$$
$$= 5.91\ \Omega$$

A Pt100 RTD (means a platinum RTD with $R_0 = 100\ \Omega$ at 0°C) has a resistance–temperature relationship described by the *Callendar–Van Dusen* equation:

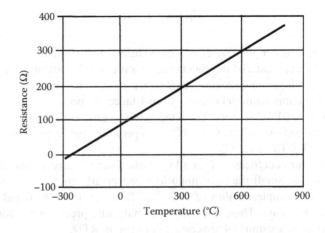

FIGURE 2.9 Resistance–temperature curve for an RTD.

TABLE 2.3
Callendar–Van Dusen Coefficients Corresponding to Standard RTDs

Standard	A'	B'	C' ($C' = 0$ for $T > 0°C$)
DIN 43760	3.9080×10^{-3}	-5.8019×10^{-7}	-4.2735×10^{-12}
American	3.9692×10^{-3}	-5.8495×10^{-7}	-4.2325×10^{-12}
ITS-90	3.9848×10^{-3}	-5.8700×10^{-7}	-4.0000×10^{-12}

Source: From Measuring temperature with RTDs—A tutorial, Application Note 046, National Instruments Corporation, Austin, Texas, USA, 1996. With permission.

$$R_T = R_0[1 + A'T + B'T^2 + C'(T - 100)T^3]_{(-200°C < T < 850°C)} \qquad (2.8)$$

where A', B', and C' are *Callendar–Van Dusen* coefficients. Their values for different ent RTD standards are listed in Table 2.3 [3].

EXAMPLE 2.5

A Pt100 sensor is used to measure the temperature of a chamber. What is its resistance under a −80°C temperature? If the chamber's temperature is increased to +80°C, what is the sensor's new resistance value? Assume the American standard Pt100.

SOLUTIONS

From the second row in Table 2.3, $A' = 3.9692 \times 10^{-3}$ and $B' = -5.8495 \times 10^{-7}$. Thus:

For $T = -80°C$, $C' = -4.2325 \times 10^{-12}$:

$$R_{-80} = R_0[1 + A'T + B'T^2 + C'(T - 100)T^3]$$

$$R_{-80} = 100[1 + (3.9692 \times 10^{-3})(-80) + (-5.8495 \times 10^{-7})(-80)^2$$
$$+ (-4.2325 \times 10^{-12})(-80 - 100)(-80)^3] = 67.83 \ \Omega$$

For $T = 80°C$, $C' = 0$:

$$R_{80} = R_0[1 + A'T + B'T^2] = 100[1 + (3.9692 \times 10^{-3})(80)$$
$$+ (-5.8495 \times 10^{-7})(80)^2] = 131.38 \ \Omega$$

These results show that as the temperature increases, the resistance of the Pt100 RTD increases.

If a sensor's coefficients A, B, and C are not available, one can measure the sensor's resistance values at a number of known temperatures, and then solve for A, B, and C or determine them using the methods shown in Example 2.6 [4].

EXAMPLE 2.6

Determine the temperature coefficients A, B, and C for a Pt100 sensor using the Callendar–Van Dusen equations.

SOLUTION

Measure resistance at the four known temperatures:

R_0 at $T_0 = 0°C$ (the freezing point of water)
R_{100} at $T_{100} = 100°C$ (the boiling point of water)
R_h at T_h—a high temperature (e.g., the melting point of zinc, 419.53°C)
R_l at T_l—a low temperature (e.g., the boiling point of oxygen, −182.96°C)

Step 1: Calculate linear parameter α
The linear parameter α is determined as the normalized slope between 0°C and 100°C:

$$\alpha = \frac{R_{100} - R_0}{100 R_0} \tag{2.9}$$

Then, the resistance at other temperatures can be calculated as

$$R_T = R_0(1 + \alpha T) \tag{2.10}$$

The temperature as a function of the resistance value is

$$T = \frac{R_T - R_0}{\alpha R_0} \tag{2.11}$$

Step 2: Calculate the Callendar constant δ
The Callendar constant, δ (introduced by Callendar), is determined based on the disparity between the actual temperature, T_h, and the temperature calculated in Equation 2.11:

$$\delta = \frac{T_h - (R_h - R_0)/(\alpha R_0)}{((T_h/100) - 1)(T_h/100)} \tag{2.12}$$

With the introduction of δ into the equation, the resistance value R_T at a positive temperature T ($T > 0°C$) can be calculated with a great accuracy:

$$R_T = R_0 + R_0 \alpha \left[T - \delta \left(\frac{T}{100} - 1 \right) \left(\frac{T}{100} \right) \right] \tag{2.13}$$

Step 3: Calculate the Van Dusen constant β
At negative temperatures ($T < 0°C$), Equation 2.12 will still give a small deviation. Van Dusen therefore introduced a term of the fourth order, β (only applicable for $T < 0°C$). β is calculated based on the disparity between the actual temperature, T_l, and the temperature that would result from employing only α and δ:

$$\beta = \frac{T_l - \left[(R_l - R_0)/(\alpha R_0) + \delta((T_l/100) - 1)(T_l/100) \right]}{((T_l/100) - 1)(T_l/100)^3} \tag{2.14}$$

With the introduction of both Callendar and van Dusen constants, the resistance value can be calculated accurately for the entire temperature range (set $\beta = 0$ for $T > 0°C$):

$$R_T = R_0 + R_0\alpha\left[T - \delta\left(\frac{T}{100} - 1\right)\left(\frac{T}{100}\right) - \beta\left(\frac{T}{100} - 1\right)\left(\frac{T}{100}\right)^3\right] \quad (2.15)$$

Step 4: Convert the results to A, B, and C
Conversion can be accomplished by simple coefficient comparison of Equations 2.8 and 2.15, resulting in

$$A = \alpha\left(1 + \frac{\delta}{100}\right) \quad (2.16)$$

$$B = -\frac{\alpha\delta}{100^2} \quad (2.17)$$

$$C = -\frac{\alpha\beta}{100^4} \quad (2.18)$$

2.3.1.2 Thermoresistive Effect for Semiconductors

In semiconductor materials, the valence electrons are bonded in covalent bonds with their neighbors. As temperature increases, thermal vibration of the atoms breaks up some of these bonds and releases electrons. These "free" electrons are able to move through the material under applied electric fields and the material appears to have a smaller resistance. Thus, electrical resistance R of semiconductor materials decreases as temperature T increases. The relationship between R and T is exponential (*Beta Equation*) [5]:

$$R_T = R_0 e^{\left[\beta\left(\frac{1}{T} - \frac{1}{T_0}\right)\right]} \quad (2.19)$$

where R_0 is the resistance at the reference temperature T_0 (in kelvin, K), usually 298 K (25°C), and β is the temperature coefficient (in K) of the material. Since resistance decreases as the temperature increases, β is a *negative temperature coefficient* (*NTC*). Most thermistors (a contraction of the words *therm*al and re*sistor*), made of semiconductor materials, are NTC sensors. A common resistance–temperature (R–T) curve for a 10 kΩ thermistor is shown in Figure 2.10 [6].

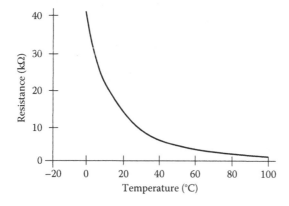

FIGURE 2.10   An NTC thermistor R–T curve.

EXAMPLE 2.7

Find β of a thermistor whose values are $R_0 = 10$ kΩ at $T_0 = 25°C$; $R_{50} = 3.3$ kΩ at $T = 50°C$.

SOLUTION

From Equation 2.19

$$\beta = \frac{\ln R_T - \ln R_0}{1/T - 1/T_0} = \frac{\ln R_{50} - \ln R_{25}}{1/(273.15 + 50) - 1/(273.15 + 25)}$$

$$= \frac{\ln(3300) - \ln(10,000)}{1/(273.15 + 50) - 1/(273.15 + 25)} = 4272.66 \text{ (K)}$$

The R–T curve of a thermistor is highly dependent upon its manufacturing process. Therefore, thermistor curves have not been standardized to the extent that RTD or thermocouple curves have been done. The thermistor curve, however, can be approximated with the *Steinhart–Hart equation*, which is an inverse and finer version of the *Beta Equation 2.19* [7]:

$$T = \frac{1}{a + b\ln(R) + c\,[\ln(R)]^3} \tag{2.20}$$

where a, b, and c are constants, normally provided by manufacturers as part of the specification for each thermistor type, or alternatively provided as the R–T tables or curves. a, b, and c can also be determined by calibrating at three different temperatures and solving three simultaneous equations based on Equation 2.20.

The Steinhart–Hart Equation 2.20 has a third-order polynomial term, which provides excellent curve fitting for temperature spans within the range of −80°C to 260°C, and it has replaced the beta equation as the most useful tool for interpolating the NTC thermistor's R–T characteristics. If the full temperature range extends beyond the −80°C to 260°C range, the Steinhart–Hart equation can be used to fit a series of narrow temperature ranges, and then splice them together to cover the full range. Table 2.4 shows a typical thermistor data sheet (from *YSI Inc.*, Yellow Springs, Ohio, USA, for its 44004 thermistor).

EXAMPLE 2.8

Three temperature points are selected for calibrating a thermistor. The resistances at each point are 7355 Ω at 0°C, 1200 Ω at 40°C, and 394.5 Ω at 70°C, respectively. (1) Find the constants a, b, and c. (2) What is the temperature if $R = 2152$ Ω?

SOLUTIONS

1. Plug the three sets of Rs and Ts into Equation 2.20:

$$(0 + 273) = \frac{1}{a + b\ln(7355) + c\,[\ln(7355)]^3}$$

$$\Rightarrow 273a + 2430.56b + 192660.03c - 1 = 0 \tag{1}$$

TABLE 2.4

Specifications of the 44004 Thermistor by *YSI Inc.*

Parameter	Specification
Resistance at 25°C	2252 Ω (100 Ω to 1 MΩ available)
Measurement range	−80°C to +120°C typical (250°C maximum)
Interchangeability (tolerance)	±0.1°C or ±0.2°C
Stability over 12 months	<0.02°C at 25°C, <0.25°C at 100°C
Time constant	<1.0 s in oil, <60 s in still air
Self-heating	0.13°C · (mW)$^{-1}$ in oil, 1.0°C · (mW)$^{-1}$ in air
Coefficients	$a = 1.4733 \times 10^{-3}, b = 2.372 \times 10^{-3}, c = 1.074 \times 10^{-7}$
Dimensions	Ellipsoid bead 2.5 mm × 4 mm

$$(40 + 273) = \frac{1}{a + b\ln(1200) + c\,[\ln(1200)]^3}$$

$$\Rightarrow 313a + 2219.19b + 111557.09c - 1 = 0 \qquad (2)$$

$$(70 + 273) = \frac{1}{a + b\ln(394.5) + c\,[\ln(394.5)]^3}$$

$$\Rightarrow 343a + 2050.32b + 73262.01c - 1 = 0 \qquad (3)$$

Solving the three simultaneous Equations (1) through (3) yields

$$a = 1.47408 \times 10^{-3} \quad b = 2.3704159 \times 10^{-4} \quad c = 1.0839894 \times 10^{-7}$$

2. When $R = 2152 \, \Omega$, the temperature is

$$T = \frac{1}{1.47408 \times 10^{-3} + 2.3704159 \times 10^{-4} \ln(2152) + 1.0839894 \times 10^{-7}\,[\ln(2152)]^3}$$

$$= 299.21 \, \text{K} \left(\text{or } 26.21°\text{C} \right)$$

2.3.2 WIEDEMANN–FRANZ LAW FOR METALS

The empirical law of Wiedemann and Franz, proposed in 1853, states that the ratio of the thermal conductivity κ (in W · m^{-1} · K^{-1}) to the electrical conductivity σ_p (in S · m^{-1}) of a metal is proportional to its absolute temperature T (in K) [8]:

$$\kappa/\sigma_p = \textit{Ł}T \qquad (2.21)$$

where $\textit{Ł}$ is the *Lorenz Number* (in W. Ω. K^{-2}). This relationship is based on the fact that both heat and electricity transport are affected by the free electrons in the metal. Thermal conductivity increases with the average particle velocity since this increases the forward transport of energy. Electrical conductivity, on the other hand, decreases when particle velocity increases because the collisions divert the electrons from forward transport of charge.

TABLE 2.5
Lorenz Numbers for Different Metals

| Metal | Lorenz Number £ (W · Ω · K⁻²) | |
	273 K (0°C)	373 K (100°C)
Copper (Cu)	2.23×10^{-8}	2.33×10^{-8}
Silver (Ag)	2.31×10^{-8}	2.37×10^{-8}
Lead (Pb)	2.47×10^{-8}	2.56×10^{-8}
Iridium (Ir)	2.49×10^{-8}	2.49×10^{-8}
Tin (Sn)	2.52×10^{-8}	2.49×10^{-8}

Source: Kittel, C., *Introduction to Solid State Physics*, 5th
Ed., New York: Wiley, 1976, p. 178.

Lorenz number can be determined theoretically by the following equation [9,10]:

$$£ = \frac{\kappa}{\sigma_\rho T} = \frac{\pi^2 k_B^2}{3q^2} = 2.44 \times 10^{-8}\, W \cdot \Omega \cdot K^{-2} \tag{2.22}$$

where k_B is Boltzmann constant (1.38065×10^{-23} J·K⁻¹) and q is the electron charge
(1.602×10^{-19} C). Lorenz number can also be determined experimentally. Some experimentally obtained Lorenz numbers for different metals are shown in Table 2.5.

2.3.3   Resistance Temperature Devices (RTDs)

2.3.3.1   RTD Characteristics

RTDs are designed based on the thermoresistive effect of metals. Although any metal
could be used to measure temperature, the metal selected should have a high melting
point and be corrosion resistant. Materials most commonly used for RTDs are platinum, copper, nickel, and molybdenum because of their chemical stability, highly
reproducible electrical properties, availability in a near pure form (which ensures
consistency in the manufacturing process), and a very predictable, near-linear R–T
relationship. Table 2.6 shows the features of these metals.

TABLE 2.6
Common RTD Sensor Materials and Their Characteristics

Metal	Temperature Range (°C)	A (Ω · Ω⁻¹ · °C⁻¹)	Comments
Platinum (Pt)	−240 ~ +850	0.00385	Good precision, broad temperature range
Nickel (Ni)	−80 ~ +260	0.00672	Low cost, limited temperature range
Copper (Cu)	−200 ~ +260	0.00427	Low cost, applied in measuring the temperature of electric motor and transformer windings
Molybdenum (Mo)	−200 ~ +200	0.00300 or 0.00385	Lower cost, alternative to platinum in the lower temperature ranges, ideal material for film-type RTDs

Platinum is by far the most common RTD material because of its long-term stability in air, broad temperature range, ease of manufacture, and reasonable cost. In fact, platinum RTDs are used to define the *International Practical Temperature Scale* (IPTS) from the melting point of hydrogen (−259.34°C) to the melting point of silver (+961.78°C). Platinum RTDs are available with alternative nominal resistance R_0 values at 0°C of 10, 25, and 100 Ω. Pt100 is dominant with the practical range from −240°C to 850°C, although special versions are available for up to 1000°C.

Nickel RTDs are preferred in cost-sensitive applications such as air conditioning and consumer goods. They are generally manufactured in higher resistance values of 1 or 2 kΩ with a simple two-wire connection (rather than the three- or four-wire connections common with platinum types). Nickel is less chemically inert than platinum and thus is less stable at higher temperatures. Nickel RTDs are normally used in the environmental temperature range and in clear air.

RTDs are recognized for their excellent linearity throughout their temperature range, while maintaining a high degree of accuracy, robustness, long-term stability, and repeatability. Some RTDs have accuracies as high as 0.01 Ω at 0°C. Common industrial RTDs drift less than 0.1°C per year, and others are stable and drift within 0.0025°C per year.

2.3.3.2 RTD Measurement

An RTD is a passive device, requiring a current to pass through to produce a measurable voltage. If the excitation current passing through an RTD is I_{ex}, and the output voltage across the RTD is V_{out}, then the measured temperature T (in °C) can be obtained by [11]

$$T = \frac{2(V_{out} - I_{ex}R_0)}{I_{ex}R_0[A + \sqrt{A^2 + 4B(V_0 - I_{ex}R_0)/(I_{ex}R_0)}]} \tag{2.23}$$

where R_0 is the resistance of the RTD at 0°C; A and B ($C = 0$ in this case when $T > 0°C$) are the *Callendar–Van Dusen Coefficients*. The excitation current also causes the RTD to heat internally (self-heating), which can result in a measurement error. Self-heating is typically specified as the amount of power that will raise the RTD's temperature by 1°C (in mW · °C^{-1}). To minimize self-heating-caused error, the smallest possible excitation current (1 mA or less) should be used in measurement. The amount of self-heating also depends greatly on the medium in which the RTD is immersed. For example, an RTD can self-heat up to 100 times higher in still air than in moving water. In addition, the temperature across the RTD must be uniform. Otherwise, an error will appear, which is directly related to the difference in temperature between the leads of the RTD.

EXAMPLE 2.9

A 0.15 mA excitation current is passed through a Pt100 RTD manufactured to a DIN 43760 standard. If the voltage output is 33 mV, what is the temperature measured (assume $T > 0°C$)?

SOLUTION

From Table 2.3, $A = 3.9080 \times 10^{-3}$ and $B = -5.8019 \times 10^{-7}$, and $C = 0$, Thus

$$T = \frac{2(V_{out} - I_{ex}R_0)}{I_{ex}R_0[A + \sqrt{A^2 + 4B(V_0 - I_{ex}R_0)/(I_{ex}R_0)}]}$$

$$= \frac{2(0.033 - 0.15 \times 10^{-3} \times 100)}{(0.15 \times 10^{-3} \times 100)\left[3.9080 \times 10^{-3} + \sqrt{\begin{array}{l}(3.9080 \times 10^{-3})^2 + 4(5.8019 \times 10^{-7})(0.033 \\ -0.15 \times 10^{-3} \times 100)/(0.15 \times 10^{-3} \times 100)\end{array}}\right]}$$

$$= 322.85\,^{\circ}C$$

RTDs can be difficult to measure due to their relatively low resistance (e.g., 100 Ω), which changes only slightly with temperature ($<0.4\ \Omega \cdot {}^{\circ}C^{-1}$). To accurately measure these small changes in resistance, special configurations should be used to minimize errors from lead wire resistance. Section 6.5.2 of Chapter 6 discusses how to correctly wire the RTD to minimize measurement error.

2.3.3.3 RTD Design

2.3.3.3.1 RTD Constructions

RTDs are constructed in two forms: *wire-wound* (Figure 2.11a) and *thin film* (Figure 2.11b). Wire-wound RTDs are made by winding a very fine strand of metal wire (platinum, typically 0.0005–0.0015 in. diameter) into a coil and packaged inside a ceramic mandrel, or wound around the outside of a ceramic housing and coated with an insulating material to prevent the sensor from shorting. Larger lead wires (typically 0.008–0.015 in. diameter) are connected to the wound wires. Wire-wound RTDs provide superior interchangeability and stability to the highest temperatures.

Thin-film RTDs are produced using *thin-film lithography* that deposits a thin film of metal (e.g., 1 μm platinum) onto a ceramic substrate through the *cathodic atomization* or *sputtering* process [11]. The metal is deposited in a specific pattern and

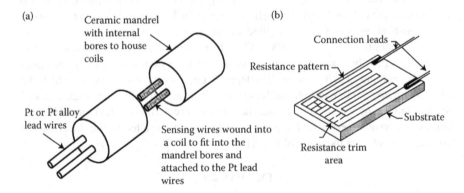

(a) Ceramic mandrel with internal bores to house coils

Pt or Pt alloy lead wires

Sensing wires wound into a coil to fit into the mandrel bores and attached to the Pt lead wires

(b) Connection leads

Resistance pattern

Substrate

Resistance trim area

FIGURE 2.11 (a) Wire-wound RTD; (b) thin-film RTD. (Courtesy of *RdF Corporation*, Hudson, New Hampshire, USA.)

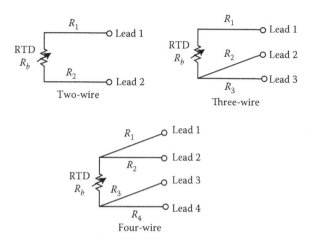

FIGURE 2.12   Two-, three-, and four-wire configurations.

trimmed by a laser to its correct resistance value. The elements are coated with a glass-like material for mechanical and moisture protection. Primary advantages of thin-film sensors are: (1) a greater resistance can be placed in a smaller area in more versatile shapes and designs. 100, 500, and 1000 Ω thin films are available; (2) they can be made much smaller than their wire-wound counterparts and can achieve higher sensitivity; (3) they have higher resolution per degree of temperature, and errors due to lead wire resistance are minimized; (4) they are relatively inexpensive; and (5) they can also function as strain gauges. However, film sensors are less accurate than wire types.

2.3.3.3.2   RTD Wiring Configurations

RTDs come with two-, three-, or four-lead wires per element (Figure 2.12). A two-wire RTD is the least expensive, but the lead wire resistance unavoidably affects the measurement results. Two-wire RTDs are mostly used with short lead wires or where high accuracy is not required. Three-wire RTDs remove the lead wire resistance from the measurement by adding a third lead wire. This works by measuring the resistance between lead 1 and lead 2 (R_{1+2}) and subtracting the resistance between lead 2 and lead 3 (R_{2+3}), which leaves just the resistance of the RTD bulb (R_b) assuming that lead wires 1, 2, and 3 are all the same resistance. Three-wire configurations are most commonly used in industrial applications. When long distances exist between the sensors and measurement instruments, significant savings can be made by using the three-wire configuration instead of the four-wire configuration [12]. Errors caused by resistance imbalance between leads can be cancelled out in a four-wire RTD circuit that is similar to a Wheatstone bridge—using wires 1 and 4 to power the circuit and wires 2 and 3 to read. Four-wire RTDs are used where superior accuracy is critical [3].

2.3.3.4   RTD Applications

Besides measuring temperature, RTDs are also used to construct other sensors, for example, bolometers and gas mass flow meters. A bolometer (Figure 2.13a) consists of a miniature RTD and a resistor to measure electromagnetic radiation range

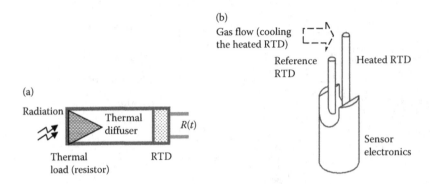

FIGURE 2.13   (a) Internal structure of a bolometer; (b) an oxygen gas flow meter.

from infrared to microwaves. When electromagnetic radiation presents, the resistor absorbs the radiation and converts it into heat. This temperature elevation is then measured by the RTD sensor, whose resistance change indicates the radiation level. Bolometers are broadly used in infrared temperature detection and imaging, high power measurement, microwave device testing, radiofrequency (RF) antenna beam profiling, and medical microwave heat monitoring. The oxygen flow meter shown in Figure 2.13b is designed by *FOX Thermal Instruments, Inc.*, Marina, California. It consists of two RTDs: a reference RTD and a heated RTD—both RTDs are constructed of reference-grade platinum wire wound around ceramic mandrels and then inserted into stainless-steel or Hastelloy tubes. The reference RTD measures the gas temperature. The other RTD is heated to a constant temperature differential (ΔT) above the gas temperature and the cooling effect of the gas flow is measured. The electrical power required to maintain a constant temperature differential ΔT between the two RTDs is directly proportional to the gas mass flow rate.

2.3.4 THERMISTORS

2.3.4.1 Thermistor Characteristics

Thermistors are either NTC type, whose resistance decreases with increasing temperature, or PTC type, whose resistance increases with increasing temperature. Each type has unique features and distinct advantages.

NTC thermistors are more commonly used than PTC type, especially in temperature measurement applications. NTC thermistors are made using basic ceramics technology and semiconductor metal oxide materials (e.g., oxides of manganese, nickel, cobalt, iron, copper, and titanium). In some thermistors, the decrease in resistance is as great as 6% for each 1°C of temperature increase although 1% changes are more typical. NTC thermistors can provide good accuracy and resolution when measuring temperatures between −100°C and +300°C. If inserted into a Wheatstone bridge, a thermistor can detect temperature changes as small as ±0.005°C.

PTC thermistors can be divided into two categories: thermally sensitive silicon resistors (*silistors*) and switching PTC thermistors. Silistors exhibit a fairly uniform PTC (about +0.0077°C^{-1}) through most of their operational range, but can also exhibit an NTC region at temperatures higher than 150°C. This type of thermistors is often used for temperature compensation of silicon semiconducting devices in the range of −60°C to +150°C. The switching PTC thermistors are made from polycrystalline ceramic materials that are normally highly resistive but become semiconductive by adding dopants. They are often manufactured using compositions of barium, lead, and strontium with additives such as yttrium, manganese, tantalum, and silica. The R–T curves of switching PTC thermistors exhibit very small NTC regions until they reach a critical temperature T_c—"*Curie*," "*Switch*" or "*Transition*" *temperature*. After T_c, the curve exhibits a rapidly increasing PTC resistance. These resistance changes can be as much as several orders of magnitude within a temperature span of a few degrees. Figure 2.14 illustrates R–T curves of both silistor and switching PTC thermistors [13].

Most PTC thermistors' applications are based on either the steady-state self-heated condition (voltage–current characteristic) or the dynamic self-heated condition (current–time characteristic) or a combination of both. For example, the dramatic rise in the resistance of a PTC at and above the transition temperature makes it ideal for over-current protection (resettable fuses). If an over-current condition occurs, the thermistor will self-heat beyond the transition temperature and its resistance rises dramatically. This causes the current in the overall circuit to be reduced. The key characteristics of NTC and PTC thermistors include [13,14]:

- *Resistance temperature (R–T) characteristics*: Describe how resistance changes as temperature changes in a thermistor. Most thermistor manufacturers provide tables, R–T curves, and coefficients for their thermistor products.
- *Resistance tolerance*: Specifies the standard tolerances available for each thermistor type.

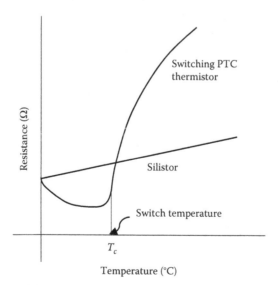

FIGURE 2.14 R–T curves of silistor and switching PTC thermistors.

- *Beta tolerance*: Is the tolerance of a thermistor's beta value β, determined by the composition and structure of the metal oxides being used in the sensor. For bead-type thermistors, beta tolerances are in the order of ±1% to ±5%; for metallized surface contact type, beta tolerances range from ±0.5% to ± 3%.
- *Heat capacity*: Is the product of the specific heat and mass of the thermistor. It represents the amount of heat required to produce a change in the body temperature of the thermistor by 1°C.
- *Dissipation constant*: Is the ratio of the change in power applied to a thermistor to the resulting change in body temperature due to self-heating. It is affected by lead wire materials, method of mounting, ambient temperature, and the shape of the thermistor.
- *Resistance range*: Defines the minimum and maximum resistance values at the reference temperature.
- *Transition temperature*: Is the Curie point at which the PTC thermistor's R–T curve begins to increase sharply. PTC manufacturers often define this temperature as the point where a specified ratio exists between the minimum resistance (or at 25°C zero-power resistance) and the transition temperature resistance. For example, *Thermometrics Inc.* specifies the point where the resistance is twice (2×) the minimum value, whereas other manufacturers might use 10 times (10×) the minimum.

The main advantages of thermistors for temperature measurement are: (1) Extremely high sensitivity. For example, a 2252 Ω thermistor has a sensitivity of −100 Ω · °C^{-1} at room temperature. Higher resistance thermistors can exhibit a sensitivity of −10 kΩ · °C^{-1} or more. In comparison, a 100 Ω platinum RTD has a sensitivity of only 0.4 Ω · °C^{-1}. (2) Very fast response to temperature changes. (3) Relatively high resistance. Thermistors are available with base resistances (at 25°C) ranging from hundreds to millions of ohms. This high resistance diminishes the effect of lead wires that can cause significant errors with low resistance devices such as RTDs. The high resistance and high sensitivity of thermistors make their measurement circuitry and signal conditioning much simpler. No special three-wire, four-wire, or Wheatstone bridge configurations are necessary, although using a Wheatstone bridge can improve linearity of thermistors.

The major disadvantages of thermistors are their high nonlinearity and limited temperature range (typically below 300°C). Figure 2.15 shows the R–T curve for a 2252 Ω thermistor. The curve of a 100 Ω RTD is also shown for comparison [15].

Table 2.7 compares the main characteristics of RTDs, NTC thermistors, and thermocouples.

2.3.4.2 Thermistor Design

Thermistors have two nonpolarized terminals. Based on the method by which these terminals are attached to the ceramic body, thermistors are classified into *bead* and *metallized surface contact* types. A bead-type thermistor has platinum alloy lead wires (about 0.5–5 mm in diameter) that are directly sintered into the ceramic body.

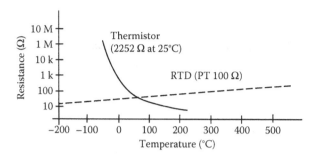

FIGURE 2.15 Comparison of R–T curves for a thermistor and an RTD. (From Potter, D., Measuring temperature with thermistors—A tutorial, Application Note 065, National Instrument Corporation, Austin, Texas, USA, 1996. With permission.)

The metallized type has metallized surface contacts (with or without radial or axial leads) for surface or spring mounting. Each type can be further characterized by differences in geometry, packaging, and/or processing techniques as shown in Table 2.8. Figure 2.16 illustrates the circuit symbol and the most common forms of thermistors.

Thermistors are often mounted in stainless-steel tubes to protect them from harsh environments during their operation. Thermoconductive grease or silicon sealant is typically used to improve the thermal contact between the sensor and the tube (see Figure 2.17). To maintain the temperature tolerance within ±0.05°C to ±1°C, thermistors are laser trimmed during the manufacturing process. Thermistors are mechanically simple and strong, providing the basis for a high reliability sensor. They are available in a large range of sizes, base resistance values, and R–T curves.

The fabrication of NTC thermistors uses basic ceramics technology: a mixture of two or more metal oxide powders combined with suitable binders and formed into a desired shape, dried, and sintered at an elevated temperature. By varying the types of oxides used, sintering temperature and atmosphere, a wide range of resistances and temperature coefficients can be obtained.

TABLE 2.7
Comparison of Characteristics for RTDs, NTC Thermistors, and Thermocouples

Characteristics	RTD	NTC Thermistor	Thermocouple
Measured parameters	Resistance	Resistance	Voltage
Resolution	Poor	Good	Moderate
Linearity	Linear	Nonlinear	Nonlinear
Temperature range	−250°C ~ 850°C	−100°C ~ 300°C	0°C ~ 1600°C
Current source	Necessary	Necessary	Not necessary
Compensation for environments	Not necessary	Not necessary	Necessary
Response	Relatively slow	Fast	Fast
Cost	Expensive	Inexpensive	Varies

Source: The data in the table are compiled based on several manufacturers' data sheets.

TABLE 2.8

Types of Bead and Metallized Surface Contact NTC Thermistors

Bead Thermistors	Metallized Surface Contact Thermistors
Bare beads	Disks
Glass-coated beads	Chips (wafers)
Ruggedized beads	Surface mounts
Miniature glass probes	Flakes
Glass probes	Rods
Glass rods	Washers
Bead-in-glass enclosures	

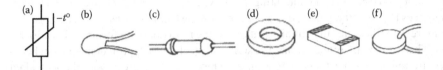

FIGURE 2.16   Thermistors: (a) circuit symbol for NTC type (with negative sign); (b) bead; (c) rod; (d) washer; (e) surface mount; (f) disk.

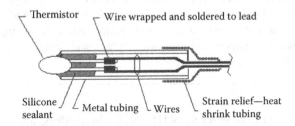

FIGURE 2.17   Internal structure of a bead-type thermistor.

2.3.4.3 Thermistor Applications

NTC thermistors have been primarily used for high-resolution temperature measurements. With high sensitivity, reliability, low price, ruggedness, and ease of use, NTC thermistors have a variety of applications such as in home appliances, automobiles (to monitor coolant or oil temperature), mobile telecommunication, computers (to monitor the temperature of battery packs while charging), overheating detection in electronic equipment, medical care, and other industrial usage. PTC thermistors are often used as current-limiting devices for circuit protection

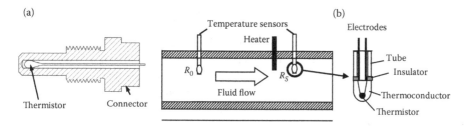

FIGURE 2.18 A thermistor for (a) water temperature measurement; (b) flow rate measurement.

as fuses. They are also used as heating elements in small temperature-controlled ovens (e.g., crystal oven).

Figure 2.18a shows a thermistor used to measure water temperature in Toyota automobile engines [16]. Figure 2.18b is a thermoanemometer for flow rate measurement. Two thermistor-type of temperature sensors, R_0 and R_S, are immersed into a moving fluid (air or liquid). R_0 measures the initial temperature of the fluid. A heater, located between R_0 and R_S, heats the fluid and its temperature is then measured by R_S. The flow rate to be measured is thus proportional to the heat loss rate.

2.4 PHOTORESISTIVE SENSORS

2.4.1 PHOTORESISTIVE EFFECT

When electromagnetic radiation, for example, infrared light, visible light, or ultraviolet (UV) light, strikes a photoconductive material, the resistance of the material decreases. This occurs because the electrons in the valence band of the photoconductive material are excited by the light and move to the conduction band, which increases the material's conductivity. The amount of the resistance change depends on the light density. Most photoresistive sensors are made of semiconductor materials, such as cadmium sulfide (CdS). The conductivity of a semiconductor, σ_p (in $S \cdot m^{-1}$, S represents siemens) is described by [17]

$$\sigma_p = q\,(\mu_e n + \mu_h p) \tag{2.24}$$

where q is the charge of an electron (1.602×10^{-19} C); μ_e and μ_h are the mobility of the free electrons and holes (in meters square per volt·second, $m^2\,V^{-1} \cdot s^{-1}$), respectively; and n and p are the density of electrons and holes or *carrier concentration* (in number of free electrons or holes per cubic meters, m^{-3}), respectively. σ_p multiplied by the number of generated carriers per second per unit of volume will result in the conductance. Both mobility and carrier concentration are temperature dependent. Thus, Equation 2.24 is actually a function of temperature:

$$\sigma_p = q[\mu_e(T)n(T) + \mu_h(T)p(T)] \tag{2.25}$$

The temperature dependence of the carriers' mobility is influenced by two basic scattering mechanisms: *lattice scattering* and *impurity scattering*. In lattice scattering, lattice vibrations cause the mobility to decrease with increasing temperature (with an approximate temperature dependence of $T^{-3/2}$), while in impurity scattering, the mobility of the carriers increases with increasing temperature (with an approximate temperature dependence of $T^{3/2}$). The total mobility μ_t is the sum of the lattice-scattering mobility μ_l and the impurity-scattering mobility μ_i governed by *Mattheisen's rule*:

$$\frac{1}{\mu_t} = \frac{1}{\mu_l} + \frac{1}{\mu_i} \tag{2.26}$$

$$\mu_l = C_l T^{-\frac{3}{2}} \tag{2.27}$$

$$\mu_i = C_i T^{\frac{3}{2}} \tag{2.28}$$

where C_l and C_i are constants. Since impurity scattering typically presents at very low temperatures only, the influence of lattice scattering alone will be considered at the normal temperature range.

In an intrinsic (pure or undoped) semiconductor, the number of electrons in the conduction band is always equal to the number of holes they leave behind in the valence band, that is, $n = p = n_i$, or

$$np = n_i^2 \tag{2.29}$$

Equation 2.29 is called the *Mass Action Law*. n_i is a function of temperature. At 300 K, silicon has a value of $n_i = 1.4 \times 10^{10}$ to 1.5×10^{10} cm^{-3}, gallium arsenide has 1.8×10^6 cm^{-3}, and germanium has 2.4×10^{13} cm^{-3}. *The mass action law* is valid for extrinsic materials as well. In an extrinsic or impure semiconductor, the material is doped with atoms from column III or V of the periodic table, resulting in a p-type or n-type semiconductor, respectively. Adding different atoms improves the conductivity of the semiconductor. If n_d is the concentration of donor atoms, and n_a is the concentration of acceptor atoms, then the majority and minority carrier concentrations can be calculated by

For $n_d < n_a$ (n-type semiconductor):

$$n = \frac{1}{2}\left[(n_d - n_a) + \sqrt{(n_d - n_a)^2 + 4n_i^2}\right]$$
$$p = n_i^2/n \tag{2.30}$$

For $n_d < n_a$ (p-type semiconductor):

$$p = \frac{1}{2}\left[(n_a - n_d) + \sqrt{(n_a - n_d)^2 + 4n_i^2}\right]$$

$$n = n_i^2/p$$

(2.31)

If $(n_d - n_a) \gg n_i$ (n-type semiconductor), then

$$n = n_d - n_a$$

$$p = n_i^2/(n_d - n_a)$$

(2.32)

If $(n_a - n_d) \gg n_i$ (p-type semiconductor), then

$$p = n_a - n_d$$

$$n = n_i^2/(n_a - n_d)$$

(2.33)

EXAMPLE 2.10

A germanium sample at $T = 300$ K has $n_d = 5 \times 10^{13}$ cm^{-3} and $n_a = 0$. If $n_i = 2.4 \times 10^{13}$ cm^{-3}, determine the electron and hole concentrations.

SOLUTION

Since $n_d > n_a$

$$n = \frac{1}{2}\left[(n_d - n_a) + \sqrt{(n_d - n_a)^2 + 4n_i^2}\right]$$

$$= \frac{1}{2}\left[(5 \times 10^{13} \text{ cm}^{-3} - 0 \text{ cm}^{-3}) + \sqrt{(5 \times 10^{13}\text{cm}^{-3} - 0 \text{ cm}^{-3})^2 + 4(2.4 \times 10^{13}\text{cm}^{-3})^2}\right]$$

$$= 5.97 \times 10^{13} \text{ cm}^{-3}$$

$$p = n_i^2/n = (2.4 \times 10^{13} \text{ cm}^{-3})^2/(5.97 \times 10^{13} \text{ cm}^{-3}) = 9.66 \times 10^{12} \text{ cm}^{-3}$$

EXAMPLE 2.11

Consider a silicon semiconductor at 300 K, $n_d = 10 \times 10^{15}$ cm^{-3} and $n_a = 3 \times 10^{15}$ cm^{-3}. Assume that $n_i = 1.5 \times 10^{10}$ cm^{-3}, determine the electron and hole concentrations in this n-type semiconductor.

SOLUTION

Since $(n_d - n_a) \gg n_i$, the majority carrier electron concentration is

$$n = n_d - n_a = 10 \times 10^{15} \text{ cm}^{-3} - 3 \times 10^{15} \text{ cm}^{-3} = 7 \times 10^{15} \text{ cm}^{-3}$$

$$p = n_i^2/(n_d - n_a) = (1.5 \times 10^{10})^2/(10 \times 10^{15} - 3 \times 10^{15}) = 3.21 \times 10^4 \text{ cm}^{-3}$$

Therefore, when $(n_d - n_a) \gg n_i$, the majority carrier (electron) concentration is approximately the difference between the donor and acceptor concentrations.

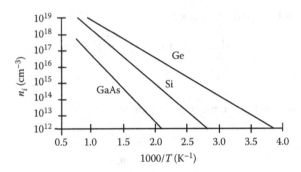

FIGURE 2.19   Temperature dependence of intrinsic carrier concentration n_i in Ge, Si, and GaAs.

The temperature dependence of the *intrinsic* carrier concentration n_i is expressed by

$$n_i = n_0 T^{3/2} \exp\left(\frac{-V_g}{2V_T}\right) \tag{2.34}$$

where n_0 is a constant, whose value depends on the material; T is the absolute temperature in K; V_g is the semiconductor bandgap voltage; and V_T is the thermal voltage. For silicon, $V_g = 1.11$ V at $T = 300$ K. The intrinsic concentration indicates the density of free electrons or holes. The thermal voltage V_T is related to temperature by

$$V_T = \frac{k_B T}{q} \tag{2.35}$$

again, k_B is the Boltzmann constant and q is the charge of an electron. Equation 2.34 can also be expressed as

$$n_i = n_0 T^{3/2} \exp\left(\frac{-V_g}{2V_T}\right) = n_s \exp\left(\frac{-E_g}{2k_B T}\right) \tag{2.36}$$

where $n_s = n_0 T^{3/2}$; $E_g(=qV_g)$ is the minimum energy gap between the bottom of the conduction band and the top of the valence band. The temperature dependence of the intrinsic carrier concentrations in Ge, Si, and GaAs is shown in Figure 2.19.

EXAMPLE 2.12

Find the thermal voltage when $T = 300$ K.

SOLUTION

$$V_T = \frac{k_B T}{q} = \frac{(1.381 \times 10^{-23}\ \text{K}^{-1})(300\ \text{K})}{1.602 \times 10^{-19}\ \text{C}} = 25.86\ \text{mV}$$

EXAMPLE 2.13

The temperature of a pure silicon strip is 300 K. After absorbing photons from a light striking it, its temperature increases to 312 K. By what factor does the intrinsic concentration n_i increase? Assume the bandgap voltage at both temperatures is $V_g = 1.11$ V.

SOLUTION

At 300 K

$$V_{300} = \frac{k_B T}{q} = \frac{(1.381 \times 10^{-23} \text{J} \cdot \text{K}^{-1})(300 \text{ K})}{1.602 \times 10^{-19} \text{C}} = 0.0259 \text{ V}$$

At 312 K

$$V_{312} = \frac{k_B T}{q} = \frac{(1.381 \times 10^{-23} \text{J} \cdot \text{K}^{-1})(312 \text{ K})}{1.602 \times 10^{-19} \text{C}} = 0.0269 \text{ V}$$

From Equation 2.34

$$\frac{(n_i)_{T=312}}{(n_i)_{T=300}} = \frac{n_0 (312)^{3/2} \exp(-1.11/2 \times 0.0269)}{n_0 (300)^{3/2} \exp(-1.11/2 \times 0.0259)} = 2.352$$

EXAMPLE 2.14

A pure silicon strip is 1 cm long and has a diameter of 1 mm. At room temperature, the intrinsic concentration in the silicon is $n_i = 1.5 \times 10^{16}$ m^{-3}. The electron and hole mobilities are $\mu_e = 0.13$ m$^2 \cdot$ V$^{-1} \cdot$ s^{-1} and $\mu_n = 0.05$ m$^2 \cdot$ V$^{-1} \cdot$ s^{-1}, respectively. Find (1) the conductivity σ_p of the silicon and (2) the resistance R of the strip.

SOLUTIONS

Since it is an intrinsic (pure) silicon, $n = p = n_i$, thus the conductivity can be calculated by Equation 2.24

$$\sigma_p = q (\mu_e n + \mu_n p) = q n_i(\mu_e + \mu_n)$$
$$= (1.602 \times 10^{-19} \text{ C})(1.5 \times 10^{16} \text{m}^{-3})(0.13 \text{ m}^2 \cdot \text{V}^{-1} \cdot \text{s}^{-1} + 0.05 \text{ m}^2 \cdot \text{V}^{-1} \cdot \text{s}^{-1})$$
$$= 4.325 \times 10^{-4} \text{ S} \cdot \text{m}^{-1} \text{ or } 4.325 \times 10^{-4} \text{ } \Omega^{-1} \cdot \text{m}^{-1}$$

The resistance can be found using Equation 2.1:

$$R = \rho \frac{L}{A} = \frac{L}{\sigma_p A} = \frac{0.01 \text{ m}}{4.325 \times 10^{-4} \text{ } \Omega^{-1} \cdot \text{m}^{-1} \times \pi(0.5 \times 10^{-3} \text{ m})^2}$$
$$= 29.44 \text{ M}\Omega$$

A photoresistor acts as a variable resistor R_2 in the circuit shown in Figure 2.2b. An increase in light intensity will correspond to a decrease in resistance. The voltage output across the photoresistor is shown in Equation 2.3. This output voltage V_{out} can be sent to an A/D (analog-to-digital) converter circuit to map V_{out} into a light intensity level ranging from 0 to 255.

2.4.2 Photoresistor Characteristics

A photoresistor is also called a *light-dependent resistor* (LDR), *photoconductor*, or *photocell* since its resistance changes as incident light intensity changes. The relationship between the resistance and light intensity can be described by the characteristic curve of a photoresistive sensor (see Figure 2.20a, for an ISL2902 CdS photoresistor from *Festo Didactic*, Hauppauge, NY). The sensor's spectral response (see Figure 2.20b) is about 550 nm (yellow to green region of visible light). When placed in the dark, its resistance is as high as 1 MΩ and then falls to 400 Ω when exposed to bright light. Table 2.9 shows the ISL2902 CdS photoresistor's datasheet. CdS sensors are of very low cost. They are often used in autodimming, darkness, or twilight detection for turning street lights ON and OFF, and for photographic exposure meters.

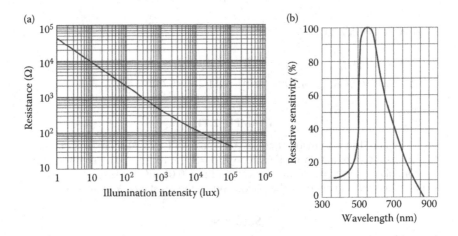

FIGURE 2.20   (a) Characteristic curve of ISL2902; (b) spectral response of ISL2902.

TABLE 2.9
ISL2902 CdS Photoresistor Datasheet

Maximum power dissipation at 30°C	250 mW
Maximum current	75 mA
Maximum peak voltage	320 V
Typical electrical resistance with illumination by tungsten lamp of 2854 K at an ambient temperature of 25°C	400 Ω (at 1000 lux) 9000 Ω (at 10 lux)
Dark resistance	>1 MΩ
Rise time: From darkness to 1000 lux	2.8 ms
From darkness to 10 lux	18 ms
Fall time: From 1000 lux to 10-fold resistance	48 ms
From 10 lux to 10-fold resistance	120 ms

Source: From *Photoresist 160495/14*, Festo Didactic, Hauppauge, NY.

Key performance characteristics of photoresistive sensors are as follows [18,19]:

- *Responsivity R_d*: the ratio of detector output to light input. It measures the effectiveness of the detector in transducing electromagnetic radiation to electrical voltage or current. If the sensor's output is voltage, R_d is the ratio of the root mean square (RMS) of the output voltage V_{RMS} to the incident radiant power Φ_e (in watts):

$$R_d = V_{RMS}/\Phi_e \tag{2.37}$$

If the sensor's output is current, R_d is the ratio of the RMS of the output current I_{RMS} to the incident radiant power Φ_e (in watts):

$$R_d = I_{RMS}/\Phi_e \tag{2.38}$$

- *Spectral response curve*: a plot of the sensitivity as a function of wavelength as shown in Figure 2.20b.
- *Noise equivalent power (NEP)*: the minimum detectable signal level defined as the radiant power that produces an output voltage equal to the noise voltage of the sensor:

$$NEP = \frac{E_e A_d}{V_s/V_n} \tag{2.39}$$

where E_e is the power density at the surface of the sensor in $W \cdot cm^{-2}$, A_d is the sensitive area of the photodetector in cm^2, and V_S/V_n is the signal-to-noise ratio. NEP has a unit watt (W).

- *Detectivity D^**: a measure of the intrinsic merit of a sensor material. It is a function of the sensitive area of the photodetector A_d (cm^2), bandwidth of the measuring system B (Hz), and NEP (W):

$$D^* = \frac{\sqrt{A_d B}}{NEP} \tag{2.40}$$

The unit of the detectivity D^* is $cm \cdot Hz^{1/2} \cdot W^{-1}$; D^* is often used to compare different types of detectors. The higher the value of D^*, the better the detector. Manufacturers often list D^* followed by three numbers in parentheses, for example, D^* (850, 900, 5), meaning that the measurement was made at a wavelength of 850 nm, with a chopping frequency of 900 Hz and a bandwidth of 5 Hz.

- *Quantum efficiency (QE)*: the effectiveness of a photodetector in producing electrical current when exposed to radiant energy. QE (in percentage) can be described by

$$QE = \frac{\text{Number of electrons ejected}}{\text{Number of incident photons}} \times 100\% \tag{2.41}$$

If over a period of time, an average of 10,000 photoelectrons are emitted as the result of the absorption of 100,000 photons of light energy, then the quantum efficiency will be 10%.

2.4.3 PHOTORESISTIVE SENSOR DESIGN

Figure 2.21a and b illustrates the typical construction of a photoresistor and its circuit symbol. To increase "dark" resistance values and reduce "dark" current, the resistive path is often designed as a zigzag pattern across the ceramic substrate.

Materials used in photoresistors include cadmium sulfide (CdS), lead sulfide (PbS), cadmium selenide (CdSe), lead selenide (PbSe), and indium antimonide (InSb). CdS is the most sensitive photoresistor to visible light. Its resistance value can change from many megaohms in the dark to several kiloohms when exposed to light. PbSe is the most efficient in near-infrared light photoresistor. CdS, PbSe, and CdSe can be made to operate at light levels of 10^{-3}–10^3 footcandles. Table 2.10 shows the main characteristics of a PGM1200 CdS photoresistor made by *Token Electronics Industry Corporation, Ltd*, Taiwan.

Photoresistors, compared to photodiodes or phototransistors, respond relatively slow to light changes. For example, a photoresistor cannot detect the characteristic blinking of fluorescent lamps (turning ON and OFF at the 60 Hz power line frequency), but a phototransistor (which has a frequency response up to 10,000 Hz) can. If both sensors are used to measure the same fluorescent light, the photoresistor would show the light to be always ON and the phototransistor would show the light to be blinking ON and

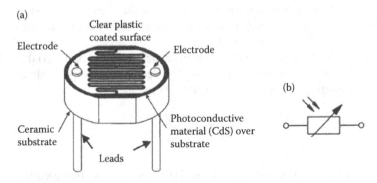

FIGURE 2.21 A typical construction (a) and circuit symbols (b) of a photoresistor.

TABLE 2.10

Characteristics of PGM1200 CdS Photoresistor

Ambient temperature	−30°C ~ +70°C
Spectral peak	560 nm
Photo (light) resistance (at 10 lux)	2 kΩ ~ 5 kΩ
Dark resistance	1.0 MΩ
Response time (rise/decay)	30 ms/40 ms
Dimension (diameter)	5 mm

OFF. Thus, phototransistors can be used to detect an incandescent lamp that acts as a timing start indicator. Photocells are commonly used to find certain objects through measuring the reflectivity of a light source such as a red LED (light-emitting diode), but they are sensitive to ambient lighting and usually need to be shielded.

2.4.4 PHOTORESISTIVE SENSOR APPLICATIONS

Photoresistors are generally low cost, small size, fast response, high sensitivity, and ease of use. They are broadly applied in light, radiation, and fire detectors; motion sensing; light intensity measurement; and inventory bar code reading. More application examples are listed in Table 2.11.

Figure 2.22 shows an electronic bagpipe [20]. Underneath each hole on the bagpipe's chanter is a photoresistor. Photoresistors were chosen, instead of capacitive contacts or piezoelectric strips, because (1) photoresistors can be mounted below the holes on the chanter to avoid direct finger contact and provide a natural feel; (2) light in this case can mimic the air traveling through the pipe; (3) the photoresistive-based electronic "keys" provide a full range of tones (such as pitch bends) when a hole is partially covered in a slurring motion.

Figure 2.23a shows a photoresistive flame detector. It contains an *anode* and a UV-sensitive photoresistive component (usually as a *cathode*). When UV light from a flame is present at the photocathode, photoelectrons are excited and emitted from the cathode and move toward the anode under the voltage provided by the batteries. A readout circuit measures charges moving toward the anode to indicate the presence of the flame. Figure 2.23b is a light-controlled LED circuit. It will turn the LED ON when the CdS photoresistor is exposed to light, or turn the LED OFF when the photoresistor is blocked from the light. The control mechanism is the variation of the transistor's base voltage V_{BE} or base current I_B. Once the transistor is ON or

---

TABLE 2.11
Application Examples of Photoresistive Sensors

Light control	Automatic gain control
Position sensor	Automatic rear view mirror
Automatic iris control in camera	Automatic headlight dimmer
Camera exposure control	Night light control
Auto slide focus	Oil burner flame out indicator
Colorimetric test equipment	Absence/presence (beam breaker)
Densitometer	Isolated circuit
Electronic scales—dual cell	Photocopy machine—density of toner

---

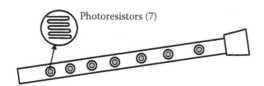

Photoresistors (7)

FIGURE 2.22   Photoresistors under each key hole of an electronic bagpipe.

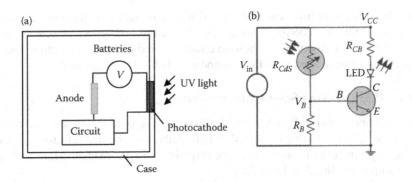

FIGURE 2.23 (a) A UV flame detector; (b) a light-controlled LED circuit.

ACTIVE, the collector current I_C controls the LED. The LED could be replaced with a relay or motor to actuate other circuits or devices.

EXAMPLE 2.15

In the light-controlled LED circuit shown in Figure 2.23b, if the supplied voltage $V_{CC} = V_{in} = 10$ V, the LED current $I_{LED} = 20$ mA, the current-limiting resistor $R_{CB} = 220$ Ω, and the base resistor $R_B = 7.5$ kΩ, find the maximum resistance of the photoresistor R_{CdS} to turn the LED ON. Assume that the LED's voltage drop is 2 V, the transistor's saturation voltage between the collector and emitter V_{CE} is 0.2 V, and the gain of the transistor $h_{fe} = 100$.

SOLUTION

To turn on the LED, the transistor must be ON, which requires a base voltage V_{BE} greater or equal to 0.7 V:

$$V_{BE} = V_B - V_E \geq 0.7 \text{ V} \quad \Rightarrow \quad V_B \geq 0.7 \text{ V (since } V_E = 0 \text{ V, grounded)}$$

$$I_C = 100 I_B \Rightarrow \frac{V_{CC} - V_{LED} - V_{CE}}{R_{CB}} = 100 \left(\frac{V_{in} - V_B}{R_{CdS}} - \frac{V_B - 0}{R_B} \right)$$

$$\frac{10 \text{ V} - 2 \text{ V} - 0.2 \text{ V}}{220 \text{ Ω}} = 100 \left(\frac{10 \text{ V} - 0.7 \text{ V}}{R_{CdS}} - \frac{0.7 \text{ V}}{7.5 \times 10^3 \text{ Ω}} \right) \Rightarrow R_{CdS} = 20.765 \text{ kΩ}$$

Thus, the maximum resistance of the photoresistor R_{CdS} must be 20.765 kΩ for the LED to be ON.

2.5 PIEZORESISTIVE SENSORS

2.5.1 PIEZORESISTIVE EFFECT IN METALS AND ALLOYS

The fundamental principle of piezoresistive sensors comes from Equation 2.1:

$$R = \rho \frac{l}{A}$$

Taking natural logarithm on both sides yields:

$$\ln(R) = \ln(\rho) + \ln(l) - \ln(A)$$

The differential of the aforementioned equation becomes:

$$\frac{dR}{R} = \frac{d\rho}{\rho} + \frac{dl}{l} - \frac{dA}{A} \tag{2.42}$$

The first term on the right side of the equation, $d\rho/\rho$, is due to changes in specific resistivity of the material. Its physical effect is governed by the electrical conduction mechanism in the solid; thus, its magnitude is very different in metals and metal alloys compared to that of semiconducting materials.

Let $A = \pi D^2/4$, where D is the diameter of a wire:

$$\ln(A) = \ln(\pi/4) + 2\ln(D)$$

The differential of this equation results:

$$\frac{dA}{A} = 2\frac{dD}{D}$$

where $dD/D = \varepsilon_D$ is the *transverse* or *lateral strain*. Since the *longitudinal strain* is $dl/l = \varepsilon$, and *Poisson's ratio* is $\upsilon = -\varepsilon_D/\varepsilon$, Equation 2.42 becomes:

$$\frac{dR}{R} = \frac{d\rho}{\rho} + (1 + 2\upsilon)\varepsilon \tag{2.43}$$

Equation 2.43 expresses the basic relationship between resistance and strain. A measure of the sensitivity of the material (i.e., its resistance change per unit of applied strain) is defined as the *gauge factor*:

$$\text{Gauge factor (GF)} = \frac{dR/R}{\varepsilon} \tag{2.44}$$

Thus, the GF of Equation 2.43 is

$$\text{GF} = (1 + 2\upsilon) + \frac{d\rho/\rho}{\varepsilon} \tag{2.45}$$

The first term $(1 + 2\upsilon)$ directly relates to the *Poisson effect*—the tendency in an elastic material (e.g., metal) to contract laterally in response to axial stretching. The second term represents the changes in specific resistivity of the material in response to the applied strain. In the absence of a direct resistivity change (i.e., $d\rho = 0$), the

GF would be $1 \leq GF \leq 2$ since theoretically υ has a range $0 \leq \upsilon \leq 0.5$. More details on GF will be discussed in Section 2.5.3.

EXAMPLE 2.16

A gauge has GF 2.0 and resistance 120 Ω. Find dR when the gauge is subjected to a strain of (1) 5 microstrain in aluminum and (2) 5000 microstrain in aluminum.

SOLUTIONS

According to Equation 2.44

1. $dR = GF \varepsilon R = 2 \ (5 \times 10^{-6}) \ (120 \ \Omega) \Rightarrow 0.0012 \ \Omega$ (0.001% change)
2. $dR = GF \varepsilon R = 2 \ (5000 \times 10^{-6}) \ (120 \ \Omega) = 1.2 \ \Omega$ (1% change)

2.5.2 PIEZORESISTIVE EFFECT IN SEMICONDUCTORS

Piezoresistive effect in silicon and germanium was discovered by Charles Smith in 1954 [21]. He found that both p-type and n-type silicon and germanium exhibited much greater piezoresistive effect than metals. Piezoresistivity of silicon arises from the deformation of the energy bands as a result of applied stress. In turn, the deformed bands affect the effective mass and the mobility of electrons and holes, hence modifying resistivity or conductivity.

To understand how the resistance change relates to the applied stress, consider an infinitesimally small cubic piezoresistive crystal element with normal stresses σ_{xx}, σ_{yy}, and σ_{zz} along the cubic crystal axes x, y, and z, respectively, and three shear stresses τ_{yz}, τ_{zx}, and τ_{xy}, as indicated in Figure 2.24. The piezoresistive effect in this case can be described by relating the resistance change ΔR to each of the six stress components using a matrix of 36 coefficients, π_{ij}, expressed in Pa^{-1}, as shown in Equation 2.46.

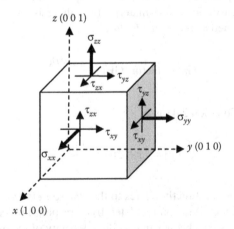

FIGURE 2.24 Definition of the normal stresses σ_i and shear stresses τ_i ($i = 1, 2, 3$).

$$
\frac{1}{R}
\begin{bmatrix}
\Delta R_{xx} \\
\Delta R_{yy} \\
\Delta R_{zz} \\
\Delta R_{yz} \\
\Delta R_{zx} \\
\Delta R_{xy}
\end{bmatrix}
=
\underbrace{
\begin{bmatrix}
\pi_{11} & \pi_{12} & \pi_{13} & \pi_{14} & \pi_{15} & \pi_{16} \\
\pi_{21} & \pi_{22} & \pi_{23} & \pi_{24} & \pi_{25} & \pi_{26} \\
\pi_{31} & \pi_{32} & \pi_{33} & \pi_{34} & \pi_{35} & \pi_{36} \\
\pi_{41} & \pi_{42} & \pi_{43} & \pi_{44} & \pi_{45} & \pi_{46} \\
\pi_{51} & \pi_{52} & \pi_{53} & \pi_{54} & \pi_{55} & \pi_{56} \\
\pi_{61} & \pi_{62} & \pi_{63} & \pi_{64} & \pi_{65} & \pi_{66}
\end{bmatrix}}_{\Pi}
\underbrace{
\begin{bmatrix}
\sigma_{xx} \\
\sigma_{yy} \\
\sigma_{zz} \\
\tau_{yz} \\
\tau_{zx} \\
\tau_{xy}
\end{bmatrix}}_{\sigma}
\tag{2.46}
$$

$$\underbrace{\phantom{\Delta R_{xx}}}_{\Delta \vec{R}}$$

where the vector $\Delta \vec{R}$ represents the change in resistance with corresponding stress components, R is the original resistance, and Π is a 6×6 *piezoresistive coefficient matrix*. If the coordinate axes coincide with the crystal axes, a cubic crystal has three independent, nonvanishing elastic components, π_{11}, π_{12}, and π_{44}, that is

$$\pi_{11} = \pi_{22} = \pi_{33}$$

$$\pi_{12} = \pi_{21} = \pi_{13} = \pi_{31} = \pi_{23} = \pi_{32}$$

$$\pi_{44} = \pi_{55} = \pi_{66}$$

Thus, the Π matrix becomes

$$
\Pi =
\begin{bmatrix}
\pi_{11} & \pi_{12} & \pi_{12} & 0 & 0 & 0 \\
\pi_{12} & \pi_{11} & \pi_{12} & 0 & 0 & 0 \\
\pi_{12} & \pi_{12} & \pi_{11} & 0 & 0 & 0 \\
0 & 0 & 0 & \pi_{44} & 0 & 0 \\
0 & 0 & 0 & 0 & \pi_{44} & 0 \\
0 & 0 & 0 & 0 & 0 & \pi_{44}
\end{bmatrix}
\tag{2.47}
$$

π_{11} and π_{12} are associated with the normal stress components, whereas the coefficient π_{44} is related to the shearing stress components. By expanding the aforementioned matrix equation (for convenience, using 1—xx, 2—yy, 3—zz, 4—yz, 5—zx, and 6—xy)

$$\Delta R_1 / R = \pi_{11}\sigma_1 + \pi_{12}(\sigma_2 + \sigma_3)$$

$$\Delta R_2 / R = \pi_{11}\sigma_2 + \pi_{12}(\sigma_1 + \sigma_3)$$

$$\Delta R_3 / R = \pi_{11}\sigma_3 + \pi_{12}(\sigma_1 + \sigma_2)$$

$$\Delta R_4 / R = \pi_{44}\tau_1$$

$$\Delta R_5 / R = \pi_{44}\tau_2$$

$$\Delta R_6 / R = \pi_{44}\tau_3$$

The actual values of these three coefficients, π_{11}, π_{12}, and π_{44}, depend on the angles of the piezoresistor with respect to silicon crystal lattice. The values of these coefficients in $\langle 100 \rangle$ orientation at room temperature (25°C) are given in Table 2.12 [22,23].

TABLE 2.12

Resistivity and Piezoresistive Coefficients of Silicon at 25°C

Materials	$\pi_{11} \cdot 10^{-11}$ Pa^{-1}	$\pi_{12} \cdot 10^{-11}$ Pa^{-1}	$\pi_{44} \cdot 10^{-11}$ Pa^{-1}
p-Silicon, $\rho = 7.8\ \Omega \cdot$ cm	+6.6	−1.1	+138.1
n-Silicon, $\rho = 11.7\ \Omega \cdot$ cm	−102.2	+53.4	−13.6
p-Germanium, $\rho = 16.6\ \Omega \cdot$ cm	−5.2	−5.5	−69.4
n-Germanium, $\rho = 9.9\ \Omega \cdot$ cm	−4.7	−5.0	−69.0

Source: From Hsu, T.R., *MEMS & Microsystems: Design and Manufacture, and Nanoscale Engineering,* 2nd ed., John Wiley Sons, Inc., Hoboken, New Jersey, 2008; Sze, S.M. (Editor), *Semiconductor Sensors,* John Wiley and Sons, Inc., Hoboken, New Jersey, 1994. With permission.

In practical applications, a thin strip of silicon is commonly used to make a strain gauge sensor, instead of a three-dimensional cube. In this case, change in resistance versus in-plane stresses in the longitudinal (parallel to the current) direction and transverse (perpendicular to the current) direction can be expressed as

$$\frac{\Delta R}{R} = \pi_L \sigma_L + \pi_T \sigma_T \tag{2.48}$$

where ΔR and R are the change in resistance and the original resistance, respectively; σ_L and σ_T are the longitudinal and transverse stress, respectively; π_L and π_T are the piezoresistive coefficient along the longitudinal and transverse direction, respectively.

EXAMPLE 2.17

A pressure sensor die contains four identical p-type piezoresistors (Figure 2.25). Resistors A and C are subjected to the longitudinal stress σ_L, and Resistors B and D are subjected to the transverse stress component σ_T. Assume that the square-shaped diaphragm is under uniform pressure loading at the top surface, $\sigma_L = \sigma_T = 186.8$ MPa, $\pi_L = \pi_T = 2.762 \times 10^{-11}$ Pa^{-1}. Estimate $\Delta R/R$.

SOLUTION

Since the square diaphragm is subjected to a uniform pressure load at the top surface, the bending moments are normal to all piezoresistors and are equal in magnitudes. Thus, for each resistor

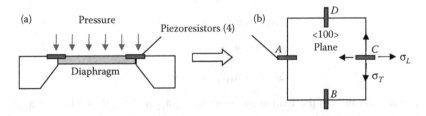

FIGURE 2.25   A semiconductor pressure sensor die: (a) side view; (b) top view.

TABLE 2.13

Longitudinal and Transverse Piezoresistive Coefficients for Various Combinations of Directions in Cubic Crystals

Longitudinal Direction, π_L		Transverse Direction, π_T	
$\langle 1\,0\,0\rangle$	π_{11}	$\langle 0\,1\,0\rangle$	π_{12}
$\langle 0\,0\,1\rangle$	π_{11}	$\langle 1\,1\,0\rangle$	π_{12}
$\langle 1\,1\,1\rangle$	$1/3(\pi_{11} + 2\pi_{12} + 2\pi_{44})$	$\langle 1\,\bar{1}\,0\rangle$	$1/3(\pi_{11} + 2\pi_{12} - \pi_{44})$
$\langle 1\,1\,\underline{0}\rangle$	$1/2(\pi_{11} + \pi_{12} + \pi_{44})$	$\langle 1\,1\,1\rangle$	$1/3(\pi_{11} + 2\pi_{12} - \pi_{44})$
$\langle 1\,1\,\underline{0}\rangle$	$1/2(\pi_{11} + \pi_{12} + \pi_{44})$	$\langle 0\,0\,1\rangle$	π_{12}
$\langle 1\,1\,0\rangle$	$1/2(\pi_{11} + \pi_{12} + \pi_{44})$	$\langle 1\,\underline{1}\,0\rangle$	$1/2(\pi_{11} + \pi_{12} - \pi_{44})$

Source: From Sze, S.M. (Editor), *Semiconductor Sensors*, John Wiley and Sons, Inc., Hoboken, New Jersey, 1994. With permission.

$$\frac{\Delta R}{R} = \pi_L \sigma_L + \pi_T \sigma_T = 2 \times (2.762 \times 10^{-11}\ Pa^{-1})(186.8 \times 10^6\ Pa)$$

$$= 0.01032\ or\ 1.032\%$$

Since the values of three piezoresistive coefficients π_{11}, π_{12}, and π_{44} (defined in a coordinate system aligned to the $\langle 100\rangle$ axis of the silicon crystal) are known, all the piezoresistance coefficients of silicon in an arbitrary Cartesian system can be determined using coordinate system transformation. Table 2.13 lists longitudinal and transverse piezoresistance coefficients for various practical directions in cubic crystals [23].

When the piezoresistors are fabricated, their orientation with respect to the silicon crystal is usually in the $\langle 110\rangle$ direction. From the last row of Table 2.13, the longitudinal piezoresistive coefficient in the $\langle 110\rangle$ direction is $\pi_L = 1/2(\pi_{11} + \pi_{12} + \pi_{44})$ and the transverse coefficient in the $\langle 110\rangle$ direction is $\pi_T = 1/2(\pi_{11} + \pi_{12} - \pi_{44})$. Based on Table 2.12, π_{44} is more significant than π_{11} and π_{12} for *p*-type silicon resistors; thus Equation 2.48 is simplified for *p*-type silicon resistors as

$$\frac{\Delta R}{R} \approx \frac{\pi_{44}}{2}(\sigma_L - \sigma_T) \tag{2.49}$$

Similarly for *n*-type silicon resistors, π_{44} can be neglected:

$$\frac{\Delta R}{R} \approx \frac{\pi_{11} + \pi_{22}}{2}(\sigma_L + \sigma_T) \tag{2.50}$$

2.5.3 CHARACTERISTICS OF PIEZORESISTIVE SENSORS

The core element of a piezoresistive sensor is the strain gauge. The characteristics of a strain gauge are mainly defined by the gauge dimensions, resistance, gauge factor, temperature coefficient, resistivity, and thermal stability.

Gauge dimensions and shape are very important in choosing a right type of strain gauge for a given application.

Gauge resistance is defined as the electrical resistance measured between two metal tabs or leads. Gauge resistance is an important design and application parameter since it determines both the output signal amplitude of the gauge ($\Delta V/V = \Delta R/R$) and the dissipation power (V^2/R).

Gauge factor or *strain sensitivity* is defined as the ratio of ($\Delta R/R$) and the strain ε (Equation 2.44). The real GF plots of common gauge materials are shown in Figure 2.26 [24], where the GF is the slope of the curve. Both Ferry alloys and Constantan alloys have relatively high and constant GF values, indicating a well-behaved and consistent pattern. The 10% rhodium–platinum alloy exhibits a desirable and high GF feature between 0% and 0.4% range of strain ε, but its performance degrades above 0.4% strain ε point. Pure nickel even demonstrates a negative GF for small strain ($\varepsilon < 0.5\%$). Table 2.14 presents the gauge factors and ultimate elongations for several materials.

The GF values of semiconductor materials are much larger than the GF values of metals. Therefore, the majority of piezoresistive strain gauges used today are made of semiconductor materials. Table 2.15 gives the typical GF range of main types of strain gauges.

Hysteresis of a strain gauge is defined as the ratio (in percent) of the difference between the output signals of the gauge (obtained with increasing and decreasing strain loading at identical strain values) divided by the maximum output signal. Figure 2.27 shows the hysteresis of a strain-gauge-type blood pressure sensor.

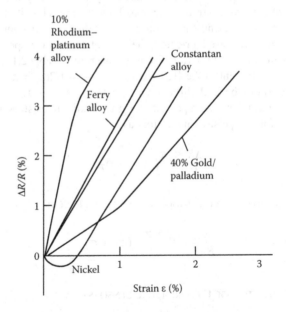

FIGURE 2.26 GF plots for various strain gauge element materials. (From Craig, J.I., *AE3145 Resistance Strain Gage Circuits*, Course Materials, Georgia Institute of Technology, Atlanta, GA, 2000. With permission.)

TABLE 2.14
Gauge Factor and Ultimate Elongation for Several Materials

Material	Gauge Factor (GF)		Ultimate Elongation (%)
	For Low Strain	For High Strain	
Copper	2.6	2.2	0.5
Constantan	2.1	1.9	1.0
Platinum	6.1	2.4	0.4
Silver	2.9	2.4	0.8
40% Gold/palladium	0.9	1.9	0.8

Source: From Craig, J.I., *AE3145 Resistance Strain Gage Circuits, Course Materials*, Georgia Institute of Technology, Atlanta, 2000. With permission.

TABLE 2.15
Typical GF Range of Main Types of Strain Gauges

Type of Strain Gauge	Gauge Factor (GF)
Metal foil	1 ~ 5
Thin-film metal	≈ 2
Bar semiconductor	80 ~ 150
Diffused semiconductor	80 ~ 200

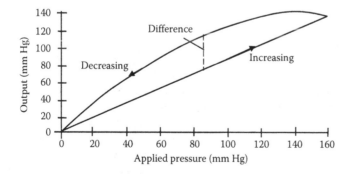

FIGURE 2.27 Hysteresis of a blood pressure sensor. (From Sensitivity of Strain Gauge Wire Materials, *efunda*. Available at (retrieved on June 12, 2013): www.efunda.com/designstandards/sensors/strain_gages/strain_gage_sensitivity.cfm. With permission.)

Creep is defined as the relative variation of the measured strain over time, Δt, when the gauge is under a constant stress:

$$\text{Creep}\,(\%) = \frac{\Delta \varepsilon / \varepsilon}{\Delta t} \times 100\% \qquad (2.51)$$

For instance, *Vishay's* Transducer-Class® strain gauge sensors have a creep less than ±0.02% of full scale (FS) during a 20-minute test. Creep is caused by the noni-deal elastic behaviors of piezoresistors and adhesive materials when bonded to the measured object. Most gauges can be adjusted in design to exhibit either a positive or a negative creep under load. Spring element materials in piezoresistive sensors exhibit only positive creep under load.

Temperature characteristics of strain gauges are often described by two coefficients:

1. *Temperature coefficient of resistance* (TCR), defined as the relative resistance variation of the measuring element per degree of temperature variation:

$$\text{TCR} = \left.\frac{\Delta R / R}{\Delta T}\right|_{\text{Free}} \qquad (2.52)$$

"Free" means the measuring element is unbounded or unembedded into any other material. TCR is an intrinsic characteristic of the gauge material; hence it is a basic design criterion.

2. *Temperature coefficient of sensitivity* (TCS), defined as the relative variation of the gauge factor GF per degree of temperature variation:

$$\text{TCS} = \frac{\Delta \text{GF} / \text{GF}}{\Delta T} \qquad (2.53)$$

Figure 2.28 shows the performance of a uniaxial metal foil strain gauge tested on a mild steel.

Other characteristics of strain gauges include *maximum permitted RMS excitation voltage* (the maximum RMS value of the applied voltage), *maximum elongation* (the strain value at which the linearity deviation exceeds ±5%), *linearity* (the maximum difference between the actual output signal and the ideal output signal under the same strain), *resolution* (the smallest strain variation that can be resolved by the gauge), *fatigue life* (the number of load cycles supported by the gauge without significantly changing its characteristics), *frequency response* (the maximum sinusoidal strain variation frequency that can be resolved by the gauge), and *smallest bending radius* (the smallest value of the radius that the gauge will withstand in bending in one direction without significant changes in its characteristics).

Other important characteristics that must be considered when selecting a strain gauge include its stability. This is because the most desirable strain gauge materials

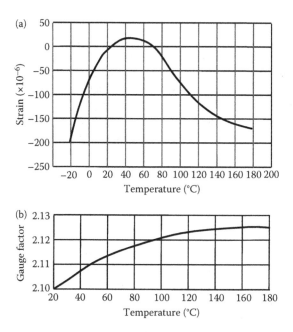

FIGURE 2.28 A foil strain gauge's (a) strain–temperature curve; (b) gauge factor–temperature curve.

are also sensitive to temperature variations and tend to change resistance as they age. Therefore, temperature and drift compensation must be included in many strain gauges.

2.5.4   Piezoresistive Sensor Design

2.5.4.1   Types and Structures of Strain Gauges

Types of strain gauges include wire strain gauges, foil strain gauges, single-crystal semiconductor strain gauges, and thin-film strain gauges.

2.5.4.1.1   Wire Strain Gauges

A wire strain gauge is the original type of resistive strain gauge. The first bonded, metallic wire-type strain gauges were developed in 1938. They are still extensively used in high-temperature environments today. The wire is made of metals or their alloys with a typical diameter of 6 ~ 30 µm. Common wire materials are CrNi alloys for standard applications and PtW alloys for high-temperature applications. The measuring grid is made by either flat winding or wrapping around a metallic wire, and is then bonded to, or completely embedded in, the substrate or carrier. The carrier or backing materials are reinforced epoxies (for standard applications), self-adhesive glass fiber-reinforced Teflon ("peel-off" backing, for high-temperature applications), and sheet brass or polycarbonate (for encapsulated strain gauges in rough ambient conditions).

2.5.4.1.2 Metal Foil Strain Gauges

A foil strain gauge (see Figure 2.29) is the most widely used type. A very thin metal foil pattern (2~5 μm thick, usually Constantan or Nichrome V) is deposited onto a thin insulating backing or carrier (10~30 μm thick, usually epoxy, polyimide, or polycarbonate). The measuring grid pattern including the metallic terminal tabs is produced by the photoetching process. The entire gauge is typically 5~15 mm long.

The main advantages of foil strain gauges over wire gauges are their better heat dissipation, low transverse sensitivity, better flexibility (the smallest bending radius is 0.3 mm), and easy creep compensation (the positive creep of the elastic sensing element can be compensated by the negative creep of the carrier material of the gauge). The disadvantages include limited working temperature due to properties of the carrier materials and adhesives, as well as technical limitations in miniaturization.

2.5.4.1.3 Single-Crystal Semiconductor Strain Gauges

This type of strain gauge is manufactured from a thin strip of semiconductor cut from a single crystal of silicon or germanium, doped with accurate amounts of impurities to obtain either *n*- or *p*-type. The output of a semiconductor gauge is very high compared to a wire or foil gauge. Semiconductor gauges can provide both positive and negative gauge factors depending on whether the gauges are *n*-type or *p*-type. The typical gauge factor of a semiconductor is −100 ~ +170, although −115 ~ +205 are achievable. The output of semiconductor gauges is usually nonlinear with strain (*p*-type gauges have better linearity in tension while *n*-type gauges are more linear in compression), but they exhibit no creep or hysteresis and have an extremely long fatigue life. Semiconductor gauges are highly sensitive to temperature; thus they require a high level of temperature compensation. They also present large TCR due to their specific resistance–temperature curves. Semiconductor gauges are widely used in small sensors such as force, acceleration, and pressure sensors since their sensing elements can be micromachined out of a single piece of silicon. Figure 2.30 illustrates a semiconductor strain gauge—a resistance strip (fabricated from a semiconductor single crystal) fixed on an insulating substrate that supports both the strip and the terminals on its surface.

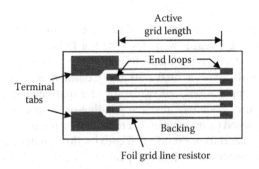

FIGURE 2.29 Metal foil strain gauge construction.

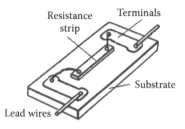

FIGURE 2.30   A single-crystal semiconductor strain gauge. (United States Patent 3,609,625.)

2.5.4.1.4   Thin-Film Strain Gauges

In thin-film strain gauges, the thin film (resistance element) is produced by sputtering or evaporating metals, alloys, or semiconductors (silicon or germanium) onto the carrier material (as a substrate) in vacuum. Several stages of sputtering and evaporation may be needed to form up to eight layers of material in thin-film polycrystalline structure. The whole surface is finally coated with an appropriate sealant. The final properties of the thin film are strongly influenced by various parameters of the deposition process (e.g., ultimate vacuum, substrate temperature, and rate of film growth). These processes are highly cost effective when strain gauges are produced in large quantities. The structure of a thin-film gauge is shown in Figure 2.31. The gauge factor of a thin-film strain gauge is higher than those of wire or foil gauges, but lower than those of single-crystal semiconductor gauges. For instance, a germanium thin-film strain gauge has a GF of 32~39.

The nominal resistance of each type of strain gauges is shown in Table 2.16.

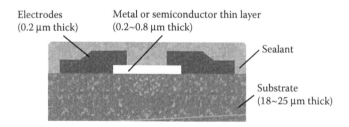

FIGURE 2.31   Structure of a thin-film strain gauge.

TABLE 2.16
Nominal Resistance of Various Strain Gauges

Type of Strain Gauge	Nominal Resistance
Wire gauge	60~350 Ω
Foil or semiconductor gauge	120 Ω~5 kΩ
Thin-film gauge	~10 kΩ

2.5.4.1.5 Gauge Spring Element

A spring element serves as a reacting element for the applied load and provides an isolated and uniform strain field, where the strain gauges are placed. It should exhibit good linearity, low hysteresis, low creep, and low relaxation. A typical sensor spring is designed for 0.1 mm or less of deflection, which requires an extremely low compliance and high-precision spring.

2.5.4.2 Strain Gauge Materials

The types of materials commonly used in strain gauges are summarized as follows [25]. Some of these materials' characteristics are listed in Table 2.17 [26].

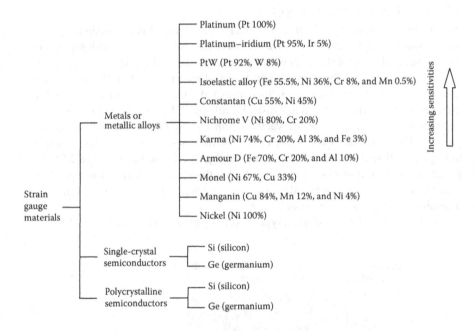

EXAMPLE 2.18

A constantan strain gauge (with a GF of 2.0 and an initial resistance of 350 Ω) is bonded to a 17-4 PH stainless-steel spring element. The spring has both tensile and compressive strains of 0.0015 in./in. at rated pressure. Find (1) the resistance change ΔR; (2) the stress in the sensing element if the elastic modulus E of the 17-4 PH is 28.5×10^6 psi.

SOLUTIONS

Since GF = $(dR/R)/\varepsilon$, $\Delta R = $ GF ε $R = (2)$ $(0.0015)(350 \ \Omega) = 1.05 \ \Omega$.

The stress σ for the 17-4 PH stainless steel spring element can be calculated using $E = \sigma/\varepsilon$:

$$\sigma = \varepsilon E = (0.0015)(28.5 \times 10^6 \text{ psi}) = 42.75 \times 10^3 \text{ psi}$$

TABLE 2.17
Characteristics of Materials Commonly Used in Strain Gauges

Material	GF	Characteristics
		Metallic Alloys
Constantan	≈2	Easy manufacturing when in an array of configurations; easy to solder and compensate with temperature; extremely linear over a wide strain range; displaying a slow but irreversible drift in resistance when over 75°C and high thermal electromotive force in contact with copper
Karma or Nichrome V	2~2.4	Exceptional linearity over a wide strain range; greater resistivity than Constantan; improved stability at elevated temperatures; longer fatigue life; adjustable negative TCR by reducing the sensors' spring modulus with temperature; more expensive to manufacture; difficult to solder
Isoelastic alloy	≈3.5	Exceptionally good fatigue life; not temperature compensable; difficult to solder
PtW	4~5	Exceptionally good fatigue life; operable at high temperatures (up to 800°C)
Armour D	2.6	Good temperature behavior; difficult to obtain in a controlled reproducible manner
		Single-Crystal Semiconductors
Ge or Si	−100~ +180	Good fatigue life; very much sensitive to temperature; narrow strain limit; brittle; high cost
		Polycrystalline Semiconductors
Ge or Si	30~45	Very much sensitive to temperature; much lower prices than single crystal

Source: From Holman, J.P., *Experimental Methods for Engineers*, 7th ed., McGraw-Hill, Boston, 2001, 477. With permission.

2.5.4.3   Supporting Structure and Bonding Methods of Strain Gauges

There are two main structures for holding piezoresistive sensing elements: membrane type (Figure 2.32a, typically in pressure and flow sensors) and cantilever beam type (Figure 2.32b, typically in acceleration sensors).

A membrane-type sensor consists of a thin silicon *membrane* or *plate* with a load (e.g., pressure) differential across the plate. The resulting deformation causes strain along the edges of the plate. Usually the plate, supported by a thicker silicon rim, is fabricated by etching away the bulk silicon in a defined region until the

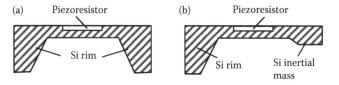

FIGURE 2.32   Cross sections of (a) membrane-type and (b) cantilever beam-type of piezoresistive sensors.

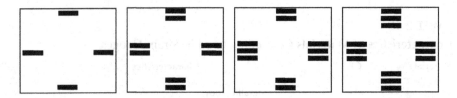

FIGURE 2.33   Various layouts of piezoresistors on a square membrane.

required thickness is reached. Both monocrystalline and polycrystalline silicon can be used for membranes. With careful design, regions along the edge of the plate can be "doped" to create resistors, which will subsequently exhibit resistance change in proportion to the applied strain. Most sensors with square membranes have four or more piezoresistors positioned at the four edges of the membrane and as close to the edge center as possible (see Figure 2.33) where the stresses are maximal. From this central point, stress drops more rapidly toward the center of the membrane than toward the corners. Thus perpendicularly placed (relative to the edge) piezoresistors are less sensitive than the parallelly placed ones.

In beam-type sensors, the stress caused by deflection of the inertial mass under a measurand (e.g., acceleration) is concentrated on the surface of the beam, where piezoresistors are placed.

EXAMPLE 2.19

An accelerometer is made by suspending a 1 mg mass at the free end of a cantilever (1 mm from the support end) and then embedding a silicon strain gauge on the upper surface of the cantilever to measure its deflection (see Figure 2.34). The strain gauge is 100 μm long and 10 μm thick. If the resonant frequency of the cantilever with the mass is 1 kHz, find (1) the deflection of the cantilever; (2) the strain ε of the strain gauge; and (3) the resistance change of the piezoresistor under the strain during 0.001g acceleration ($g = 9.8$ m.s^{-2}). The gauge factor for the silicon strain gauge is 100.

SOLUTIONS

1. At resonant frequency $\omega_n = 1$ kHz, a 1 mg mass with an acceleration $\ddot{x} = 0.001g$ would cause a deflection x of:

$$x = \frac{\ddot{x}}{\omega_n^2} = \frac{0.001 \times 9.8\,\text{m} \cdot \text{s}^{-2}}{\left[2\pi(1 \times 10^3)\,\text{rad} \cdot \text{s}^{-1}\right]^{-2}} = 2.485 \times 10^{-10}\,(\text{m})$$

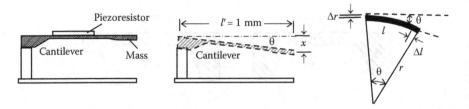

FIGURE 2.34   Acceleration measurement using a cantilever beam-type sensor.

2. The cantilever's deflection angle θ can be found using the trigonometric relation (see Figure 2.34b):

$$\theta = \arcsin\left(\frac{x}{l'}\right) \approx \frac{x}{l'} \text{ (for a small } \theta) = \left(\frac{2.485 \times 10^{-10} \text{ m}}{1 \times 10^{-3} \text{ m}}\right) = 2.485 \times 10^{-7} \text{ rad}$$

Therefore, the radius of curvature of the 100 μm length of cantilever is found by

$$r = \frac{l}{\theta} = \left(\frac{100 \times 10^{-6} \text{ m}}{2.485 \times 10^{-7}}\right) = 4.024 \times 10^2 \text{ (m)}$$

Because of this radius of curvature, the length of the upper surface of the cantilever is greater than the length of the lower surface of the cantilever. The strain ε is then calculated by (note Δl is taken at the middle line of thickness):

$$\varepsilon = \frac{\Delta l}{l} = \frac{\Delta r}{r} = \frac{(10/2) \times 10^{-6}}{4.024 \times 10^2} = 1.243 \times 10^{-8}$$

3. The resistance change can be found by applying Equation 2.44:

$$\frac{\Delta R}{R} = GF \times \varepsilon = (100)(1.243 \times 10^{-8}) = 1.243 \times 10^{-6}$$

This change in resistance is very small, but it is measurable with a resistance bridge circuit.

Figure 2.35 shows four bonding methods for strain gauge: (a) adhesive bonding with backing, (b) adhesive bonding without backing, (c) deposited molecular bonding, and (d) diffused molecular bonding.

In Figure 2.35a, a metallic foil strain gauge is directly bonded to a strained surface through a thin layer of epoxy resin. The backing and the adhesive agent work together to transmit strain. The adhesive also serves as an electrical insulator between the foil grid and the strained surface. A semiconductor gauge usually has no

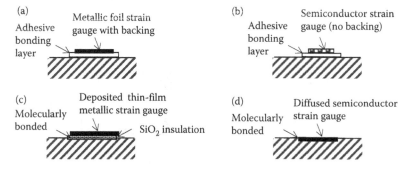

(a) Adhesive bonding layer — Metallic foil strain gauge with backing

(b) Adhesive bonding layer — Semiconductor strain gauge (no backing)

(c) Molecularly bonded — Deposited thin-film metallic strain gauge — SiO₂ insulation

(d) Molecularly bonded — Diffused semiconductor strain gauge

FIGURE 2.35 Four strain gauge-bonding methods: (a) adhesive bonding; (b) diffused bonding; (c) molecular bonding; (d) embedded bonding.

backing; therefore, it is diffused to the substrate and is then bonded to the strained surface with a thin layer of epoxy (Figure 2.35b). It is smaller and lower in cost than a metallic foil sensor. The epoxy used to attach foil gauges can also be used to bond semiconductor gauges. A thin-film metallic strain gauge is molecularly bonded to the specimen (Figure 2.35c) by first depositing an electrical insulation (typically a ceramic) onto the stressed metal surface, and then depositing the strain gauge onto this insulation layer. This results in a much more stable installation with less resistance drift. The errors due to creep and hysteresis therefore are also eliminated. The diffused (or embedded) semiconductor strain gauge also eliminates the need for an adhesive bonding agent (Figure 2.35d). It uses photolithography masking techniques and solid-state diffusion of boron to molecularly bond the resistance elements.

The resistance change of piezoresistive gauges is typically measured using a balanced Wheatstone bridge. This allows a small change in resistance to be measured relative to an initial zero resistance value in a balanced bridge, rather than to a large initial resistance value, which greatly improves sensitivity, accuracy, and resolution. Chapter 6 gives the details of a piezoresistive sensor measurement using bridge circuits.

2.5.5 PIEZORESISTIVE SENSOR APPLICATIONS

Piezoresistive strain gauges are the most widely used sensors among all types of strain gauges. Capacitive and inductive strain gauges' sensitivity to vibration, the special mounting requirements, and circuit complexity have limited their applications. Photoelectric strain gauges are costly and delicate although they can be made as small as 1/16 inch in length. Piezoresistive strain gauges are applied in measuring acceleration, force, torque, pressure, and vibration as discussed below.

2.5.5.1 Piezoresistive Accelerometers

A piezoresistive accelerometer usually contains a mass-spring system designed so that the force exerted by the spring exactly equals the force required to accelerate the mass, and the displacement of the mass (deflection) is directly proportional to the acceleration measured by the strain gauge. Other common methods for measuring the deflection of the inertial mass are capacitive (measuring the gap between a movable electrode and a fixed electrode) and piezoelectric methods (measuring the produced voltage). Figure 2.36 shows two piezoresistive accelerometers developed by *Honeywell Sensotec*, Columbus,

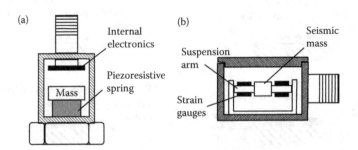

FIGURE 2.36 (a) Piezoresistive spring type accelerometer; (b) suspension arm type accelerometers. (Courtesy of *Honeywell Sensotec*, Columbus, Ohio.)

Ohio. The first one (Figure 2.36a) uses a piezoresistive spring, where the force exerted by the mass changes the resistance of the spring. The second one (Figure 2.36b) measures the deflection of a seismic mass using silicon or foil strain gauges placed on the suspension arm of the mass. Both accelerometers contain a built-in Wheatstone bridge circuit to measure the resistance change and produce an electric output.

Piezoresistive accelerometers have advantages over piezoelectric accelerometers in that they can measure both dynamic and static (or zero Hz) accelerations. This is due to the fact that a resistive sensor is a nonenergy storage component (a *zero-order* system) and there is no energy stored within the component; therefore, its output is immediately available resulting in a near zero response time.

2.5.5.2 Piezoresistive Pressure Sensor

Piezoresistive pressure sensors are critical devices in a variety of control and automobile applications. Figure 2.37 shows an internal combustion engine sensor designed by *Kulite Semiconductor Products Inc.*, Leonia, New Jersey. It uses four piezoresistors to measure the stress in a silicon diaphragm caused by the force or pressure of the media. These four piezoresistors are connected electrically to form a Wheatstone bridge. At the corners of the diaphragm, five 0.024-mm-diameter gold bond wires (ultrasonically ball bonded to the sensor) allow electrical connections to the bridge. The sensor has a resonant frequency above 150 kHz, which also meets the stringent combustion requirements [27]. This sensor can withstand the engine's harsh environment—extreme operating temperature of 500°C and high vibration.

2.5.5.3 Piezoresistive Flow Rate Sensor

Flow rate can be measured using a variety of methods. One technique is to take the differential pressure across two points in a flowing medium (e.g., one at a static point and one in the flow stream). *Pitot tubes* operate based on this principle and have long been used to measure flow rates. Another way is to use the *Venturi effect* by placing a restriction in the flow path and measuring the pressure difference. Figure 2.38 shows a third means to measure flow rate—a bending vane with an attached piezoresistive strain gauge, whose resistance change is proportional to the flow rate and is measured using a Wheatstone bridge circuit. The advantages of a piezoresistive flow rate sensor are: (1) it can measure both air or water flow rate in one, two, or three dimensions and (2) it can detect sporadic, multidimensional, or turbulent flow.

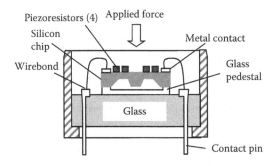

FIGURE 2.37 A piezoresistive internal combustion engine sensor.

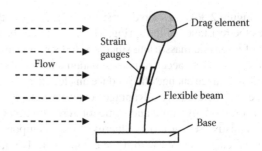

FIGURE 2.38   A piezoresistive flow rate sensor.

2.5.5.4   Piezoresistive Blood Pressure Sensor

A silicon piezoresistive sensor can also be used for intravascular blood pressure monitoring [28]. When the silicon piezoresistive element (directly attached to a stainless-steel plunger) is under an applied force or pressure, the resistance of the silicon element increases, providing a low-cost, extremely compact, and fast-response blood measurement method.

2.5.5.5   Piezoresistive Force Sensor

Many force, torque, and tactile sensors are piezoresistive types. Table 2.18 shows the general specifications of a compact LPM 560 piezoresistive micro force sensor made by *Cooper Instruments & Systems*, Warrenton, Virginia, USA.

2.5.5.6   Piezoresistive Imaging Sensor

Carnegie-Mellon University developed an implantable MEMS stress imager to directly measure bone strength *in situ* on a microlevel scale. This imaging sensor is composed of an array of piezoresistive sensors ("ohm-pixels") with a resolution of 100 Pa and an average measurement time of 1 s. It also includes a coil antenna for RF power and remote sensing. The sensor array is embedded into a textured surface to accommodate sensor integration into the bone. The array of piezoresistive elements offers a safe and convenient way to measure bone strength *in situ* and provides improved and timely information for bone treatment, including prescription of drugs, fixation adjustments, rehabilitation regiments, and preemptive surgical intervention.

TABLE 2.18

Specifications of LPM 560 Micro Force Sensor

Load range: 0–1500 grams	Linearity: 22.5–25 grams
Hysteresis: 45–180 grams	Repeatability: 30–120 grams
Temperature range: −40 to 185°F	Output: 290–430 mV/FS
Bridge resistance: 5 kΩ	Excitation voltage: 10 VDC

Source: LPM 560 Micro Force Sensor, *Cooper Instruments and Systems*, Warrenton, Virginia, USA.

2.6 CHEMORESISTIVE SENSORS

2.6.1 CHEMORESISTIVE EFFECT

In 1938, Wagner and Hauffe discovered that the interaction of certain atoms and molecules with semiconductor materials caused the material's surface properties (e.g., conductivity and voltage potential) to change [29]. In 1953, Brattain and Bardeen studied the phenomenon of large changes in the electrical resistance of metal oxide semiconductors due to the adsorption of gases into their surfaces [30, 31]. In 1962, the first chemoresistive semiconductor gas sensor was developed by Seiyama et al. [32].

When a target gas makes contact with a chemoresistive sensor, it interacts with the sensor both physically and chemically, causing charge exchanges between the adsorbate layer of the target gas and the sensor material, resulting in a variation in conduction or resistance.

There are two main types of chemoresistive sensors: *conductometric sensors* (operate based on conductivity change) and *potentiometric sensors* (operate based on voltage potential change). This section primarily focuses on conductometric sensors due to their broad application in gas sensing, relatively simple structure, and low cost. Conductometric sensors are further divided into *bulk conductance* and *surface conductance* sensors. In a bulk conductance sensor, the entire volume of the material is involved in the chemical reaction; thus, the bulk chemistry defect of the sensor is most important. Bulk sensors are often used in combustion process control (e.g., to measure oxygen partial pressure). In a surface conductance sensor, only the surface of the material is involved in the reaction. Surface conductivity changes are primarily due to changes in free electron concentration caused by charge exchanges between the sensing material and the adsorbed target during the chemosorption and heterogeneous reactions occurred on the sensor surface. Most chemoresistive sensors are surface conductance sensors. The following kinetic scheme describes the complex, temperature-dependent process of oxygen adsorption that involves charge (electron) exchange [33]:

$$O_2^{gas} \rightleftharpoons O_2^{ads} \xrightleftharpoons{e} O_2^- \xrightleftharpoons{e} O^- \xrightleftharpoons{e} O^{2-} \quad \begin{array}{c} 2OH^- \\ \uparrow\downarrow H_2O \\ \\ \uparrow\downarrow X \\ XO \end{array} \tag{2.54}$$

where "ads" means adsorb, e is the electron, and X represents reducing gas. The law of mass action applied to each step of the aforementioned scheme yields the steady-state occupancy for the different oxygen surface states, which contributes to the variation in the surface resistance. By reacting with oxygen ion species at the surface, the reducing gas X generates conduction electrons, resulting in a decrease in the surface resistance. Thus, the electric resistance is directly related to the surface state's (oxygen) occupancy. Using an n-type semiconductor as an example, the two main processes are involved:

Oxidation process:

$$O_2 + 2e^- \rightarrow 2O^-_{ads} \tag{2.55}$$

In this reaction, the surface conductivity decreases (resistance increases) because two electrons are taken away during the oxidation process.

Reduction process:

$$X + O^-_{ads} \rightarrow XO + \text{ }^- \tag{2.56}$$

In this reaction, the surface conductivity increases (resistance decreases) because one electron is returned during the reduction process.

If the oxide is a p-type semiconductor, oxidation will reduce the resistance, while reduction will increase the resistance. The reaction between the gas and the oxide surface depends on the sensor temperature, the gases involved, and the sensor material.

The following generic power law describes the variation of sensor resistance, R, in the presence of a gas. This model is suitable for carbon black polymer composite films as well as other film materials [34]:

$$R = R_0 \left[1 + k_G C_G^\gamma \exp\left(\frac{K_T}{T} \right) \right] \tag{2.57}$$

where R_0 is the sensor's base resistance measured in a reference gas (e.g., clean air); k_G is a sensitivity coefficient for a target gas, and it can be positive or negative depending on the nature of the gas, the sensing material used, an increase or decrease in sensor resistance when the gas is introduced; C_G is the gas concentration in ppm (parts per million); γ is a low power exponent; T is the temperature in kelvin; and K_T is the temperature coefficient of the gas. If K_T the positive, an increase in temperature will reduce the resistance of the sensor.

The aforementioned resistance model can be applied to a mixture of gases by adding the effect of each individual gas as a separate input assuming no interaction between them. This assumption is valid for low concentrations of volatile organic compounds. The above resistance model can also be expanded to include the independent additive effect of humidity [35]:

$$R = R_0 \left[1 + k_G C_G^\gamma \exp\left(\frac{K_T}{T} \right) + k_H (C_H)^{\gamma_H} \exp\left(\frac{K_{TH}}{T} \right) \right] \tag{2.58}$$

where k_H is the sensitivity coefficient for humidity (water vapor), C_H is the water vapor concentration in ppm, γ_H is the low power exponent, and K_{TH} is the temperature coefficient of the water vapor.

EXAMPLE 2.20

A gas has a sensitivity coefficient k_G of 0.701, a power component γ of -0.5, a concentration C_G of 200 ppm, and a temperature coefficient K_T of 7.4. If the air resistance R_0 is 180 Ω, what is the resistance change when the gas sensor is exposed to 620 K?

SOLUTION

From Equation 2.57, the resistance change ΔR is

$$\Delta R = R - R_0 = R_0 k_G C_G^{\gamma} \exp\left(\frac{K_T}{T}\right)$$

$$= (180)(0.701)(200)^{-0.5} \exp\left(\frac{7.4}{620}\right) = 9.03 \ \Omega$$

EXAMPLE 2.21

A chemoresistive sensor's resistance changed from $R_0 = 200$ Ω to $R = 210.63$ Ω when exposed to a gas. Knowing that $k_G = 0.64$, $\gamma = -0.5$, $K_T = 12$, and $T = 710$ K, what is the concentration C_G of the gas in ppm?

SOLUTION

From Equation 2.57, the concentration C_G of the gas is expressed by

$$C_G = \left(\frac{R - R_0}{R_0 k_G \exp(K_T/T)}\right)^{1/\gamma} = \left(\frac{210.63 - 200}{(200)(0.64)\exp(12/710)}\right)^{-1/0.5} = 149.98 \ \text{ppm}$$

2.6.2 CHARACTERISTICS OF CHEMORESISTIVE SENSORS

Chemoresistive sensors (or *chemoresistors*) can be high-temperature chemoresistors ($200\sim600°C$) with semiconductor metal oxide coatings or low-temperature chemoresistors (room temperature) with polymeric and organic coatings.

2.6.2.1 Characteristics of Mixed Metal Oxide Semiconductor Sensors

In mixed metal oxide semiconductor (MMOS) sensors, the large resistance change is caused by a loss or a gain of surface electrons as a result of adsorbed oxygen reacting with the target gas. If the oxide is an n-type, there is either a donation (reducing gas) or subtraction (oxidizing gas) of electrons from the conduction band. When oxidizing gases (e.g., NO_2 and O_3) are present, the n-type oxides increase their resistance; if reducing gases (e.g., CO, CH_4, and C_2H_5OH) are present, the n-type oxides reduce their resistance. Commercially, only a few oxides are available due to the requirement for a unique combination of resistivity, sensitivity (magnitude of resistance change in target concentration), thermal and humidity effects, and wide bandgap semiconductors. The commonly used materials for MMOS sensors include ZnO, TiO_2, Cr_2TiO_3, WO_3, SnO_2, and In_2O_3. These materials have higher sensitivity, quicker response, and enhanced capability to detect gases at low concentrations compared to thin-film

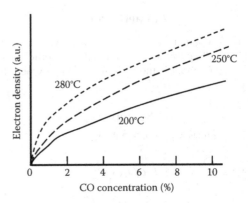

FIGURE 2.39   Change in conductance of an SnO_2 nanowire sensor as a function of CO concentration.

materials [36,37]. Furthermore, doping with noble metals (Au, Pd, Pt) can effectively improve the sensitivity, selectivity, and response time of the nanomaterial gas sensors at low operating temperatures [38,39]. These metal oxides can be deposited onto thin or thick films of semiconductor substrates. In recent years, novel nanostructures such as nanowires, nanotubes, nanorods, nanonails, nanocages, nonosheets, nonocables, and nanobelts, are increasingly used in chemoresistive sensor designs due to their large surface-to-volume ratios, single crystalline structures, and great surface activities [40,41]. Figure 2.39 shows the change in conductance (in terms of electron density) of an individual SnO_2 nanowires as a function of CO concentration at three different temperatures, where the electron density (in arbitrary units—a.u.) is proportional to CO partial pressure [42].

Semiconductor metal oxide sensors usually respond to almost any analyte (carbon monoxide, nitrogen oxide, hydrogen, or hydrocarbon), but they are not very selective. One way to modify the selectivity pattern is surface doping the metal oxide with catalytic metals such as platinum, palladium, gold, and iridium. Metal oxide sensors are more stable at higher operating temperatures. Under lower temperatures (e.g., 200°C or below), polymer sensors work better.

2.6.2.2   Characteristics of Polymer or Organic Material Sensors

Conducting polymers, such as polypyrroles, polyaniline, and polythiophene, change their electrical resistance when exposed to certain targets. They are broadly used in making chemoresistive sensors to monitor a variety of polar organic volatiles such as ethanol and methanol. Other conducting polymers, such as carbon black, can be dispersed in nonconducting polymers. When the polymer absorbs vapor molecules, it swells and the particles are pushed further apart, causing a decrease in conductivity [43]. Figure 2.40 shows how the conductance of an H_2O sensor changes as ethanol concentration changes at constant humidity and constant temperature [44]. For instance, if the measured conductance change is 8%, the ethanol concentration is 24 ppm.

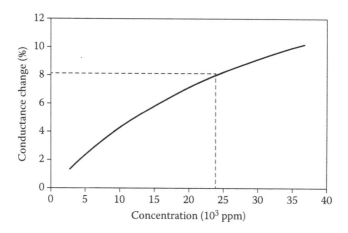

FIGURE 2.40 A Pan/DSA/H$_2$O sensor's conductance change versus ethanol concentration.

Conducting polymer-based chemoresistors are most commonly used in vapor or odor sensing due to the wide range of available polymer combinations and their ease of deposition, ability to operate at room temperature, low power consumption, and sensitivity to a broad range of volatile organic compounds and organic solvents (e.g., hydrocarbons, chlorinated compounds, and alcohols). The characteristics of polymer chemoresistive sensors depend upon the polymer material, electrode geometry, temperature, and ambient humidity. Some disadvantages of polymeric chemoresistors include their batch-to-batch variation in baseline resistance, large temperature and humidity coefficients, long-term drift, and small signal-to-noise ratio (common for most types of gas sensors).

2.6.3 CHEMORESISTIVE SENSOR DESIGN

Typical configurations of chemoresistive sensors include tube, volume, chip, and liquid electrolyte types [45]. Figure 2.41 shows a tubular chemoresistive sensor developed by the University of Kentucky, USA [46]. It uses nanotube structure in its sensor chip.

A planar chemoresistive sensor is shown in Figure 2.42, the typical configuration for metal oxide sensors. The sensor has three main components: a sensing element, electrodes, and a heater [47,48]. The sensing element, often made of a metal oxide material such as SnO$_2$, WO$_3$, or In$_2$O$_3$, reacts with the target molecules, causing changes in resistance. The electrodes, connected to the sensing element, measure the resistance of the sensing material. A heating element mounted under the substrate is used to regulate the sensor temperature (since metal oxide sensors operate at high temperature –200~600°C, and exhibit different gas response characteristics at different temperatures). An additional power supply is needed for the heater. In order to achieve high sensitivity for target detection, the sensing element is designed to have as high *specific area* (surface-area-to-volume ratio) as possible.

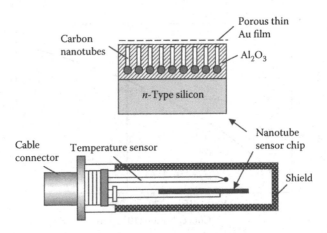

FIGURE 2.41   A carbon nanotube chemoresistor.

This is achieved in practice through using either thin films or thick, porous layers of partially sintered materials.

A typical polymer chemoresistor has a chip configuration. Figure 2.43 shows a carbon black polymer sensor [35]. The sensor electrodes are the two square metal pads with two 96×96 μm open areas (indicated by the crosses "X") to allow the sensing element, carbon black polymer, to be deposited onto the metal pads. The "nitride" functions as an insulator. The sensor has a baseline resistance of 10 kΩ, achieved by using the minimum pad dimensions, a 80 μm gap between the electrodes, and the proper polymer thickness with the right fraction of carbon black in the composite. If the sensing element is made of organic materials, the electrode spacing will be typically 5~100 μm with an applied voltage of 1~5 V.

The liquid electrolyte structure is commonly used in fuel cell CO sensors as shown in Figure 2.44. The sensor has three electrodes: working (sensing), reference, and counter (output) electrodes [49]. The gas molecules react at the sensing electrode as follows:

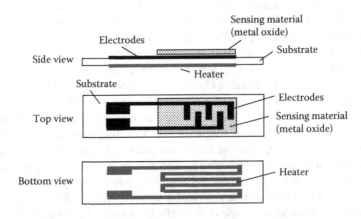

FIGURE 2.42   A typical metal oxide chemoresistive sensor.

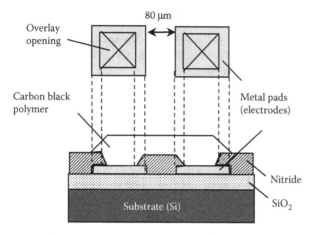

FIGURE 2.43 A typical polymer chemoresistor.

$$CO + H_2O \Rightarrow CO_2 + 2H^+ + 2e^-$$

The generated CO_2 diffuses away into the air while the positively charged ions (H^+) migrate into the electrolyte. The electrolyte facilitates H^+ moving to the counter electrode, resulting in current flow between the sensing and the counter electrodes. The electrons generated (e^-) are conducted through the counter electrode to the external measuring circuit. Thus, the change in conductivity is directly proportional to the concentration of CO. The aforementioned oxidation reaction is balanced by a corresponding reduction reaction at the counter electrode:

$$O_2 + 4H^+ + 4e^- \Rightarrow 2H_2O$$

The reference electrode is introduced to improve the performance of the sensor by maintaining a constant voltage at the sensing electrode. No current flows to or from the reference electrode. The CO diffuses into the cell through the barrier (capillary). A filter is installed in front of the sensing electrode to block unwanted gases. The most commonly used filter medium is activated charcoal. Properly selecting the filter medium can make a sensor more selective to its target gases.

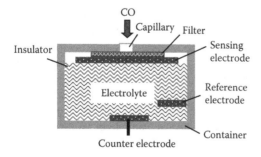

FIGURE 2.44 Typical layout of a three-electrode fuel cell sensor.

2.6.4 Chemoresistive Sensor Applications

Chemoresistors and their arrays have been widely used for gas or vapor detection of substances such as H_2, O_2, CO, CO_2, NO, NO_2, CH_4, pH, and H_2O. Following are a few examples of their applications.

2.6.4.1 Hygristor

A *hygristor* (from the words "*hygro-*" and "*resistor*") is a humidity sensor that uses a moisture-dependent resistor to measure humidity. It is made of hygroscopic materials (e.g., conductive polymer, silica gel, paper, or treated substrate) whose specific resistivity depends largely on the concentration of absorbed water molecules. The resistance changes inversely and exponentially as humidity changes. A typical hygristor contains a substrate and two silkscreen-printed conductive electrodes, which are usually covered with a hygroscopic semiconductive gel.

The most important attributes of resistive humidity sensors—small size, low cost, interchangeability, and long-term stability—have made them suitable for industrial, commercial, and residential applications. Some resistive humidity sensors also resist chemical and physical contaminants and have the ability to recover from repeating condensations. The life expectancy of humidity sensors is about 5 years for residential and commercial applications, but exposure to chemical vapors and other contaminants (e.g., oil mist) may shorten the sensors' life span. Resistive humidity sensors are also significantly influenced by temperature. If they are used in environments with large temperature fluctuations (>10°F), temperature compensation should be incorporated into their design to ensure accurate measurement.

2.6.4.2 Groundwater Monitoring System

A four-chemoresistor array has been developed for groundwater monitoring wells at *Edwards Air Force Base* and for the chemical waste landfill at *Sandia National Laboratories* [50]. The sensor is fitted with a GORE-TEX® membrane to allow chemical vapors to go through to the sensor while protecting internal electronics and wiring from contaminants. The chemoresistors are incorporated into a single die with the circuitry to measure sensor resistance.

2.6.4.3 Electronic Nose

An *electronic nose* (*e-nose*) has been widely used to analyze volatile organic compounds, monitor vehicle emissions, detect explosives, and perform clinical diagnoses. Many chemoresistors can be integrated in an array to form an e-nose, including metal oxide semiconductors, conducting polymers, nanotubes or nanowires, etc. [51]. These sensing elements or materials form a sensitive layer and allow a wide variety of chemical compounds to be detected. An e-nose also includes a pattern-recognition mechanism (chemometrics) that compares the patterns from the measurements to the known patterns for identification. An example of an e-nose's response to different gases is shown in Figure 2.45. The different output pattern from this eight-sensor array e-nose indicates a different gas. If the array is "trained" properly, it can recognize an individual gas in mixtures.

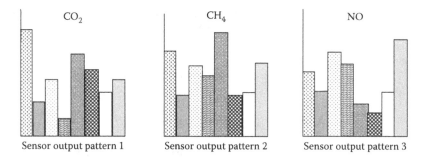

CO₂ CH₄ NO

Sensor output pattern 1 Sensor output pattern 2 Sensor output pattern 3

FIGURE 2.45 An eight-sensor array e-nose's output patterns corresponding to different gases.

An increasing research interest is to integrate chemoresistors into robotic systems for radiation detection or toxin sensing without risking human exposure. If chemical sensors are mounted in multirobot systems (e.g., swarm robots), a mobile sensor network can be established to perform more powerful functions.

2.7 BIORESISTANCE/BIOIMPEDANCE SENSORS

2.7.1 SENSING PRINCIPLES

Biosensors convert the analyte–receptor reactions into a quantitative electrical, electrochemical (amperometric, potentiometric, conductimetric), piezoelectric, calorimetric, acoustic, mechanical, magnetic, or optical signals (fluorescence absorbance).

2.7.1.1 Types of Bioresistance/Bioimpedance Sensors

All living cells, tissues, or organs are conductive materials. For example, blood has approximately 0.67 S · m⁻¹ conductivity (or 1.50 Ω · m resistivity), urine 3.33 S · m⁻¹ (or 0.30 Ω · m resistivity), and fatty tissue 0.04 S · m⁻¹ (or 25 Ω · m resistivity). Living organisms are also composed of positive and negative ions in various quantities and concentrations. Bioresistance and bioimpedance sensors extract physiological and pathological information of living organisms through measuring a sample's conductivity, ion migration, electron increase/decrease, resistance change, electrical potential difference, and impedance variation. Thus, bioresistive sensors can be classified into five categories: (1) Sensing based on the migration of ions through a region that causes an electrical potential difference or current flow between two points. (2) Sensing based on electrochemical reactions that generate ions or electrons, which alters the overall conductivity of a living sample. (3) Sensing based on employing other resistive sensors (e.g., potentiometers, piezoresistive sensors, or chemoresistive sensors) to perform a biosensing task. For example, in an enzyme–substrate reaction (a biometabolic process), instead of measuring how many ions and electrons are produced, the released products, such as O_2, pH, CO_2, or NH_3, are measured using chemoresistive gas sensors. (4) Sensing based on bioelectrodes that convert ionic conduction into electronic conduction (e.g., Ag/AgCl electrodes) so that a signal can be processed in an electronic circuit. (5) Sensing based on measuring bioimpedance instead of resistance only. Bioimpedance includes electrical resistance R, capacitive reactance X_C (in Ω), and inductive reactance X_L (in Ω) expressed by

$$X_C = \frac{1}{2\pi f C} \quad (C\text{—capacitance in farad}) \tag{2.59}$$

$$X_L = 2\pi f L \quad (L\text{—inductance in henry}) \tag{2.60}$$

where f is the frequency (in Hz) of the measuring signal. Indeed, many of the impedance measurements conducted in physiology laboratories are actually AC (alternating current) resistance measurements. This is because a direct conductance measurement often has relatively low sensitivity, while using a sinusoidal current or voltage with a frequency f in the measurement can minimize undesirable effects such as *Faradaic processes* (electrons transfer at the interface between the electrode and the chemical solution), double-layer charging, and concentration polarization in living samples. Proper selection of the AC frequency f is important since impedance is frequency dependent. For example, the impedance of skin is about 100 Ω with a 40 Hz AC current and about 200 Ω with a 20 Hz AC current. Besides frequency f, the current level is also important since higher current stimulates cells and causes measurement artifacts and discomfort in the subject. For DC (direct current) or low-frequency AC, the current level can be 1 mA, while for high-frequency AC (e.g., 25 kHz), the RMS value of the current can be 0.1 mA.

2.7.1.2 Modeling of Bioresistance/Bioimpedance Sensors

Bioimpedance Z is defined as the total opposition to the flow of an AC current. The magnitude of Z can be calculated by

For a series connection:

$$|Z| = \sqrt{R^2 + (X_L - X_C)^2}$$

$$\tag{2.61}$$

For a parallel connection:

$$|Z| = \frac{R(X_L - X_C)}{\sqrt{R^2 + (X_L - X_C)^2}}$$

$$\tag{2.62}$$

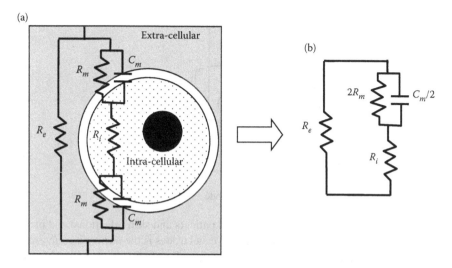

FIGURE 2.46 The electric model (a) and equivalent circuit (b) for a cell.

A cell, tissue, or the entire human body is often modeled as a resistor and a capacitor in parallel, since the inductance tends to be very small relative to the capacitance (i.e., $X_L \ll X_C$). Figure 2.46 shows a cell membrane modeled as a capacitor (C_m) and a resistor (R_m) in parallel, while its intra-cellular and extra-cellular regions function as resistors (R_i and R_e, respectively). The capacitance value of a cell membrane is about 10 μF per cm².

2.7.2 SENSOR MATERIALS AND CHARACTERISTICS

The most important characteristics of biosensors include *selectivity* (or *specificity*), *sensitivity, response time, regenerability*, and *simplicity*. Selectivity, meaning detection selectivity, depends entirely on the inherent binding capability of the bioreceptor molecule, whereas sensitivity depends on both the nature of the biological element and the sensing material. The biomaterials that can be used as biological sensing elements include organisms, tissues, cells, organelles, membranes, enzymes, receptors, antibodies, and nucleic acids. The bioagents that can be recognized by the bio-recognition elements are metabolic chemicals (oxygen, methane, ethanol), enzyme substrates (glucose, penicillin, urea), ligands (hormones, pheromones, neurotransmitters), antigens and antibodies (human immunoglobulin, antihuman immunoglobulin), and nucleic acids (DNA, RNA).

A biosensor should be (1) highly specific for the analyte of interest; (2) responsive in the appropriate concentration range and have a moderately fast response time; (3) reliable, reproducible, accurate, and sensitive; and (4) miniaturizable.

Electrodes used for bioelectrical recording and stimulation can be classified as *noninvasive* or *invasive* types. Noninvasive (mostly *surface* type) electrodes are used outside the body (e.g., on the skin), while invasive electrodes use a needle to punch through the skin and are inserted beneath the tissue. An invasive electrode is only

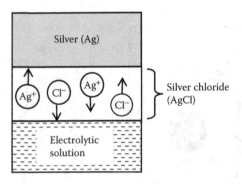

FIGURE 2.47   Structure of an Ag–AgCl electrode.

used for specialized cases (e.g., anesthetized patients and skin infections) and in veterinary applications. Another type of invasive electrodes is the *indwelling electrode*. An indwelling electrode, inserted into the body rather than into the layers beneath the skin, is typically a tiny exposed metallic contact at the end of a long insulated catheter (e.g., the catheter for veins or arteries). There are two categories of electrode materials: *perfectly polarized* (or called *perfectly nonreversible*) and *perfectly nonpolarized* (or called *perfectly reversible*). In the first type, there is no net transfer of charge across the metal–electrolyte interface, while in the second type, there is a free transfer of charge between the metal and the electrolyte. Silver–silver chloride (Ag–AgCl) is an ideal polarized (Ag) and nonpolarized (AgCl) pair, which becomes excellent standard, and is by far the most common reference electrode used today due to its simplicity, low cost, stability, reproducibility, and toxin-free character. Figure 2.47 shows an Ag–AgCl electrode consisting of a silver (Ag) substrate with a thin layer of silver chloride (AgCl) deposited onto its surface. The AgCl provides a free two-way exchange of Ag^+ and Cl^- ions between the metal Ag and the electrolytic solution.

Besides silver–silver chloride, other noble metals and alloys, for example, gold and platinum, tungsten alloys, and platinum–platinum black, are also used to make practical bioelectrodes, not only because of their ideal features but also their resistance to corrosive biofluids inside the human body. A conductive gel or paste is also applied when using bioelectrodes to reduce the impedance between skin and electrodes. One of the commonly used bioelectrode gels is a 0.5% saline-based electrode gel. It has high conductivity and is hypoallergenic, bacteriostatic, water soluble, and nongritty. Note that not all electrodes are acceptable for bioapplications. For example, zinc–zinc sulfate, although it has ideal features, is toxic to living tissues.

2.7.3   DESIGNS AND APPLICATIONS OF BIORESISTANCE/BIOIMPEDANCE SENSORS

Several electrode configurations are available for clinical applications. Figure 2.48a shows an Ag–AgCl disk electrode with a flexible lead wire attached to the back. It can be used as a direct-contact skin electrode for electrocardiogram recording.

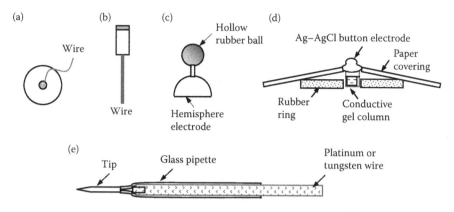

FIGURE 2.48 Typical clinical electrode designs: (a) disk electrode; (b) pellet electrode; (c) suction-cup electrode; (d) button electrode with gel-filled column; (e) needle electrode.

Figure 2.48b is a pellet electrode that has a cylindrical AgCl pellet embedded on a wire. A 0.8–2.0 mm diameter Ag–AgCl matrix is at the tip. Pellet electrodes are suitable for probing tissues. Figure 2.48c is a suction cup electrode (chest electrode). Figure 2.48d shows a column electrode. In this configuration, an Ag–AgCl metal contact button is placed at the top of a conductive gel-filled column. The assembly is held in place by an adhesive-coated rubber ring. This design reduces movement artifacts since the electrode is lifted off the surface and it is often used for long-term recording and monitoring. Figure 2.48e shows a typical needle electrode. A very fine platinum or tungsten wire is slip-fit through a 1.5–2.0 mm glass pipette. The tip is etched and then fire-formed into the shallow-angle taper.

EXAMPLE 2.22

Figure 2.49 shows a KCl-filled microelectrode (KCL—potassium chloride). Its spreading resistance R_s is a function of the tip diameter, d, and can be approximated by (ρ is the resistivity of the solution in $\Omega \cdot$ m)

$$R_s = \frac{\rho}{2\pi d} \tag{2.63}$$

If ρ for physiological saline is 0.70 $\Omega \cdot$ m and tip diameter is 1.2 μm, find the tip spreading resistance of the microelectrode.

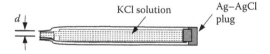

FIGURE 2.49 A KCl-filled microelectrode.

SOLUTION

$$R_s = \frac{\rho}{2\pi d} = \frac{0.70\ \Omega \cdot m}{2\pi\ (1.2 \times 10^{-6} m)} = 92.89\ k\Omega$$

Some applications of bioresistance/bioimpedance sensors are shown as follows.

2.7.3.1 Body Composition Monitor

Commercial body composition monitoring systems, such as *BIA 310e* [52], use measured body resistance to estimate the percentage of body fat, fat weight, lean weight, basal metabolic rate, and hydration status. In BIA 310e, sensor pads are placed on the subject's right wrist and ankle. To measure the body's resistance, the system emits a low-level high-frequency electrical current (<1 mA, 50 kHz) that flows through the body. Lean tissue is more conductive than fat tissue due to the lean tissue's high water content. Thus, the resistance is an indication of how well the body conducts an electrical current. For individuals who have the same height and weight, the lower the resistance, the lower the percentage of body fat. A typical measurement report with automatically generated target recommendations (based on optimal body fat values stored in the analyzer's memory) is shown in Table 2.19.

2.7.3.2 Resistance Measurement at Acupuncture Points

The electrical characteristics (especially resistance) of the acupuncture meridians in Chinese medicine have been extensively studied since 1950 to establish their existence by scientific methods. It is now generally accepted that both meridian and acupuncture points have lower electrical resistance or impedance than nearby surrounding points [53]. These low resistance points are thought to be the result of sensory and motor nerves emerging from deep tissue to superficial layers of the skin

---

TABLE 2.19

Test Results from a BIA 310e Body Composition Monitor

Data:

Sex: Male	Height: 73.0 in.	Age: 28	Weight: 200 lb

Test results:

Bioresistance: 450 Ω	Percent body fat: 19.3%	Fat body weight: 38.6 lb
Lean body weight: 161.4 lb	Total body water: 52.4 liter	

Basal metabolic rate:
2227 calorie per day

Target recommendations:

Percent body fat: 16.0%	Total weight: 192.1 lb	Fat body weight: 30.7 lb
Lean body weight: 161.4 lb		Weight to lose: 7.9 lb

Source: From BIA 310e bioimpedance analyzer datasheet, *Biodynamics Corporation*, Seattle, Washington, USA. www.biodyncorp.com. With permission.

---

[54]. The estimation of resistance at acupuncture points can be made through skin resistance measurement, which is a function of the cutaneous region, skin humidity, and pressure of the measurement electrode. Measurement methods include using an excitation current or measuring the current (0.01–5 μA DC current) produced by the electromotive force resulting from differences in these tiny points' potentials using appropriate electrodes.

2.7.3.3   Ovulation Predictor

The *CUE Fertility Monitor*, developed by *Zetek Inc.* (Aurora, Colorado, USA), consists of a hand-held digital monitor with an oral and a vaginal sensor. The system detects and records the changes in electrical resistance and ionic concentration of saliva and vaginal secretions, in response to the cyclical changes in estrogen. Based on the measurement results, the CUE monitor is able to predict and confirm an ovulation. The peak electrical resistance in the saliva occurs 5~7 days before ovulation, and the lowest electrical resistance in cervical secretions occurs about a day before ovulation.

2.7.3.4   Venous Blood Volume Measurement

Medis Medizinische Messtechnik GmbH in Germany has used an impedance method to measure changes in venous blood volume as well as pulsation of the arteries. As blood volume changes, the electrical impedance also changes proportionally. This measurement requires passing a small magnitude of high-frequency AC current through the body using four electrodes. The two middle electrodes measure the voltage, and their positions define the measurement region. The outer two electrodes are used to emit the small AC current mentioned earlier. The locations of these outer electrodes are not critical. This method allows doctors to detect blood flow disorders, early stage arterioscleroses, functional blood flow disturbances, deep venous thromboses, migraines, and general arterial blood flow disturbances. This bioimpedance sensing method is safe, noninvasive, inexpensive, and easy to operate.

EXERCISES

Part I: Concept Problems

1. Potentiometric sensors are often designed based on the change of a conductor's
   A. Electrical conductivity, σ_p
   B. Resistivity, ρ
   C. Cross-sectional area, A
   D. Effective length, l
2. In the International Annealed Copper Standard (IACS), electrical materials are rated relative to copper (in percent). For example, platinum has a conductivity of $0.943 \times 10^7 \ \Omega^{-1} \cdot m^{-1}$ and copper has a conductivity of $5.95 \times 10^7 \ \Omega^{-1} \cdot m^{-1}$. Thus, platinum has 15.85 IACS percent, while copper has 100 IACS percent.
   i. Which of the following common electrical materials listed in Table 2.20 has the highest IACS rating?

TABLE 2.20

Conductivity of Common Electrical Materials

Material	Conductivity σ_p ($\Omega^{-1} \cdot m^{-1}$)
Silver	6.29×10^7
Gold	4.17×10^7
Aluminum	3.77×10^7
Tungsten	1.79×10^7
Iron	1.03×10^7
Lead	0.45×10^7

---

   ii. If using one of these materials to make a metal wire, which material requires the largest size to achieve the same or equivalent conductivity?

3. Which of the following statements is true?

   A. The electrical conductivity of a metal increases as temperature rises.

   B. The electrical resistance of a semiconductor increases as temperature rises.

   C. All resistance temperature detectors (RTDs) are PTC sensors.

   D. All semiconductor temperature sensors are NTC sensors.

4. A temperature sensor, made of semiconductor metal oxide materials, is a typical

   A. PTC thermistor

   B. NTC thermistor

   C. PTC RTD

   D. NTC RTD

5. The resistance and temperature relationship of an RTD can be described by which of the following equations?

   A. $R(T) = R_0[1 + A(T - T_0) + B(T - T_0)^2 + C(T - T_0)^3 + \cdots]$

   B. $R(T) = R_0 e^{\left[\beta\left(\frac{1}{T} - \frac{1}{T_0}\right)\right]_a}$

   C. $T = \dfrac{1}{a + b\ln(R) + c\,[\ln(R)]^3}$

   D. $R(T) = \text{Ł}lT/(\kappa A)$

6. Which following excitation current is most likely used in an RTD measurement?

   A. 1 A

   B. 0.1 A

   C. 0.01 A

   D. 0.001 A

7. Based on your knowledge on *solid mechanics*, explain why a cubic crystal has only three independent, nonvanishing elastic components, π_{11}, π_{12}, and π_{44}, when the coordinate axes coincide with the crystal axes, making a 6×6 *piezoresistive coefficient matrix* become

$$\begin{bmatrix} \pi_{11} & \pi_{12} & \pi_{13} & \pi_{14} & \pi_{15} & \pi_{16} \\ \pi_{21} & \pi_{22} & \pi_{23} & \pi_{24} & \pi_{25} & \pi_{26} \\ \pi_{31} & \pi_{32} & \pi_{33} & \pi_{34} & \pi_{35} & \pi_{36} \\ \pi_{41} & \pi_{42} & \pi_{43} & \pi_{44} & \pi_{45} & \pi_{46} \\ \pi_{51} & \pi_{52} & \pi_{53} & \pi_{54} & \pi_{55} & \pi_{56} \\ \pi_{61} & \pi_{62} & \pi_{63} & \pi_{64} & \pi_{65} & \pi_{66} \end{bmatrix} \Rightarrow \begin{bmatrix} \pi_{11} & \pi_{12} & \pi_{12} & 0 & 0 & 0 \\ \pi_{12} & \pi_{11} & \pi_{12} & 0 & 0 & 0 \\ \pi_{12} & \pi_{12} & \pi_{11} & 0 & 0 & 0 \\ 0 & 0 & 0 & \pi_{44} & 0 & 0 \\ 0 & 0 & 0 & 0 & \pi_{44} & 0 \\ 0 & 0 & 0 & 0 & 0 & \pi_{44} \end{bmatrix}$$

8. Explain why many resistive sensors (e.g., piezoresistive or photoresistive sensors) have a zigzag pattern.

9. Which of the following statements about gauge factors is *not* true:
   A. A typical semiconductor strain gauge has a gauge factor of 100.
   B. A typical metal or thin-film metal strain gauge has a gauge factor of 1.
   C. Gauge factor is not always positive.
   D. Gauge factor increases as the temperature increases.

10. Refer to the visible light spectrum, indicate to which color that CdS and CdSe sensors shown in Figure 2.50 are the most sensitive?

11. Among RTDs, thermistors, and thermocouples, which temperature sensor(s) can
    A. Measure a temperature that is below 0°C: _____
    B. Measure a temperature that is above 1000°C: _____
    C. Be made the smallest: _____
    D. Have a direct voltage output (without using a voltage divider circuit): _____

12. Match the materials in the left column with sensors in the right column.

Materials	Sensors
CdS	RTD
Pt100	Piezoresistive sensor
Ferry alloy	Photoresistive sensor

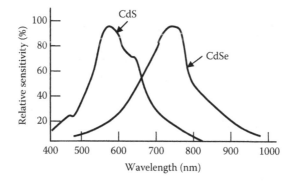

FIGURE 2.50 Relative sensitivity versus wavelength curves of a CdS and a CdSe sensor.

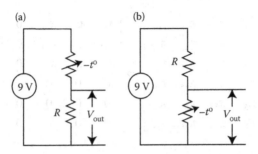

FIGURE 2.51   Two sensor circuits.

13. A fire sensor circuit is designed to deliver a high voltage to trigger an alarm when a hot condition is detected. Which of the circuits in Figure 2.51 should be chosen for this purpose? (Note: " $-t^o$ " in Figure 2.51 indicates an NTC sensor).

14. Which of the following chemical reactions will cause an increase in resistivity?
    A. $O_2 + 4H^+ + 4e^- \Rightarrow 2H_2O$
    B. $CO + H_2O \Rightarrow CO_2 + 2H^+ + 2e^-$
    C. $Zn + 2HCl \Rightarrow ZnCl_2 + H_2$
    D. $AgNO_3 + NaCl \Rightarrow AgCl + NaNO_3$

15. A chemoresistor is chosen to detect a gas at 500°C operating temperature. Which coating should be used for the sensor?
    A. Semiconductor metal oxide
    B. Polymer
    C. Organic material
    D. Glass

16. Which of the following features is unique for chemical and biosensors?
    A. Hysteresis
    B. Linearity
    C. Selectivity
    D. Repeatability

17. An electronic nose can detect different gases because of the following facts, *except*
    A. It uses several different sensors that are sensitive to different gases.
    B. The sensors in an e-nose are pretrained to recognize different gases.
    C. A "lookup" table or a pattern recognizer is used to compare e-nose's output pattern with the patterns stored in the recognizer.
    D. A single sensing element can recognize different gases.

18. Why do most biosensors use Ag–AgCl electrodes?

Part II: Calculation Problems

19. In a light intensity measurement circuit shown in Figure 2.52, V_S is +5 V, R_1 is 47 kΩ, and V_{out} is 2.7 V. What is the resistance of the photoresistor R_2?

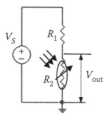

FIGURE 2.52 A photoresistor circuit.

20. The circuit shown in Figure 2.52 can be used as a "dark sensor" to turn ON a lighting system automatically in the evening. If $R_1 = 10$ kΩ, $V_{in} = 9$ V, and R_2 has a resistance of 500 Ω in bright light and 200 kΩ in the dark, find R_2's output voltage when (1) R_2 is in the bright light; (2) R_2 is in the dark.
21. In the circuit shown in Figure 2.53, what should the value of resistor R_1 be to turn on the light? The LDR has a resistance of 500 Ω in bright light and 200 kΩ in the dark. (*Hint:* For the light to be ON, the transistor has to be in the ON status, which means the voltage V_{BE} must be larger than 0.7 V.)
22. A 120 Ω strain gauge with GF = 2.1 is used to measure a strain of 0.105. What is the resistance change of the gauge from the unloaded state to the loaded state?
23. Compare the resistance change produced by a 150 mm · m^{-1} strain in a metallic gauge with GF = 2.13 and a semiconductor gauge with GF = −151. Assume the nominal resistances for both gauges are 120 Ω.
24. A nickel RTD manufacturer recommends the following polynomial:

$$R_T = R_o (1 + AT + BT^2 + DT^4 + FT^6)$$

where $R_0 = 78$ Ω at 25°C, $A = 5.485 \times 10^{-3}$, $B = 6.650 \times 10^{-6}$, $D = 2.805 \times 10^{-11}$, and $F = -2 \times 10^{-17}$. Find the resistance value of this nickel RTD sensor when it is exposed to 39°C environment.
25. Given a Pt100 platinum RTD with α = 0.003911, find its resistance value at 100°C.
26. A pure silicon strip is 8 mm long, 2 mm wide, and 1 mm thick. At room temperature, the intrinsic concentration in the silicon is 1.2×10^{16} m^{-3}. The electron and hole mobilities are 0.11 m^2 · V^{-1} · s^{-1} and 0.048 m^2 · V^{-1} · s^{-1},

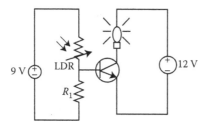

FIGURE 2.53 A light control circuit.

respectively. Find (1) the conductivity σ_p of the silicon; (2) the resistance R of the strip.

27. Calculate the hole and electron densities in a piece of p-type silicon that has been doped with 5×10^{16} cm^{-3} acceptor atoms, knowing that $n_i = 1.4 \times 10^{10}$ cm^{-3} at room temperature (300 K). (*Note*: For a p-type silicon, $p \gg n$ and $n \ll n_i$.)

28. An n-type silicon semiconductor has $N_d = 10^{16}$ cm^{-3} and $N_a = 0$ at $T = 300$ K. If the intrinsic carrier concentration is $n_i = 1.5 \times 10^{10}$ cm^{-3}, determine the electron and hole concentrations for the given doping concentration.

29. A chemoresistive CO gas sensor has a resistance change from $R_0 = 645$ Ω (in air) to $R = 693$ Ω when CO (with concentration of 60 ppm) is present. Knowing that $K_G = 0.61$, $K_T = 15$, and $T = 590°C$, find the power coefficient γ.

REFERENCES

1. Bunce, N., Industrial applications and sensors, *CHEM7234/CHEM 720 Fundamentals of Electrochemistry Lecture Notes*, University of Guelph, Ontario, Canada, Spring 2003.
2. Temperature coefficient of resistance, Creative Commons, Stanford, California, USA, 2008.
3. Measuring temperature with RTDs—A tutorial, Application Note 046, National Instruments Corporation, Austin, Texas, USA, 1996.
4. The Callendar–van Dusen coefficients, Unitek Systems, 1999.
5. Alciatore, D.G. and Histand, M.B., *Introduction to Mechatronics and Measurement Systems*, 3rd ed., McGraw-Hill, New York, 371, chap. 9, 2007.
6. Thermistor calibration and the Steinhart-Hart equation, Application Note, ILX Lightware, Bozeman, Montana, USA, 2003.
7. Steinhart-Hart equation, Cornerston Sensors, Vista, California, USA, 2007.
8. Ashcroft, N.W. and Mermin, N.D., *Solid State Physics*, Saunders College Publishing, Philadelphia, Pennsylvania, USA, 1976, p. 20.
9. Kumar, G.S., Prasad, G., and Pohl, R.O., Experimental determinations of the Lorenz number, *J. Mater. Sci.*, 28, 4261, 1992.
10. Nave, R., Thermal conductivity, Georgia State University website, 2010. Available at: http://hyperphysics.phy-astr.gsu.edu/Hbase/thermo/thercond.html.
11. Garvey, D., So what is Platinum RTD? RdF Corporation, Hudson, New Hampshire, USA, 2010.
12. RTD Theory, Pyromation, Inc., 2013.
13. PTC Thermistors, Thermometrics, Inc., Edison, New Jersey, USA, 1999.
14. NTC Thermistors, Thermometrics, Inc., Edison, New Jersey, USA, 1999.
15. Potter, D., Measuring temperature with thermistors—A tutorial, Application Note 065, National Instrument Corporation, Austin, Texas, USA, 1996.
16. Sensors and actuators, Toyota Motor Sales, USA, Inc., Torrance, California, USA, p. 3.
17. Callister, W.D. and Rethwisch, D.G., *Materials Science and Engineering: An Introduction*, 7th ed., John Wiley & Sons, Inc., New York, USA, 2007.
18. Module 4 photodetector characteristics, LEOT laser tutorial—applications in photonics and telecommunications, 2006. www.csun.edu/~dchoudhary/optexp/ex9-photodetector.pdf.
19. Hossain, Q.D., Design and characterization of a current assisted photo mixing demodulator for TOF based 3d CMOS image sensor, Ph.D. Dissertation, University of Trento, Trento, Italy, pp. 16–20. 2010.

20. Kirk, T., The FrankenPipe: A bagpipe MIDI controller with RC car control. Available at: https://ccrma.stanford.edu/~turner/website/projects/frankenpipepaper.pdf
21. Smith, C.S., Peizoresistive effect in germanium and silicon, *Phys. Rev.*, 94, 42, 1954.
22. Hsu, T.R., *MEMS & Microsystems: Design and Manufacture, and Nanoscale Engineering*, 2nd ed., John Wiley & Sons, Inc. Hoboken, New Jersey, 2008.
23. Sze, S.M. (Editor), *Semiconductor Sensors*, John Wiley and Sons, Inc., Hoboken, New Jersey, 1994.
24. Craig, J.I., *AE3145 Resistance Strain Gage Circuits,* Course Materials, Georgia Institute of Technology, Atlanta, 2000.
25. Sensitivity of strain gauge wire materials, *efunda Inc.* Available at (retrieved on June 12, 2013): www.efunda.com/designstandards/sensors/strain_gages/strain_gage_sensitivity.cfm.
26. Holman, J.P., *Experimental Methods for Engineers*, 7th ed., McGraw-Hill, Boston, 2001, 477.
27. Kurtz, A.D. et al., High accuracy piezoresistive internal combustion engine transducers, Kulite Semiconductor Products, Inc., Leonia, New Jersey, USA.
28. Harsanyi, C., Sensor in biomedical applications: may they change the quality of life? *Sensor Rev.*, 21, 259, 2001.
29. Wagner, C. and Haufffe, K., Untersuchungen über den stationären zustand von. katalysatoren bei heterogenen reaktionen, *Ztschr. Elektrochem.*, 44, 172, 1938.
30. Brattain, W.H. and Bardeen, J., Surface properties of germanium, *Bell Syst. Tech. J.*, 32, 1, 1953.
31. Garrett, C.G.B. and Brattain, W.H., Physical theory of semiconductor surfaces, *Phys. Rev.*, 99, 376, 1955.
32. Seiyama, T. et al., A new detector for gaseous components using semiconductive thin films, *Anal. Chem.*, 34, 1502–1503, 1962.
33. McAleer, J.F. et al., Tin dioxide gas sensors, *J. Chem. Soc. Faraday Trans.*, 183, 1323, 1987.
34. Covington, J.A., CMOS and SOI CMOS FET-based gas sensors, Ph.D. Thesis, University of Warwick, Coventry, CV4 7AL, UK, 2001.
35. Colea, M. et al., Parametric model of a polymeric chemoresistor for use in smart sensor design and simulation, *Microelectr. J.*, 34, 865, 2003.
36. Liu, Z.F. et al., O_2 and CO sensing of Ga_2O_3 multiple nanowire gas sensors, *Sens. Actuat. B*, 129, 666, 2008.
37. Wang, H.T. et al., Hydrogen-selective sensing at room temperature with ZnO nanorods, *Appl. Phys. Lett.*, 86, 243503, 2005.
38. Tien, L.C. et al., Hydrogen sensing at room temperature with Pt-coated ZnO thin films and nanorods, *Appl. Phys. Lett.*, 87, 222106, 2005.
39. Neri, G. et al., In_2O_3 and $Pt–In_2O_3$ nanopowders for low temperature oxygen sensors, *Sens. Actuat. B*, 127, 455, 2007.
40. Pan, Z.W., Dai, Z.R., and Wang, Z.L., Nanobelts of semiconducting oxides, *Science*, 291, 1947, 2001.
41. Cui, Y. and Lieber, C.M., Functional nanoscale electronic devices assembled using silicon nanowire building blocks, *Science*, 291, 851, 2001.
42. Kolmakov, A. et al., Detection of CO and O_2 using tin oxide nanowire sensors. *Adv. Mater.*,15, 997, 2003.
43. Severin, E.J., Doleman, B.J., and Lewis, N.S., An investigation of the concentration dependence and response to analyte mixtures of carbon black/insulating organic polymer composite vapor detectors, *Anal. Chem.*, 72, 658, 2000.
44. Ingleby, P., Gardner, J.W., and Bartlett, P.N., Effect of micro-electrode geometry on response of thin-film polypyrrole and polyaniline chemoresistive sensors, *Sens. Actuat.*, B 57, 17, 1999.

45. Stetter, J.R., Penrose, W.R., and Yao, S., Sensors, chemical sensors, electrochemical sensors, and ECS, *J. Electrochem. Soc.*, 150, S11, 2003.
46. Dr. Zhi Chen's research website at: www.engr.uky.edu/~zhichen/RESEARCH/research.html (retrieved on April 23, 2010).
47. Hsu, L.C. et al., Evaluation of commercial metal-oxide based NO_2 sensors, *Sens. Rev.*, 27, 121, 2007.
48. Arshak, K. et al., A review of gas sensors employed in electronic nose applications, *Sens. Rev.*, 24, 181, 2004.
49. Electrochemical sensors, International Sensor Technology Company, Irvine, California, USA. Available at: www.intlsensor.com/pdf/electrochemical.pdf.
50. Ho, C.K. McGrath, L.K., and May, J., FY02 Field Evaluations of an In-Situ Chemiresistor Sensor at Edwards Air Force Base, California, Sandia National Laboratories, Albuquerque, New Mexico, USA, 2002.
51. James, D. et al., Review chemical sensors for electronic nose systems, *Microchim. Acta*, 149, 1, 2005.
52. BIA 310e bioimpedance analyzer datasheet, Biodynamics Corporation, Seattle, Washington, USA. www.biodyncorp.com.
53. Boccaletti, C. et al., A non-invasive biopotential electrode for the correct detection of bioelectrical currents, *Proc. 6th IASTED Int. Conf. Biomed. Eng.*, Innsbruck, Australia, 2008, 353.
54. Heine, H., Manuale di medicina biologica, Guna Editore, Milano, 207, 1999.

3 Capacitive Sensors

3.1 INTRODUCTION

Capacitive sensors operate based on changes in electrical capacitance. Capacitive sensors are the most precise of all electrical sensors (including resistive and inductive sensors) and are known for their extremely high sensitivity, high resolution (e.g., 0.01 nm), broad bandwidth (e.g., 1 ~ 100 kHz), robustness, long-term stability and durability, drift-free character, simple structures, low cost, and noncontact detection features. Most capacitive sensors are immune to humidity, temperature, target material, and stray electric field variations. Some can be integrated into a printed circuit board (PCB) or embedded into a microchip or a nanodevice to provide excellent accuracy and nearly infinite resolution, higher reliability, less weight, and lower power consumption.

The earliest capacitive sensing can be traced back to 1600 when William Gilbert experimented with frictional electrical charges on objects [1]. He found that electrical charges cause objects to attract or repel each other. This attracting or repelling force was greatly affected by the distance between the objects. In 1745, the first capacitor, the Leyden jar, was invented independently by Ewald Georg von Kleist and Pieter van Musschenbroek [2].

Capacitive sensors, also called *detecting probes*, are traditionally divided into *passive* or *active* types, depending on whether or not there are any electronic components in the probe. Passive sensors do not come with any electronics; thus the sensor sizes can be minimized. They are also more flexible in probe design, more stable, and less expensive. Their disadvantages include cable length restrictions (≤3 m), narrow bandwidth, and lower drive frequency. The active sensors, on the other hand, have electronics in the probe. The electronics can be as simple as a few diodes, or as complex as an integrated circuit board (typically enclosed within the guard shield). Active sensors are not restricted by the cable length. They operate at much higher frequencies with broader bandwidths, and are particularly suitable to applications that involve stray electrical noise on the target. The disadvantages of active sensors are their higher cost and lesser design flexibility.

Capacitive sensors can also be classified based on their configurations (e.g., parallel, cylindrical, and spherical), dielectric materials (e.g., polymeric and fluidic), measurants (e.g., acceleration and CO_2), mechanisms that cause capacitance change (e.g., space variation and area variation), or manufacturing means (e.g., micromachined and macroscopic) [3]. Micromachined capacitive sensors are built directly on a silicon wafer, usually integrated with an application-specific integrated circuit (ASIC) for high reliability and low production cost. A macroscopic capacitive sensor is usually manufactured as an individual component or is packaged on a PCB.

Capacitive sensors find their applications in precision motion detection, coating thickness gauging, liquid level and flow rate monitoring, pressure or force measurement, diamond turning, occupancy identification, fingerprint acquisition, chemical element selection, biocell recognition, engine rotational alignment, as well as on keyswitches, touchpads, and touchscreens.

This chapter will explore capacitive sensors based on their configurations (flat/parallel, cylindrical/coaxial, spherical/concentric, and array) and their sensing mechanisms that cause capacitance changes through

- *Spacing variation*: varying the space or distance between plates
- *Area variation*: varying the overlap area between plates
- *Electrode property change*: changing conductivity, charges, mass, or other physical or chemical properties of the electrodes
- *Dielectric material property change*: changing properties of the dielectric media

In each configuration, the sensing principles will be discussed first, followed by their features, designs, and applications. To assist readers understand these principles better, an overview of capacitors, capacitance, and associated physical laws are presented as follows.

3.2 CAPACITORS AND CAPACITANCE

A capacitor is a passive electrical or electronic component that can store energy in the form of an electric field. Capacitance, typified by a parallel-plate arrangement, is defined in terms of charge storage:

$$C = \frac{Q}{V} \tag{3.1}$$

where C is the capacitance (in farads, F), Q is the charge (in coulomb, C), and V is the voltage difference between the two plates (in volts, V). V can be expressed in terms of the work done on a positive test charge Q when it moves from the positive to the negative plate:

$$V = \frac{\text{work done}}{\text{charge}} = \frac{Fd}{Q} = Ed \tag{3.2}$$

where E is the electric field (in volt per meter, $V \cdot m^{-1}$ or newton per coulomb, $N \cdot C^{-1}$) between two parallel plates, F is the force (in newton, N), and d is the distance between two plates.

Commonly used capacitors have a capacitance range from 1 pF to 1000 µF. The voltage–current relationship of a capacitor is expressed by

$$V(t) = \frac{1}{C} \int I(t)\,dt \tag{3.3}$$

Capacitors in series:

$$\frac{1}{C_{eq}} = \sum \frac{1}{C_i} \tag{3.4}$$

Capacitors in parallel:

$$C_{eq} = \sum C_i \tag{3.5}$$

Thus, if two capacitors are connected in series, the total capacitance is $(C_1 C_2)/(C_1 + C_2)$; if they are connected in parallel, the total capacitance is $C_1 + C_2$. Table 3.1 summarizes the types of capacitors. Figure 3.1 shows the circuit symbols and certain types of capacitors.

Capacitive sensors are variable capacitors, that is, their capacitance changes during the sensing process. Capacitance is a function of geometry, dielectric constant, plate materials, and plate configuration. The capacitance calculations for different configurations are summarized in Table 3.2.

TABLE 3.1

Types of Capacitors

Based on Structures		**Based on Electrodes**
Parallel (flat) configuration and their array		Electrostatic
Cylindrical (coaxial) configuration		Electrolytic
Spherical (concentric) configuration		Multielectrode

Based on Dielectric Materials		**Based on Applications**
Air gap	Paper	Energy storage
Ceramic	Polyester	Power conditioning
Mica	Polystyrene	Signal processing
Mylar	Polycarbonate	Sensing
Glass	Teflon	Memory

Based on Capacitance Values	**Based on Packaging**
Fixed (constant)	Through-hole
Variable (adjustable)	Surface mount
Super	

Based on Polarization
Polarized
Nonpolarized

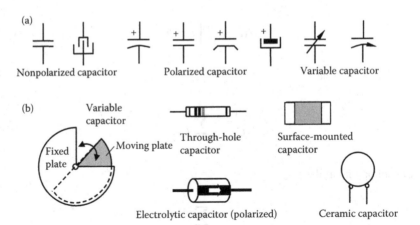

FIGURE 3.1   (a) Circuit symbols of capacitors; (b) certain types of capacitors.

TABLE 3.2
Capacitance Calculations for Certain Configurations

Parallel-(flat) plate capacitor

$$C = \frac{\varepsilon_0 \varepsilon_r A}{d} \tag{3.6}$$

Ratio A/d is called the *geometry factor* for a parallel-plate capacitor.

Cylindrical (coaxial) capacitor

$$C = \frac{2\pi\varepsilon_0\varepsilon_r h}{\ln(r_2/r_1)} \quad (h \gg r_2) \tag{3.7}$$

Ratio $2\pi h/\ln(r_2/r_1)$ is the *geometry factor* for a cylindrical capacitor.

Spherical (concentric) capacitor

$$C = \frac{4\pi\varepsilon_0\varepsilon_r r_1 r_2}{r_2 - r_1} \tag{3.8}$$

Ratio $4\pi r_1 r_2/(r_2 - r_1)$ is the *geometry factor* for a spherical capacitor.

TABLE 3.2 (continued)
Capacitance Calculations for Certain Configurations

Two spheres [4]

$$C = \frac{4\pi\varepsilon_0\varepsilon_r}{\dfrac{r_1+r_2}{r_1 r_2} - \dfrac{1}{d}}$$

(3.9)

Two parallel cylinders [4]

$$C = \frac{\pi\varepsilon_0\varepsilon_r h}{\ln\left(\dfrac{d + \sqrt{d^2 - 4r^2}}{2r}\right)}$$

(3.10)

One cylinder and one plate [4]

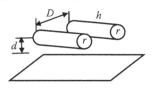

$$X = \frac{2\pi\varepsilon_0\varepsilon_p \eta}{\ln\left(\dfrac{\delta + \sqrt{\delta^2 - \rho^2}}{\rho}\right)}$$

(3.11)

Two cylinders and one plate [4]

$$C = \frac{\pi\varepsilon_0\varepsilon_r h \cdot \ln\left(1 + \dfrac{2d}{D}\right)}{\ln\left(\dfrac{2d}{r}\right)^2} \quad (2d \gg r)$$

(3.12)

Note: A is the area of the flat plate (in meter square, m²), d is the spacing (in meter, m) between the plates; ε_0 is the electric permittivity in vacuum (8.85 × 10⁻¹² F · m⁻¹), ε_r is the *relative dielectric permittivity* (RDP, unitless) or *dielectric constant* of the media between the plates. r_1, r_2, and r are the radii of the cylindrical or spherical plates (in meter, m), and h is the length or height of the cylindrical plates (in meter, m).

Source: From Baxter, L.K., *Capacitive Sensors: Design and Applications*, 1st ed., IEEE Press, New York, USA, 1997. With permission.

Dielectric constant ε_r (usually determined empirically) is the ratio of the permittivity of a substance ε_p to the permittivity of free space ε_0 (i.e., $\varepsilon_r = \varepsilon_p/\varepsilon_0$). It is an expression of the extent to which a material concentrates electric flux, or the ability of a material to store a charge under an applied electric field. ε_r has the range of 2 ~ 20 for most dry solid materials and often higher for liquids (see Table 3.3).

TABLE 3.3

RDP of Common Materials at Room Temperature

Material	RDP, ε_r	Material	RDP, ε_r	Material	RDP, ε_r
Air, vacuum	1	Paper	3.5	Mica	6
Clay	1.8–2.8	Silica glass	3.7	Rubber	7
Teflon	2	Silicon dioxide	3.9	Marble	8
Soft rubber	2.5	Nylon	4–5	Silicone	11–12
Wood	2.7	Porcelain	4.4	Alcohol	16–31
Silicone rubber	2.8	Diamond	5.5–10	Fresh water	80
Ice	3–4	Glass	5	Sea water	81–88

Source: *Dielectric Constants of Materials*, Clipper Controls, Inc., San Francisco, California, USA, 2011. With permission.

The ratio of ε_r/t_h is called *sensitivity factor S* (in m^{-1}), where t_h is the thickness of the dielectric layer. S is an important factor in touch sensor design. In the case of a stack of dielectric materials, the combined S_i for the whole stack S_{Stack} is given by

$$\frac{1}{S_{Stack}} = \sum_{i=1}^{n} \frac{1}{S_i} \tag{3.13}$$

The electric field E between two parallel plates is

$$E = \frac{Q}{\varepsilon_0 A} \tag{3.14}$$

The electric field E on a cylindrically symmetric surface around a rod contactor is

$$E = \frac{Q}{2\pi\varepsilon_0 r} \tag{3.15}$$

EXAMPLE 3.1

The plates of a parallel capacitor have a separation of 2.85 mm, and each has an area of 10.2 cm^2. If a charge of 3.95×10^{-8} C is carried by each plate and the plates are in vacuum, find (1) the capacitance C, (2) the potential difference V between the plates, and (3) the magnitude of the electric field E between the plates.

SOLUTIONS

1. The capacitance C is

$$C = \frac{\varepsilon_0 \varepsilon_r A}{d} = \frac{(8.85 \times 10^{-12}\, F \cdot m^{-1})\,(1)(10.2 \times 10^{-4}\, m^2)}{2.85 \times 10^{-3}\, m} = 3.17\, pF$$

2. The potential difference V between the plates is

$$V = \frac{Q}{C} = \frac{3.95 \times 10^{-8}\, C}{3.17 \times 10^{-12}\, F} = 12.46 \times 10^3\, V$$

3. The magnitude of the electric field E between the plates is

$$E = \frac{V}{d} = \frac{12.46 \times 10^3\, V}{0.00285\, m} = 4.37 \times 10^6\, V \cdot m^{-1}$$

EXAMPLE 3.2

A Geiger counter (a capacitive sensor used to detect ionizing radiation) consists of a thin straight wire surrounded by a coaxial conducting shell and filled with low-pressure inert neon gas ($\varepsilon_r \approx 1$). The diameter of the wire is 0.025 mm and that of the shell is 25 mm. If the length of the tube is 120 mm, what is the capacitance of the Geiger counter?

SOLUTION

Since the Geiger counter tube has a cylindrical configuration, thus its capacitance can be calculated using Equation 3.7:

$$C = \frac{2\pi \varepsilon_0 \varepsilon_r h}{\ln(r_2/r_1)} = \frac{2(3.14)(8.854 \times 10^{-12}\, F \cdot m^{-1})(1)(0.12\, m)}{\ln[(25 \times 10^{-3}/2)/(25 \times 10^{-6}/2)]} = 0.97\, pF$$

EXAMPLE 3.3

The diameters of the internal and external plates of a spherical capacitor are 38.0 and 40.0 mm, respectively. The dielectric media between the plates is air ($\varepsilon_r = 1$). (1) Calculate its capacitance. (2) What would be the plate area of a parallel capacitor with the same plate separation, capacitance, and dielectric media?

SOLUTIONS

1. Apply Equation 3.8 for the spherical capacitor:

$$C = \frac{4\pi \varepsilon_0 \varepsilon_r r_1 r_2}{r_2 - r_1} = \frac{4(3.14)(8.85 \times 10^{-12}\, F \cdot m^{-1})(1)(19 \times 10^{-3}\, m)(20 \times 10^{-3}\, m)}{(20 \times 10^{-3}\, m - 19 \times 10^{-3}\, m)}$$
$$= 42.24\, pF$$

2. For a parallel-plate capacitor with the same plate separation:

$$d = r_2 - r_1 = 20 \times 10^{-3} \text{ m} - 19 \times 10^{-3} \text{ m} = 0.001 \text{ m}$$

and with the same capacitance: $C = 42.24$ pF, the plate area is (from Equation 3.6)

$$A = \frac{Cd}{\varepsilon_0 \varepsilon_r} = \frac{(42.24 \times 10^{-12} \text{ F})(1 \times 10^{-3} \text{ m})}{(8.85 \times 10^{-12} \text{ F} \cdot \text{m}^{-1})(1)} = 4.77 \times 10^{-3} \text{ m}^2 \text{ or } 47.70 \text{ cm}^2$$

3.3 PHYSICAL LAWS AND EFFECTS GOVERNING CAPACITIVE SENSORS

3.3.1 COULOMB'S LAW

Coulomb's law describes the force F between any two charges Q_1 and Q_2 separated by a distance d:

$$F = \frac{Q_1 Q_2}{4\pi\varepsilon_0\varepsilon_r d^2} \tag{3.16}$$

EXAMPLE 3.4

A 2 C charge is 1.5 m away from a −3 C charge with paper between them. Determine the force they exert on each other.

SOLUTION

Apply Equation 3.16 and $\varepsilon_r = 3.5$ from Table 3.2:

$$F = \frac{Q_1 Q_2}{4\pi\varepsilon_0\varepsilon_r d^2} = \frac{(2\text{C})(-3\text{C})}{4(3.14)(8.85 \times 10^{-12} \text{ F} \cdot \text{m}^{-1})(3.5)(1.5\text{m})^2} = -6.85 \times 10^9 \text{ N}$$

The negative sign indicates an attractive force between the charges.

EXAMPLE 3.5

Three charges in air are arranged in a triangle as shown in Figure 3.2a. Determine the force on Q_2.

SOLUTION

The force exerted on Q_2 by Q_1 is (the arrow indicates the direction of the force)

$$F_{1-2} = \frac{Q_1 Q_2}{4\pi\varepsilon_0\varepsilon_r (d_{1-2})^2} = \frac{(-2\text{C})(1\text{C})}{4(3.14)(8.85 \times 10^{-12} \text{ F} \cdot \text{m}^{-1})(1)(3\text{m})^2} = -2.0 \times 10^9 \text{ N}\uparrow$$

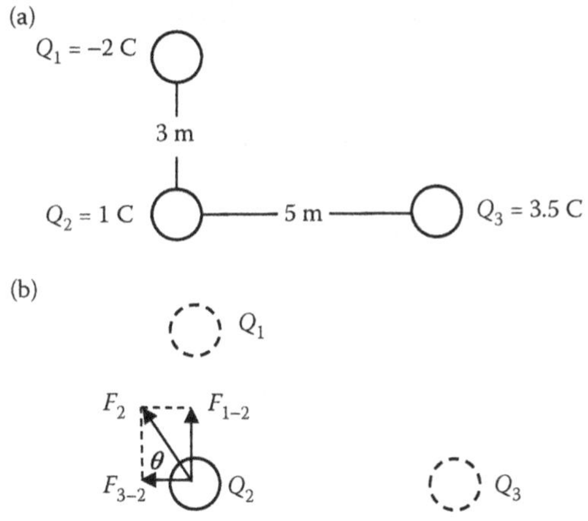

FIGURE 3.2 (a) Three charge diagram; (b) force diagram for Q_2.

The force exerted on Q_2 by Q_3 is

$$F_{3-2} = \frac{Q_3 Q_2}{4\pi\varepsilon_0\varepsilon_r(d_{3-2})^2} = \frac{(3.5\,C)(1\,C)}{4(3.14)(8.85 \times 10^{-12}\,F \cdot m^{-1})(1)(5\,m)^2} = 1.26 \times 10^9\,N \leftarrow$$

The net force acting on Q_2 is

$$F_2 = \sqrt{(F_{1-2})^2 + (F_{3-2})^2}\sqrt{(-2.0 \times 10^9\,N)^2 + (1.26 \times 10^9\,N)^2} = 2.36 \times 10^9\,N$$

The direction of the net force is (see Figure 3.2b)

$$\theta = \tan^{-1}\left(\frac{F_{1-2}}{F_{3-2}}\right) = \tan^{-1}\left(\frac{-2.0 \times 10^9}{1.26 \times 10^9}\right) = 57.79°$$

3.3.2 Gauss's Law for Electric Field

Gauss's law is one of Maxwell's four fundamental equations for electricity and magnetism. It states that the total electric flux out of a closed surface, Φ_E, is equal to the total charge enclosed ΣQ_i divided by the permittivity in free space ε_0 [4]:

$$\Phi_E = \frac{\Sigma Q_i}{\varepsilon_0} \tag{3.17}$$

The electric flux Φ_E (in volt · meter, V · m; or in newton meter square per coulomb, N · m² · C⁻¹) is defined as a surface integral of the electric field E over any closed surface A:

$$\Phi_E = \oint \vec{E} \cdot d\vec{A} \tag{3.18}$$

Gauss's law can be used to find the electric field in situations with a high degree of symmetry.

EXAMPLE 3.6

(1) Find the electric field E at distance r from a point charge Q. (2) Find the electric field E at distance r from the charge Q that is uniformly distributed on a thin spherical surface (centered at O) with a radius of R.

SOLUTIONS

1. In order to apply Gauss's law, an imaginary sphere (Gaussian sphere) around the charge Q (i.e., Q is at the center of the sphere) with a radius of r is established. E is uniformly distributed on the sphere's surface (i.e., E is constant) as shown in Figure 3.3a.

   Apply Equation 3.18 with constant E and total surface area of a sphere $4\pi r^2$:

$$\oint \vec{E} \cdot d\vec{A} = \frac{\Sigma Q_i}{\varepsilon_0} \Rightarrow E 4\pi r^2 = \frac{Q}{\varepsilon_0} \Rightarrow E = \frac{Q}{4\pi\varepsilon_0 r^2}$$

2. Create a Gaussian sphere centered at O with a diameter of r as shown in Figure 3.3b

   If $r < R$, $Q = 0$ (since no charge exists inside the thin sphere), thus

$$\oint \vec{E} \cdot d\vec{A} = \frac{\Sigma Q_i}{\varepsilon_0} \Rightarrow E 4\pi r^2 = \frac{0}{\varepsilon_0} \Rightarrow E = 0$$

If $r > R$

$$\oint \vec{E} \cdot d\vec{A} = \frac{\Sigma Q_i}{\varepsilon_0} \Rightarrow E 4\pi r^2 = \frac{Q}{\varepsilon_0} \Rightarrow E = \frac{Q}{4\pi\varepsilon_0 r^2}$$

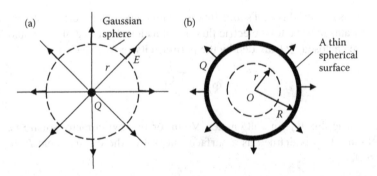

FIGURE 3.3 (a) The electrical field around a point charge Q; (b) the electrical field around a charge Q that is uniformly distributed on a thin spherical surface.

The electric field for parallel (Equation 3.6) and cylindrical plates (Equation 3.7) can also be found using Gauss's law.

EXAMPLE 3.7

Derive the capacitance equation for a cylindrical capacitor (Equation 3.7) using Gauss's law for $h \gg r_2$.

SOLUTION

Choose an arbitrary cylindrical surface (Gaussian surface) distant r from the center point (see Figure 3.4). Using Gauss's Law, the total electric flux out of the closed cylindrical surface is

$$\Phi_E = \oint \vec{E} \cdot d\vec{A} = E(2\pi rh) \Rightarrow$$

$$E(2\pi rh) = \frac{Q}{\varepsilon_0} \Rightarrow E = \frac{Q}{2\pi\varepsilon_0 rh}$$

Voltage, V, between cylindrical plates is

$$V = V(r_1) - V(r_2) = \int_{r_1}^{r_2} Edr = \int_{r_1}^{r_2} \frac{Q}{2\pi\varepsilon_0 rh} dr = \frac{Q}{2\pi\varepsilon_0 h} \int_{r_1}^{r_2} \frac{dr}{r} = \frac{Q}{2\pi\varepsilon_0 h} \ln(r_2/r_1)$$

Since $C = Q/V$, the capacitance for the cylindrical geometry is

$$C = Q/V = \frac{2\pi\varepsilon_0 h}{\ln(r_2/r_1)}$$

If the dielectric media is not in free space, then ε_r should be added into the above equation.

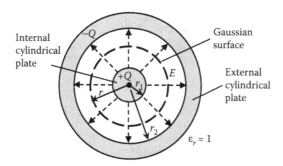

FIGURE 3.4 The cross section of a cylindrical capacitor.

3.3.3 PIEZOELECTRIC EFFECT

A piezoelectric element will produce an electric charge when a mechanical stress (longitudinal, transverse, or shear) is applied. Conversely, mechanical deformation (e.g., shrinkage or expansion) occurs when an electric field is applied to a piezoelectric element. This effect is called *Piezoelectric effect*, discovered by the Curie brothers more than 100 years ago. They found that quartz changed its dimensions when subjected to an electrical field, and conversely, generated electrical charge when mechanically deformed. "Piezo," derived from the Greek word *piezein*, means to squeeze or press. The first piezoelectric sensor was published by J. Thomson in 1919 for pressure measurement [5]. In the 1920s, Langevin developed a quartz transmitter and receiver using *Piezoelectric effect* for underwater sound detection the first SONAR (SOund Navigation And Ranging).

Some solid materials (e.g., crystals and certain ceramics) and biological matters (e.g., bones, tendons, DNA, and various proteins) display the piezoelectric effect. In 1969, Kawai also found very high piezoelectric activity in polarized fluoropolymer and polyvinylidene fluoride (PVDF). The most commonly used piezoelectric materials are single crystal (or monocrystalline, quartz), tourmaline (a crystal boron silicate mineral), gallium orthophosphate ($GaPO_4$), lithium niobate ($LiNbO_3$), and lithium tantalite ($LiTaO_3$). In crystals, each crystal molecule is polarized (called a *dipole*, i.e., one end is negatively charged and the other end is positively charged). In a monocrystalline, all dipoles lie in one direction (symmetrical crystal), while in a polycrystal, there are different regions within the material that have different polar axes (asymmetrical crystals), as shown in arrows in Figure 3.5.

For a cylindrical shape of piezoelectric element with a radius r, if an axial force F is applied normal to the ends of the element (assuming that the force is evenly distributed over the end of face), the stress σ (in newton per square meter, $N \cdot m^{-2}$) is

$$\sigma = \frac{F}{A} = \frac{F}{\pi r^2} \tag{3.19}$$

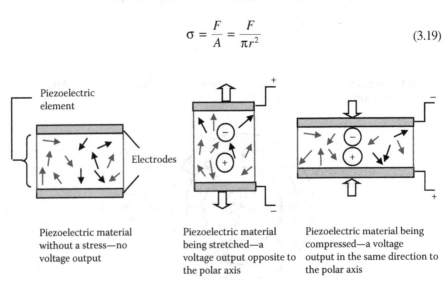

Piezoelectric material without a stress—no voltage output

Piezoelectric material being stretched—a voltage output opposite to the polar axis

Piezoelectric material being compressed—a voltage output in the same direction to the polar axis

FIGURE 3.5 Piezoelectric effect in polycrystals.

The electric flux density D_Φ (in coulomb per square meter, $C \cdot m^{-2}$) in the piezo-electric element is

$$D_\Phi = d_{11}\sigma = \frac{d_{11}F}{\pi r^2} \tag{3.20}$$

where d_{11} is the *piezoelectric charge coefficient* (in coulomb per newton, $C \cdot N^{-1}$). If the piezoelectric element is fitted between two metal plates (electrodes), the charge Q accumulated on the metal plates can be calculated by multiplying the above equation with the area of the piezoelectric element:

$$Q = D_\Phi A = D_\Phi (\pi r^2) = d_{11}F \tag{3.21}$$

The above equation indicates that the polarization charge Q yielded in a piezoelectric element is proportional to the applied force F. Piezoelectric sensors are primarily used for measuring force, strain, pressure, acceleration, and acoustic emission.

EXAMPLE 3.8

A piezoelectric sensor is made by fitting a 2.5 cm × 1 cm rectangular quartz between two metal plates. The plates can transmit the force to the piezoelectric element and also serve as electrodes to collect the electric charges yielded for output. The quartz has a piezoelectric charge coefficient of 2.3×10^{-12} C · N^{-1} at 20°C. If a 100 N force is applied, find (1) the stress σ of the sensor, (2) the electric flux density D_Φ, and (3) the output charge Q.

SOLUTIONS

1. Assume that the force is evenly distributed over the ends of faces, the stress is then

$$\sigma = \frac{F}{A} = \frac{100\,N}{(0.025\,m)(0.01\,m)} = 400\,kPa$$

2. The electric flux density D_Φ is

$$D_\Phi = d_{11}\sigma = (2.3 \times 10^{-12}\,C \cdot N^{-1})(400 \times 10^3\,N \cdot m^{-2}) = 9.2 \times 10^{-7}\,C \cdot m^{-2}$$

3. The polarization charge Q is

$$Q = d_{11}F = (2.3 \times 10^{-12}\,C \cdot N^{-1})(100N) = 2.3 \times 10^{-10}\,C$$

3.3.4 EFFECT OF EXCITATION FREQUENCIES

A capacitive sensor may require a high-frequency excitation, close to the resonance frequency of the circuit, so that the electrode impedance reaches its minimum and the sensitivity of the sensor reaches its maximum. The excitation frequency also affects the operating range of a capacitive sensor. For example, a *ChenYang Technologies'* (in Germany) capacitive displacement sensor has a measurement range of 1 mm at the excitation frequency 28 kHz, but its measurement range can reach 4 mm at the resonance frequency of 30.4 kHz [6]. The measurement range of a capacitive sensor also depends on the size of the active sensor area. Once the excitation frequency passes the resonance frequency, the sensor's sensitivity starts to decrease. For the same displacement sensor, at the resonance frequency of 30.4 kHz, its sensitivity is 10.25 V·mm^{-1}, while at 32 kHz, its sensitivity is decreased to 3.41 V·mm^{-1} [6]. Some capacitive motion detectors can achieve a resolution of 0.5 nm using an excitation frequency that is closer to the resonance frequency. The linearity of a capacitive sensor can also be improved by optimizing the excitation frequency. Piezoelectric polymer film sensors have a wide frequency range from 0.001 to 10^9 Hz and capacitance in the order of a couple of hundred pF to a few nF [7].

3.4 PARALLEL-PLATE (FLAT-PLATE) CAPACITIVE SENSORS

The simplest capacitive sensor consists of two parallel metal plates separated by a distance d (Figure 3.6). To eliminate the fringe field or edge effect of the electrodes, the guard electrodes are added and kept at the same potential as the sensing electrode.

Most capacitive sensors have the parallel-plate or configuration. They are widely used to detect motion, distance, pressure, acceleration, fluid level, chemicals, and biocells. The sensing principles are well explained by Equation 3.6: any variation in the dielectric constants (ε_r), the plate area (A), or the plate spacing (d) will cause a change in capacitance (C). Figure 3.7 shows the four sensing principles used in parallel capacitive sensors. Each type is discussed in the following sections.

3.4.1 SPACING-VARIATION-BASED SENSORS

3.4.1.1 Sensing Principle and Characteristics

In this design type, spacing d between the two parallel plates varies, causing the capacitance C to change. An inverse relationship exists between the spacing d and the capacitance C. This gives a large change in capacitance value with a small spacing

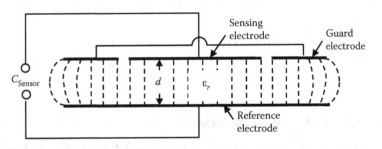

FIGURE 3.6 A parallel capacitive sensor with a guard electrode.

(a) Spacing variation   (b) Area variation   (c) Dielectric constant   (d) Electrode property
                                                    variation   variation

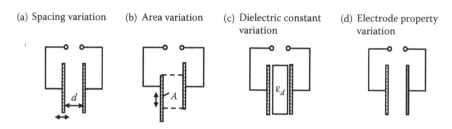

FIGURE 3.7 Four basic principles underneath the parallel capacitive sensor design.

change, but it also displays the nonlinearity. A bridge or signal conditioning circuit is often required to compensate for this nonlinearity.

Compared to other capacitive sensors, spacing-variation-based sensors are generally more sensitive, but only suitable for a small displacement range (usually the spacing variation is less than the electrode size and in the range of micrometers) [8]. They are used for industry standards for ultra-high-precision measurements. Area-variation-based capacitive sensors, on the other hand, are less sensitive, but suitable for a larger displacement range [9].

3.4.1.2 Sensor Design

Spacing-variation-based sensors can have single-plate, dual-plate, or multiplate designs. In the single-plate design, the sensing plate (or probe) functions as one electrode, while the target, made of a conductive material, functions as the second electrode. Note that in this case, the sensor is not affected by the specific target material as long as it is conductive. This is because the sensing electric field stops at the surface of the conductive target and the thickness of the target does not affect the measurement [10]. Figure 3.8 shows the application examples of single-plate sensors [11]. Traditionally, a capacitive sensor system has the sensing plate (or probe)

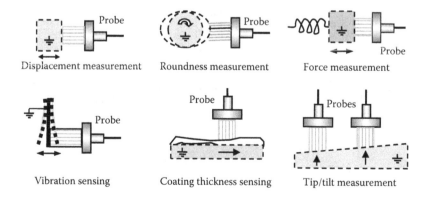

FIGURE 3.8 Application examples of single-plate, spacing-variation-based sensors. (From Capacitive position sensors/controllers for nanometrology, Physik Instrumente, GmbH & Co. KG, Germany. www.capacitance-sensors.com/index.htm. With permission.)

driven and the target grounded. A poorly grounded target will be more susceptible to noise pickup. Target resistance of less than 100 kΩ will unlikely cause measurement errors. High-performance displacement sensors usually use small sensing surfaces and they are positioned close to the targets (0.25 ~ 2 mm).

The important characteristics of spacing-variation-based sensors include their high resolution and high accuracy (often better than 1 nm) at bandwidths between 1 and 100 kHz. The performance of a capacitive sensor also depends on its size—in general, the larger the sensor, the larger the sensing range. However, doubling a sensor's size will also increase its RMS noise level; larger sensing range also increases the sensor's nonlinearity and decreases its spatial resolution. Although there is no specific constraint on a sensor's shape, typical sensing probes are round. As a general rule for a round sensor, the diameter of the sensor should be no less than 4 times the total detection range and about 0.5–10 mm in diameter.

A target's surface may be curved or irregular; capacitive sensors usually measure the average distance to the target. Smaller sensors have better abilities to distinguish small features on a target, as shown in Figure 3.9. Thus, a sensor's size is often chosen based on the size of the target and is usually chosen smaller than the target. Figure 3.10 gives some guidelines for determining a maximum sensor size.

Another principal advantage of noncontact capacitive sensors is their ability to measure a rapid motion. Applications such as servo control, spindle analysis, and vibration evaluation often require a sensor to measure very small and fast motions. Active sensor systems are designed to meet most of these application requirements with a bandwidth up to 100 kHz, while passive systems can work well at a bandwidth of 20 kHz or less. The resolution of a system is usually proportional to the square root of the bandwidth.

In a dual-plate design shown in Figure 3.11, the sensor has two parallel-plate electrodes. When the moving plate experiences a force, pressure, or vibration, the distance d will change, causing a change in capacitance.

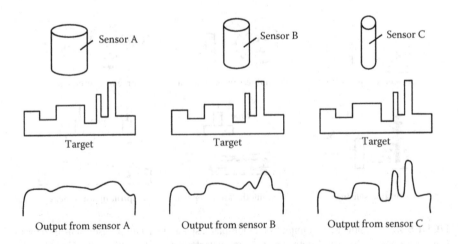

FIGURE 3.9 Sensor size and its spatial resolution.

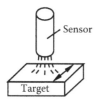

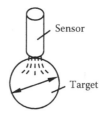

Flat surface:
Maximum sensor diameter =
0.6 × Minimum target
detecting surface dimension

Cylindrical surface:
Maximum sensor diameter =
0.25 × Minimum target dimension

Spherical surface:
Maximum sensor diameter =
0.2 × Minimum target dimension

FIGURE 3.10   Selection guidelines for the maximum sensor size.

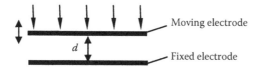

FIGURE 3.11   A dual-plate capacitive sensor design.

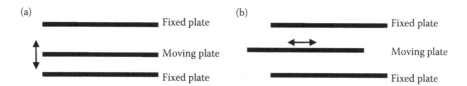

FIGURE 3.12   Three-plate configurations for (a) spacing-variation- and (b) area-variation-based sensors.

Sometimes, temperature change can also affect d due to the thermal expansion of the plates and/or dielectric media. This can be compensated by applying the differential capacitance technique that uses three plates as shown in Figure 3.12. By adding the third electrode, twin capacitors are formed: C_1 (between the top and middle plates) and C_2 (between the middle and bottom plates). Proper wiring and adding an amplifier circuit can make a sensor's output voltage proportional to $C_1 - C_2$, C_1/C_2, or $(C_1 - C_2)/(C_1 + C_2)$ to compensate for temperature/humidity variation and small changes in dielectric constants. Another advantage of the twin capacitor system is that it can be incorporated into two arms of an AC bridge to linearize the output signal [12]. Both spacing-variation- and area-variation-based sensors have the three-plate configurations as shown in Figure 3.12a and b, respectively.

EXAMPLE 3.9

Compare the capacitance values of (1) a two-plate capacitor with a spacing d, (2) a three-plate capacitor (in series) with a spacing $d/2$ between two adjacent plates, and (3) a three-plate capacitor (in parallel) with a spacing $d/2$ between two adjacent plates. Assume that the plate area A and dielectric constants ε_r maintain the same in each case.

SOLUTIONS

1. The capacitance for a two-plate capacitor is

$$C = \frac{\varepsilon_0 \varepsilon_r A}{d}$$

2. The capacitance of a three-plate capacitor in series is

$$\frac{1}{C} = \frac{1}{C_1} + \frac{1}{C_2} \Rightarrow C = \frac{C_1 C_2}{C_1 + C_2} = \frac{\left(\varepsilon_0 \varepsilon_r A/(d/2)\right)\left(\varepsilon_0 \varepsilon_r A/(d/2)\right)}{\left(\varepsilon_0 \varepsilon_r A/(d/2)\right) + \left(\varepsilon_0 \varepsilon_r A/(d/2)\right)} = \frac{\varepsilon_0 \varepsilon_r A}{d}$$

3. The capacitance of a three-plate capacitor in parallel is

$$C = C_1 + C_2 = \frac{\varepsilon_0 \varepsilon_r A}{d/2} + \frac{\varepsilon_0 \varepsilon_r A}{d/2} = \frac{4\varepsilon_0 \varepsilon_r A}{d}$$

The above results show that a serially connected three-plate capacitor (with its middle plate located at the midpoint) has the same capacitance as a two-plate capacitor, while a parallelly connected three-plate capacitor has four times as much capacitance as a two-plate capacitor.

To improve a sensor's measurement range, accuracy, sensitivity, stability, and signal-to-noise ratio, a multiplate configuration can be used, which could be a replication of the single-plate system many times or a replication of the three-plate system many times. This technique works well for both linear and rotary designs. An additional circuit is often required to count plates. Figure 3.13 shows a multiplate capacitive sensor that is often seen in capacitive acceleration, vibration, and displacement measurements.

EXAMPLE 3.10

Figure 3.14a and b show a simplified version of a capacitive accelerometer at rest and under an applied acceleration, respectively. The actual sensor consists of 42 silicon finger cells (as fixed plates) and a common beam (as moving plates),

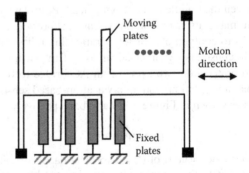

FIGURE 3.13   A multiplate capacitive sensor.

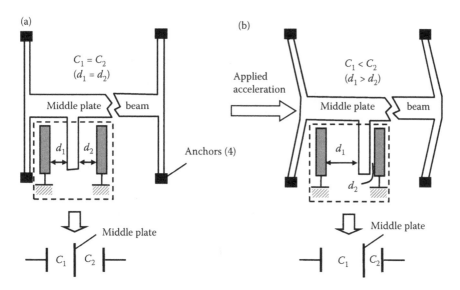

FIGURE 3.14 A capacitive accelerometer (a) at rest and (b) under an applied acceleration.

together forming 42 groups of twin capacitors. The twin capacitors in each group, C_1 and C_2, are serially connected. If no acceleration is applied, $C_1 = C_2$; if an acceleration is applied, $C_1 \neq C_2$. Assume that the sensor is under an acceleration causing $C_1 = 0.004$ pF, $C_2 = 0.006$ pF, what is the total capacitance of the entire sensor?

SOLUTION

Since the two capacitors are serially connected, the resultant capacitance in each finger unit is

$$C = \frac{C_1 C_2}{C_1 + C_2} = \frac{(0.004\,\text{pF})(0.006\,\text{pF})}{(0.004\,\text{pF} + 0.006\,\text{pF})} = 0.0024\,\text{pF}$$

The total capacitance for 42 finger cells is

$$0.0024 \text{ pF} \times 42 = 0.10 \text{ pF}$$

3.4.1.3 Sensor Applications

3.4.1.3.1 Capacitive Defect Detection

Because capacitive sensors have much higher sensitivities to conductors than non-conductors, they can be used to detect the presence or absence of metallic parts in completed assemblies. Figure 3.15 shows a connector assembly. It requires a metallic cap to be assembled that is invisible in the final assembly, but can be "seen" by a capacitive sensor. Thus, if a cap is missing, the capacitive sensor can detect the defective connector and signal the system to remove the part from the line [10].

3.4.1.3.2 Capacitive Distance/Presence Sensing

Figure 3.16 shows a capacitive sensor formed by a human (a conductive target) and a single sensing plate. The capacitance increases when the human gets closer to the

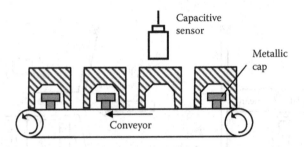

FIGURE 3.15   A capacitive detector detecting a missing internal metallic cap.

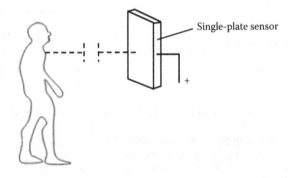

FIGURE 3.16   A capacitive distance sensor formed by a human and a metal plate.

metal plate. By measuring this capacitance, the distance between the human and the sensing plate can be estimated. Ideally, a conductive target should be grounded, as the earth provides a quasi endless charge reservoir. However, in this case, the human body itself, as a massive conductive target, provides a sufficient charge reservoir. Thus, there is no need for the human to be grounded in order to act as a good capacitor plate.

3.4.1.3.3 Capacitive Pressure Sensor

Over half of the sensors used in the chemical industry are capacitive pressure sensors. These sensors measure absolute, relative, and differential pressures, ranging from 1 psi (pounds per square inch) to 5000 psi. Accuracies of 0.1–0.5% of full scale are common for capacitive pressure sensors. A typical spacing-variation-based capacitive pressure sensor consists of a fixed plate and a flexible plate (diaphragm). Both plates are usually constructed from metal or metal-coated quartz materials. When liquid or gas enters the chamber, it presses the diaphragm, causing the spacing d and capacitance C to change. Thus, the capacitance variation directly relates to the pressure variation. The change in capacitance in this type of sensors is typically a lesser percentage of the total capacitance.

Although a resistor–capacitor (RC) circuit can be used to detect plate spacing for less critical applications, a low-impedance amplifier circuit provides better performance (see Figure 3.17). This is because the stray capacitance is grounded and

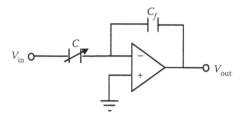

FIGURE 3.17 A two-plate capacitive sensor measurement circuit.

will not affect the measurement. A grounded shield can also protect the device from unwanted local electrostatic fields. The transfer function of spacing d to output voltage V_{out} in Figure 3.17 can be expressed by (C_f–feedback capacitor)

$$V_{out} = -\frac{\varepsilon_0 \varepsilon_r A}{C_f d} V_{in} \tag{3.22}$$

The advantages of spacing-variation-based capacitive pressure sensors are compactness, low drift, and broad bandwidth. However, they are sensitive to vibrations.

3.4.1.3.4 Capacitive Accelerometer

If one plate of a parallel capacitor moves subject to acceleration or vibration, a capacitive accelerometer or vibration sensor is formed. Figure 3.18 shows a capacitive accelerometer designed by *PCB Piezotronics*, Depew, New York [13]. It contains a diaphragm (acting as a mass undergoing deflection in the presence of acceleration), and two plates that sandwich the diaphragm (creating two capacitors, each with an individual fixed plate and sharing the diaphragm as a movable common plate). The deflection causes a capacitance shift between the plates. The two capacitance values are sent to a bridge circuit, and the voltage output is proportional to the input acceleration. This design can achieve high performance for uniform acceleration and low-frequency vibration measurement.

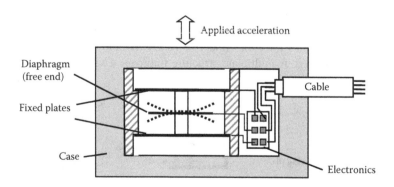

FIGURE 3.18 A capacitive accelerometer. (After Capacitive Accelerometers: Series 3701, 3702, 3703, and 3801, PCB Piezotronics, Inc., Depew, New York, USA, 2004.)

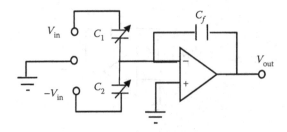

FIGURE 3.19　A three-plate sensor's measurement circuit.

The circuit for this three-plate sensor is shown in Figure 3.19. The transfer function of the center plate deflection x to the output voltage V_{out} is

$$V_{\text{out}} = -\frac{C_1 - C_2}{C_f d} V_{\text{in}} = \frac{2x}{d_0^2 - x^2} \frac{C_0}{C_f} V_{\text{in}} \tag{3.23}$$

where C_1 and C_2 are the capacitances between the plates, with an average value of C_0 and undeflected spacing d_0.

Capacitive accelerometers generally have higher sensitivities than piezoresistive accelerometers.

3.4.2　Area-Variation-Based Sensors

3.4.2.1　Sensing Principle and Characteristics

From Equation 3.6, the capacitance of a flat plate capacitor is proportional to the area of the plate A.

As one of the plates slides transversely, the overlapped area changes, causing the capacitance changes linearly. This change is usually converted into a voltage change. Area-variation-based sensors are often used for noncontact measurements of motion, displacement, pressure, and electrical field.

3.4.2.2　Sensor Design

An area-variation-based capacitive sensor can be designed in one-pair or multi-pair plate configurations for either linear or rotary motion detection as shown in Figure 3.20a through c. A multiplate structure improves a sensor's sensitivity and

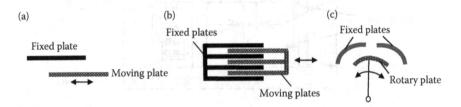

FIGURE 3.20　One-pair or multipair, linear or rotary configurations.

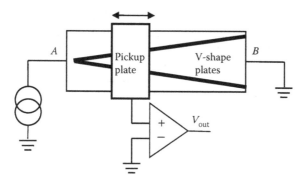

FIGURE 3.21 Chevron-shaped driven plate reducing tilt. (From Baxter, L.K., *Capacitive Sensors: Design and Applications*, 1st ed., IEEE Press, New York, 1997. With permission.)

measurement accuracy. However, it requires larger space and also is more expensive. Challenges in making area-variation-based capacitive sensors include how to (1) avoid tilt in any axis of a plate, especially when area A is large in order to achieve a big capacitance C; (2) maintain a small and constant gap between plates all the time to minimize the effect of spacing variation in measurement and maximize capacitance C; and (3) eliminate the coupling from the back of the plate, especially in compact, multiple-plate sensors.

Figure 3.21 shows a method to reduce the tilt sensitivity by making one plate in a chevron shape [14]. The second plate, a small rectangular plate, moves laterally above the chevron-shaped plate with a 0.5 mm gap. If the moving plate tilts about any axis, the total vertical distance between the rectangular plate and each arm of the chevron-shaped plate will maintain the same. The chevron plate is driven by a 5 V voltage at 100 kHz, and the sensor signal is picked up by the rectangular plate and fed into a high-impedance unit amplifier. The sensor output V_{out} varies linearly from 5 V at the left side to 0 V at the right side, and V_{out} is not sensitive to the pickup plate's tilt.

Tilt can also be compensated by adding a third electrode between the two plates. This structure also doubles the sensor capacitance and makes shielding easier.

EXAMPLE 3.11

An air capacitor consists of two flat plates, each with area A, separated by a distance d. Then a metal slab—with a thickness a ($<d$) and the same shape and size as the plates—is inserted between the two flat plates (at the middle) and parallel to the plates without touching either plate. Express the capacitance C of this three-plate capacitor in terms of capacitance C_0 when the metal slab is not inserted.

SOLUTION

This three-plate capacitor forms the two equal capacitors ($C_1 = C_2$) in series (see Figure 3.22), each with a separation of $(d - a)/2$. Thus

$$C = \frac{C_1 C_1}{C_1 + C_1} = \frac{C_1}{2} = \frac{1}{2} \frac{\varepsilon_0 A}{(d-a)/2} = \frac{\varepsilon_0 A}{d-a} = \frac{\varepsilon_0 A}{d} \frac{d}{d-a} = \frac{d}{d-a} C_0$$

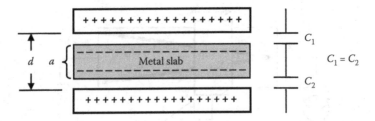

FIGURE 3.22 A three-plate capacitor.

So the resultant capacitance C is larger than the capacitance without the metal slab C_0. As $a \to 0$, $C \to C_0$, and as $a \to d$, $C \to \infty$.

3.4.2.3 Sensor Applications

Area-variation-based capacitive sensors are often used for noncontact displacement measurement, especially when the movement is larger than the electrode dimension (if the movement is less than the electrode size, the spacing-variation-based sensors are most suitable). As the plates slide transversely, the capacitance changes linearly with motion. Using an impedance–voltage converter, the capacitance change can be converted into a voltage change for further signal processing. Some application examples of the area-variation-based capacitive sensors are given as follows.

3.4.2.3.1 Pressure Sensor

Figure 3.23 shows an area-variation-based capacitive pressure sensor designed by *VEGA Technique* (France). It consists of multiple plates and is more robust than other types of pressure sensors. One group of the electrodes is connected to the sensing diaphragm whose displacement is proportional to the pressure, causing the overlap area changes and therefore the capacitance change. The unit forms a capacitor whose variation in plate area is determined by the movement of the diaphragm. The multiplate design improves the sensitivity and provides a larger output.

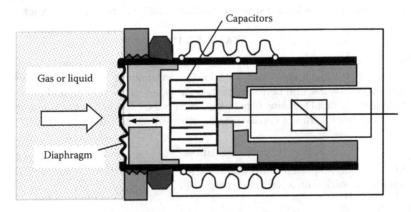

FIGURE 3.23 Capacitive pressure sensor with area variation. (Modified based on original drawing from *VEGA Technique*, France.)

3.4.2.3.2 Accelerometer

Figure 3.24 shows a z-axis torsional capacitive accelerometer [15,16]. It consists of a 12-μm-thick boron-doped silicon inertial mass, which has long fingers at the end, and a fixed multifinger plate. The mass is suspended 7.5 μm above the glass substrate by two torsion beams anchored to the glass substrate. The moving fingers of the mass and the fixed fingers (anchored to the substrate) form a group of parallel capacitors. When the mass experiences an acceleration, a torque T is produced on the suspension beams that cause the mass fingers to rotate an angle θ. As a result, the overlap area of capacitors is reduced (see Figure 3.24b), causing a reduction in capacitance (note that the air gap between the fingers remains constant). The high-aspect-ratio geometry and the number of finger electrodes provide a high capacitance change. Table 3.4 lists the main geometry parameters of this accelerometer [15].

The capacitance change ΔC, in terms of the finger length l_f, the mass length l_m, the total number of moving fingers n, the finger gap d_f, and the relative permittivity ε_r, can be derived from a trapezoidal overlap area using the parallel plate and approximated by (with an assumption of small θ) [16]

$$\Delta C \approx \frac{n\varepsilon_0\varepsilon_r l_f}{d_f}(2l_m + l_f)\theta \tag{3.24}$$

(a)

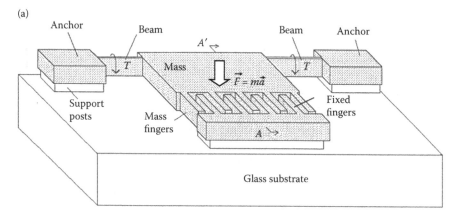

(b)

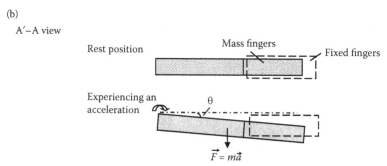

FIGURE 3.24 An area-variation-based z-axis torsional capacitive accelerometer.

TABLE 3.4

Geometry Parameters of a Torsional Accelerometer

Mass length l_m: 1000 µm	Beam length l_b: 150 µm
Mass width w_m: 750 µm	Beam width w_m: 5 µm
Mass thickness t_m: 12 µm	Finger length l_f: 500 µm
Total moving mass M: 2.446×10^{-3} g	Finger width w_m: 10 µm
Etch pit depth d': 15 µm	Inter finger gap d_f: 5 µm
	Total number of moving fingers n: 25

EXAMPLE 3.12

Find the capacitance change ΔC using the parameters listed in Table 3.3, given $\theta = 1.6°$, $\varepsilon_r = 1$.

SOLUTION

$$\Delta C \approx \frac{n\varepsilon_0\varepsilon_r l_f}{d_f}(2l_m + l_f)\theta$$

$$= \frac{(25)(8.85 \times 10^{-12}\,\text{F} \cdot \text{m}^{-1})(1)(500 \times 10^{-6}\,\text{m})}{(5 \times 10^{-6}\,\text{m})}\Big[2(1000 \times 10^{-6}\,\text{m}) + (500 \times 10^{-6}\,\text{m})\Big]$$

$$\times \left(\frac{(1.6°)(3.14)}{180°}\right)$$

$$= 1.54\,\text{pF}$$

3.4.2.3.3 *Position Sensor*

Figure 3.25 shows an area-varying capacitive position sensor [17]. It consists of two fixed parallel plates (the top and the bottom plates) and a sliding plate between them. All the plates have periodic geometries (e.g., rectangular shapes as shown in the figure; they can also be triangular, sinusoidal, or of any other shape). The periodically repeated patterns (1) allow the sensor to cover a large measurement range by increasing the number of periods and (2) provide much finer and much more accurate

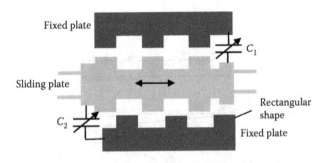

FIGURE 3.25   An area-variation-based capacitive position sensor.

position measurement by evaluating each small capacitor formed by two finger tips within one period. This unique combination makes it possible to measure with nanometer accuracy over the entire measurement range. Simulation results have shown that for two arrays of 50 rectangular finger pairs, the maximum capacitance change $\Delta C_{max} \approx 7$ fF can be achieved.

3.4.3 Dielectric-Constant-Variation-Based Sensors

3.4.3.1 Sensing Principle and Characteristics

Capacitive sensors that are designed based on the dielectric constant change can also be explained by Equation 3.6: a linear relationship exists between dielectric constant ε_r and the capacitance C: the larger the ε_r, the bigger the C.

In some sensors, capacitance change depends on the sensor's dielectric constant; in other sensors, capacitance change depends on the target material's dielectric constant. The larger the dielectric constant of the target material, the bigger the capacitance change, and the easier it is to detect the target. Figure 3.26 shows the relationship between the dielectric constant ε_r of a target material and the detectable distance of a Siemens capacitive sensor [18], in terms of the percentage of the rated sensing distance.

EXAMPLE 3.13

A Siemens capacitive sensor has a rated sensing distance of 9 mm and the target material is marble. What is the effective sensing distance?

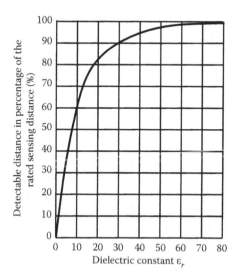

FIGURE 3.26 Relationship between a target material's dielectric constant and the rated sensing distance. (After Capacitive proximity sensors: Theory of operation, Siemens online course. www.automationmedia.com/Port1050%5CSiemensFreeCourses%5Csnrs_3.pdf.)

SOLUTION

From Table 3.3, marble has a dielectric constant of 8, which corresponds to ~50% of the rated distance according to Figure 3.26; thus, the effective sensing distance for marble is (9 mm) × 50% = 4.5 mm.

EXAMPLE 3.14

A Siemens capacitive sensor has a normal (rated) sensing range S_n of 20 mm. Can this capacitive sensor be used to sense the presence of ammonia from behind a 3 mm glass panel? The dielectric constant for ammonia is 20, and for glass is 10.

SOLUTION

From Figure 3.26, $\varepsilon_r = 20$ corresponds to 82% normal sensing range; $\varepsilon_r = 10$ corresponds to 60% normal sensing range.

Since $S_n = 20$ mm, the effect sensing distance for glass is (20 mm) × 60% = 12 mm, and the effect sensing distance for ammonia behind the glass is (12 mm) × 82% = 9.84 mm. The glass is only 3 mm thick, which results in 6.84 mm operating margin. Thus, the sensor is able to sense the ammonia behind the 3-mm-thick glass panel.

Usually a change in a sensor's dielectric constant is due to: (1) the dielectric material alters its density when a force or pressure is applied to the dielectric media, or (2) the dielectric material absorbs moisture, or (3) the dielectric media reacts with a target material. Capacitive humidity sensors, for example, require a dielectric material that easily absorbs and releases moisture based on the surrounding RH (relative humidity). The commonly used dielectric materials for humidity sensors are glass, ceramic, or silicone [19]. Figure 3.27 shows the RH–capacitance curve of a capacitive humidity sensor made by *Measurement Specialties*, Hampton, Virginia [20].

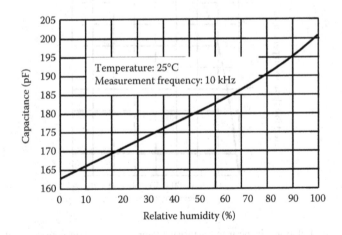

FIGURE 3.27 Typical response curve of an HS1100 capacitive humidity sensor. (After HS1100/HS1101 relative humidity sensor technical data, *Measurement Specialties*, Hampton, Virginia, USA, 2002.)

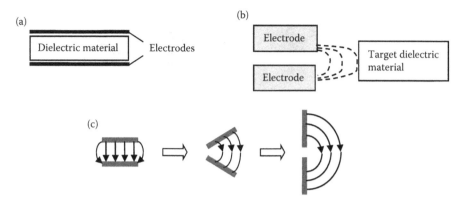

FIGURE 3.28 Designs of dielectric media-based capacitive sensors.

3.4.3.2 Sensor Design

Dielectric-based capacitive sensors have two basic designs. One is to have two electrodes and one dielectric media in the sensor as shown in Figure 3.28a. This type of design is often seen in capacitive force, pressure, chemical, and humidity sensors where the permittivity of the dielectric material changes when a sensor is subjected to a force, pressure, chemical gas, or water vapor. The other design is to have two electrodes but no dielectric media, and it relies on the target's material to interfere with the electric field of the sensor to change capacitance (see Figure 3.28b). The two metallic electrodes can be positioned like an "opened" capacitor as shown in Figure 3.28c. The change in capacitance depends on the target material and the distance to the target. This type of sensor is often used to detect motion or proximity of nonconductive materials.

Most capacitive proximity sensors are probe types (Figure 3.29a) with a typical internal layout shown in Figure 3.29b. The sensing surface is often formed by two concentric ring electrodes. When an object nears the sensing surface and enters the electrostatic field of the electrodes, it alters the amount of electric flux reaching the second electrode (see Figure 3.29c) and causes the capacitance to increase. As the target moves away from the sensor, the capacitance decreases, switching the sensor output back to its original state (see Figure 3.29d).

Capacitive proximity sensors could be affected by humidity or contaminants, causing false outputs. Therefore, some sensors add a compensating electrode E_c (see Figure 3.30). When contaminants are on the sensor face, they affect both the sensor's main field and the compensation field. A sensor circuit can then detect the increase in both fields, and filter out the effects of the contaminants. When a target nears the sensor, the sensor circuit can distinguish the different responses between these two fields and generate an output.

3.4.3.3 Sensor Applications

Dielectric-constant-variation-based capacitive sensors are widely used to detect motion or proximity of nonconductive objects (e.g., dust, paper, plastics, and cloth), chemical-substance, and biocells. They can also measure RH, force, and pressure.

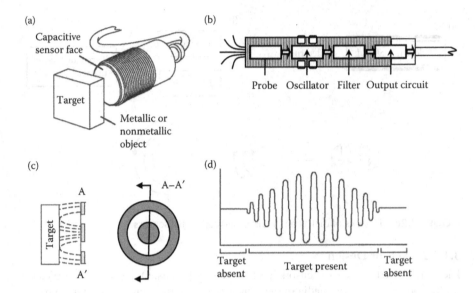

FIGURE 3.29   Example of a capacitive proximity sensor design and sensing. (After Capacitive proximity sensors: Theory of operation, Siemens online course. www.automationmedia.com/ Port1050%5CSiemensFreeCourses%5Csnrs_3.pdf.)

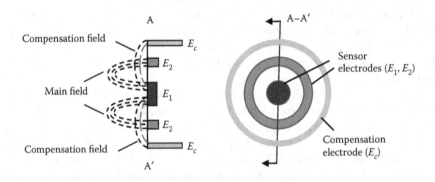

FIGURE 3.30   Sensing face design of a capacitive proximity sensor. (After Capacitive proximity sensors: Theory of operation, Siemens online course. www.automationmedia.com/ Port1050%5CSiemensFreeCourses%5Csnrs_3.pdf.)

3.4.3.3.1   Water Level Detection

One application of ε_r-based capacitive sensors is to detect water level through a barrier (see Figure 3.31). Water is a much higher dielectric constant than plastic, giving the sensor the ability to "see through" the plastic and detect the water level.

3.4.3.3.2   Pressure Measurement

Figure 3.32 illustrates the structure of a pellicular pressure sensor (about 80 μm thin) designed by *ONERA French Aerospace Lab*, France. The sensor can detect

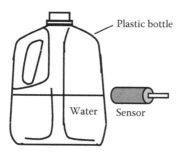

FIGURE 3.31 A ε_r-based capacitive water level detector.

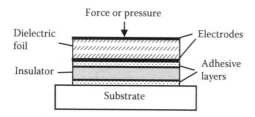

FIGURE 3.32 Structure of a pellicular sensor.

the change in ε_r between two electrodes when a dynamic force or pressure is exerted on the plates. The advantages of such pellicular sensors are their compactness, ability to detect micro pressure, resistance to vibration, and high bandwidth (e.g., 50 ~ 200 kHz). The primary disadvantage is that they are temperature sensitive.

To eliminate usage of an external power supply, one can use a diaphragm to preserve the constant electric polarization (*electret effect*). The electret effect is also broadly used in microphones, which are actually sensitive pressure sensors.

3.4.3.3.3   Chemical Sensing

A capacitive chemical sensor, as shown in Figure 3.33, is often composed of two electrodes separated by a chemically sensitive polymer that can absorb specific chemicals (analytes) [21]. Upon analyte absorption, the polymer swells and increases not only the distance between the two electrodes but also the polymer's dielectric permittivity. Both change the capacitance of the sensor that can be electrically detected and measured.

3.4.3.3.4   Level Measurement

The capacitive level sensor shown in Figure 3.34 evaluates the capacitance changes $(C - C_0)$ due to the space between two planar electrodes being filled with a liquid

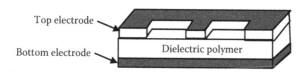

FIGURE 3.33 A capacitive chemical sensor.

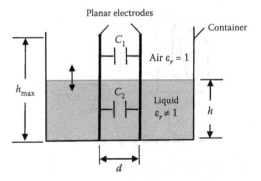

FIGURE 3.34   A capacitive level sensor.

$(\varepsilon_r > 1)$. $(C - C_0)$ is approximately proportional to the filling height h. The total capacitance C is the sum of the two parallelly connected capacitances C_1 and C_2:

$$C = C_1 + C_2$$

Plug in $C_1 = \varepsilon_0 b(h_{max} - h)/d$ ($\varepsilon_r = 1$ for air; b is the width of the sensor's planar electrodes) and $C_2 = \varepsilon_r \varepsilon_0 bh/d$ (for filled liquid) into the above equation and use C_0 for the capacitance of empty container, that is, $C_0 = \varepsilon_0 bh_{max}/d$, resulting in capacitance C that depends on level h:

$$C = C_0 + \varepsilon_0(\varepsilon_r - 1)\frac{bh}{d} \tag{3.25}$$

or the level h can be expressed by

$$h = \frac{(C - C_0)d}{\varepsilon_0(\varepsilon_r - 1)b} = \frac{\Delta Cd}{\varepsilon_0(\varepsilon_r - 1)b} \tag{3.26}$$

Several factors could influence capacitive level sensors: density of bulk material, concentration (mixture proportion of the filling material), temperature, and humidity.

EXAMPLE 3.15

The capacitive level sensor in Figure 3.34 is chosen to measure a water level $(\varepsilon_r = 80)$ that can rise 300 mm high. If the capacitance reading is 503.4 pF, what is the water level? The width of the sensor's planar electrodes b is 20 mm, and the distance between the two plates d is 1 mm.

SOLUTION

First find C_0:

$$C_0 = \frac{\varepsilon_0 bh_{max}}{d} = \frac{(8.85 \times 10^{-12}\,\text{F} \cdot \text{m}^{-1})(0.02\,\text{m})(0.3\,\text{m})}{0.001\,\text{m}} = 53.1\,\text{pF}$$

From Equation 3.26

$$h = \frac{(C - C_0)d}{\varepsilon_0 b(\varepsilon_r - 1)} = \frac{[(503.4 - 53.1) \times 10^{-12}\,\mathrm{F}](0.001\,\mathrm{m})}{(8.85 \times 10^{-12}\,\mathrm{F \cdot m^{-1}})(0.02\,\mathrm{m})(80 - 1)} = 0.0322\,\mathrm{m\ or\ 32.2\,mm}$$

3.4.3.3.5 *Droplet Detector*

Figure 3.35 shows a capacitive droplet detector developed by the *University of Freiburg* in Germany [22]. A free droplet is introduced between two plates of an open capacitor. The presence of the droplet changes the dielectric property between the plates, causing a significant change in capacitance [23,24]. If the droplet's dielectric constant is fixed (or known) and its velocity can be controlled (or known), then the sensor can directly relate the droplet's volume to the capacitance change. The sensor can determine the volume of dispensed droplets in the range of 20–65 nL with a resolution of <2 nL. This capacitive droplet detection method is reliable and noncontact (preventing contamination), and well suited for droplet presence detection and high-precision volume measurement for microdispensing processes in various applications.

3.4.3.3.6 *Angular Speed Sensor*

The capacitive angular speed sensor shown in Figure 3.36 [25] consists of two coaxial stator plates (form a capacitor) and a rotor between them (functions as a dielectric media). One of the stator plates, called the *transmitter*, is divided into four segments:

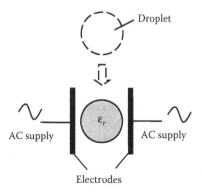

FIGURE 3.35 A capacitive droplet detector.

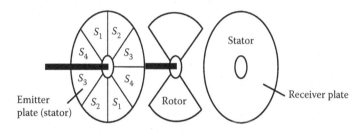

FIGURE 3.36 A capacitive angular speed sensor for rotor speed measurement.

S_1–S_4. The number of the segments is determined by the resolution requirements—more segments provide higher resolution. To measure an angular position, a constant voltage is applied to one of the segments (a zero voltage is applied to all other segments), then the induced charge or signal is received by the other stator plate (called *receiver*). The position of the rotor will affect the capacitance between the active segment electrode and the receiver electrode. The angular position change is then converted to a speed. The rotor is usually grounded and used as a shield. The symmetrical structure and cross-quad connection of the sections significantly reduce the effects of many mechanical and systematic nonidealities such as eccentricities, nonflatness, and obliqueness of the electrodes. Fulmek et al. [26] improved the above design by introducing a second rotor, allowing both relative movement and rotating direction to be measured. Li et al. [27] used six segments and an additional inner window to provide a broader and finer angular measurement.

3.4.4 ELECTRODE-PROPERTY-VARIATION-BASED SENSORS

3.4.4.1 Sensing Principle and Characteristics

There are four main sensing principles that are based on electrode property variation:

1. The sensor has only one electrode, while the conductive target (e.g., copper, aluminum, or conductive fluid) serves as the second electrode. A complete capacitor is formed between the sensor and the target. Current usually flows from the probe face to the target and back to the sensor amplifier to complete the circuit. Standard capacitive sensors often require the target to be electrically grounded.
2. The sensor has two electrodes. A voltage is applied across the electrodes and an electric field is established between the electrodes. When a metallic target approaches the sensor (serves as the third electrode), it alters the electrostatic field between the sensor's electrodes.
3. The sensor has two electrodes: one is a sensing or working electrode, the other is a reference electrode. The sensing electrode can be made of a material sensitive to chemicals, biocells, humidity, force, pressure, or acceleration. When a force, pressure, vapor, dirt, or chemical or biological substance is applied or present, the properties of the working electrode (e.g., conductivity, mass, and ability to accumulate charges) change, causing the capacitance or the voltage across the capacitor to change. In this case, the change in capacitance depends on the state of the working electrode.
4. The sensor has two electrodes and the charge Q on the electrodes change during the sensing process. For example, a piezoelectric element is fitted between two electrodes. When the piezoelectric element is under strain, the electric charges are generated or changed. The accumulated charge Q on the sensor's electrodes can be measured by a form of voltage. Thus, piezoelectric type of sensors usually require no external power and are often used for dynamic force measurements. Most accelerometers and microphones today are piezoelectric because of their high natural frequency, wide frequency

response (e.g., 200 kHz), small size, and low noise level [28]. Piezoelectric pressure sensors can measure very fast pressure changes (e.g., shock waves from explosions or a very fast pressure spike).

3.4.4.2 Electrode Materials and Design

Electrodes are critical parts of all capacitive sensors. There are several considerations when choosing electrode materials: low resistivity (or high conductivity), easy fabrication, good RC (R-resistance; C-capacitance) time constant, and low cost. Common electrode materials include copper, silver, degenerately doped polysilicon, Orgacon™, and indium tin oxide (ITO).

3.4.4.3 Sensor Applications

Electrode-variation-based capacitive sensors have many applications. The following are just a few examples.

3.4.4.3.1 Capacitive Touch Sensing

Figure 3.37 shows a capacitor formed by a finger and a copper with a glass overlay as the insulating dielectric. Charges exist in the finger as conductive electrolytes and in the copper trace from a constant-current source. When a finger is placed over the glass overlay, a capacitance change is detected. Capacitive touch sensors are effective and popular methods of input and control. Their primary applications are switches, touch-screen inputs, and consumer electronic devices such as tablet PCs or iPhones.

3.4.4.3.2 Cell Adhesion Strength Characterization

Figure 3.38 illustrates a CMOS capacitive sensor for characterizing cell adhesion strength [29]. The coupling capacitance C is the combination of the two capacitances:

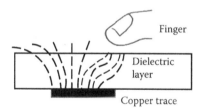

FIGURE 3.37 A parallel-plate capacitive touch sensor.

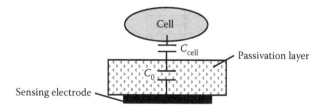

FIGURE 3.38 A capacitive sensor for cell adhesion strength evaluation.

between the cell and the passivation layer C_{cell}, and between the passivation layer and the metal electrode. The metal electrode functions as the sensing electrode. C_{cell} varies inversely with the distance of the cell from the chip surface, which is an indication of the cell adhesion strength. As C_{cell} increases with increasing cell proximity to the electrode, the output voltage also increases.

3.4.4.3.3 Capacitive Tactile Sensor

A collapsible capacitive tactile sensor constructed at the *University of Illinois at Urbana-Champaign* [30] used soft conductive elastomers as electrodes, instead of traditional solid metals, for better flexibility and robustness. To increase the capacitance of the sensor, the design incorporated a large electrode area and a small electrode gap (2.4 μm), achieved by inserting a thin polydimethylsiloxane sheet with a regular array of small pillars between the electrodes. The pillars not only define the air gap but also supply the restoring force to separate the electrodes.

3.4.4.3.4 Piezoelectric Pressure Sensor

Figure 3.39 shows a pressure sensor that uses a stack of thin piezoelectric elements (crystal wafers) to improve its sensitivity [28]. The crystals are stacked in series, that is, negative surface of one crystal is in contact with the positive surface of the next crystal. The electrode running through a connector within the housing acts as the signal return. The pressure is applied to the diaphragm, which applies a compression force to the piezoelectric stack. As a result, a charge proportional to the pressure is generated and collected by the electrode.

3.4.4.3.5 Capacitive DNA Biosensors

Well-designed electrodes also allow capacitive sensors to analyze deoxyribonucleic acid (DNA) for medical and biological applications. The *University of Bologna* in Italy has developed a capacitive DNA sensor [31], where DNA targets in the solution bond with the probe molecules and change the state of the probe surface, causing a change in the interface capacitance. Advantages of this label-free method over a conventional optical marker are (1) improved sensitivity, (2) no need for an expensive optical reading device, and (3) real-time detection.

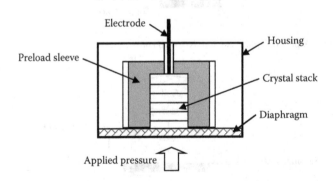

FIGURE 3.39   A piezoelectric pressure sensor construction.

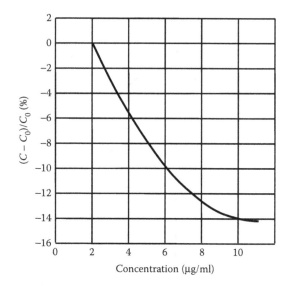

FIGURE 3.40 Ratio of capacitance change vs. the fenvalerate concentration. (After Gong, J.L. et al., *Anal. Bioanal. Chem.*, 379, 302, 2004.)

3.4.4.3.6 Electropolymerized Capacitive Chemical Sensor

A capacitive chemical sensor was developed for fenvalerate, one of the most widely used synthetic pyrethroid insecticides [32]. The sensing layer is made of 2-mercapto-benzimidazole (2-MBI) molecularly imprinted onto a 10 mm^2 surface of a gold electrode (this process is called *electropolymerization*). In the presence of fenvalerate, the capacitance of the imprinted electrode decreases due to the interaction with fenvalerate. The sensor can be restored by immersing it into an ethanol–water (4:1) alkaline solution containing 0.2 mol · L^{-1} KOH (pH = 13) for one hour. The values of $(C - C_0)/C_0$ as a function of fenvalerate concentration are shown in Figure 3.40. By using the ratio $(C - C_0)/C_0$, one can eliminate the deviation caused by the variation in the capacitance measurement.

3.5 CYLINDRICAL CAPACITIVE SENSORS

The cylindrical or coaxial capacitive sensor configuration is another popular design in capacitive sensors. In fact, one of the earliest capacitors, the *Leyden jar*, had a cylindrical shape. Cylindrical capacitive sensors (CCSs) are broadly used in proximity sensing, fluid level gauging, displacement/temperature/humidity measurement, and chemical substance detection. The CCS was originally introduced by P.D. Chapman for its advantages of insensitivity to geometric errors (by the averaging effect) and the high resolution with a large sensing area [33]. Figure 3.41 shows the basic structures of CCSs.

Based on Equation 3.7, the capacitance C of a cylindrical capacitor has a linear relationship with height h and dielectric constant ε_r. In practical design, r_1 and r_2 are usually fixed, and any variation in height h or dielectric constants ε_r will cause

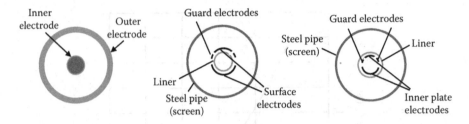

FIGURE 3.41   Basic structures of cylindrical capacitive sensors.

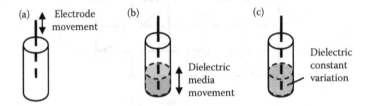

FIGURE 3.42   Three underlying principles of the cylindrical capacitive sensor designs.

a change in capacitance C. Figure 3.42 shows the three design principles of CCSs: (a) inner electrode movement, (b) dielectric media movement, and (c) dielectric constant variation. These designs are discussed as follows.

3.5.1   ELECTRODE-MOVEMENT-BASED SENSORS

3.5.1.1   Sensing Principle and Characteristics
In this type of design, the moving electrode (usually the inner one) causes the overlap height h of the cylindrical plates to change, resulting in a change in capacitance C. A linear relationship exists between the overlap height h and the capacitance C as indicated in Equation 3.7: the bigger the height h, the larger the capacitance C.

3.5.1.2   Sensor Design and Applications
A simple displacement sensor can be easily constructed using a cylindrical configuration if the inner conductor can move relative to the outer conductor. More complex design examples are shown in the following sections.

3.5.1.2.1   Capacitive Touch Transducer
The cross section of a touch transducer is shown in Figure 3.43. This sensor uses a coaxial capacitor structure and a high dielectric polymer (*PVDF*) to maximize the capacitance change when a force is applied. The movement of the inner electrode is caused by the applied force, resulting in a capacitance change. The coaxial design is better than a flat plate design as it will provide greater capacitance increase for a given force.

3.5.1.2.2   Concentric Capacitive Displacement Sensor
Thomas et al. [34] invented a cylindrical capacitive displacement sensor for measuring the displacement of a stylus. The sensor comprises three capacitance cylinders (see Figure 3.44a). The inner and the outer cylinders, S_1 and S_3, are fixed; the middle cylinder

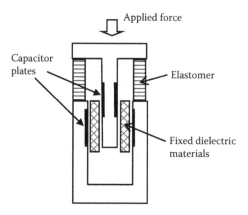

FIGURE 3.43 Schematic of a coaxial capacitor touch sensor.

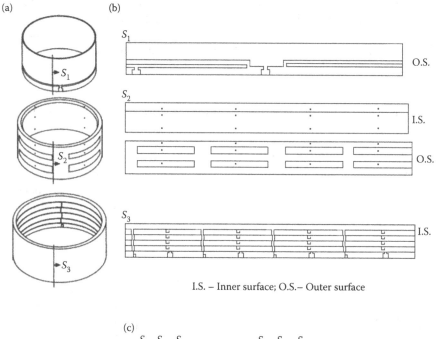

I.S. – Inner surface; O.S.– Outer surface

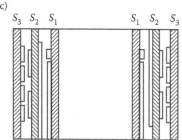

FIGURE 3.44 A concentric capacitive displacement sensor.

S_2 is movable as the stylus moves with respect to the two static cylinders. S_1 and S_3 are spaced from the inner and outer surfaces of the middle cylinder by an air gap. When S_2 moves, the capacitance changes. Each cylinder is made of a ceramic material metallized with copper and coated with gold (about 2 μm thickness). This coating is then through-etched so that discrete capacitive areas are formed in the coating, some of which have electrical connections incorporated.

Other unique features of this capacitive sensor are as follows:

1. It can measure a 3-DOF (degree of freedom) movement. The middle cylinder can move in the z (axial) direction (which causes a large capacitance change), and can also pivot about the center (in the x and y directions). The pivot movement also alters the capacitance due to the changes in the air gap between the cylinders, although the capacitance change is not as significant as in the z direction movement. This feature makes it possible to measure movement for devices such as joysticks.
2. The inner surface of the outer cylinder and the outer surface of the middle cylinder are divided into capacitor segments (quadrants, as shown in Figure 3.44b). Thus, a greater range and better accuracy for displacement measurement can be achieved. This quadrant design of the cylindrical surface also enable the signal circuit to determine in which direction the middle cylinder (and hence the stylus) is moving.

Figure 3.44c shows the cross section of this cylindrical sensor assembly with a clearer display of the capacitor's layout between the cylinders.

3.5.1.2.3 Radial Displacement Measurement

A CCS for the precision radial displacement measurement of active magnetic bearing (AMB) spindles is shown in Figure 3.45 [35]. The rotating spindle (rotor) functions as the inner cylindrical electrode; the outer (fixed) cylindrical electrode, instead of one piece, consists of eight segments. These eight segments, each with arbitrary arc length

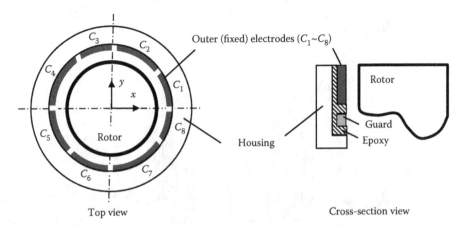

Top view   Cross-section view

FIGURE 3.45   An eight-segment cylindrical sensor with arbitrary arc length.

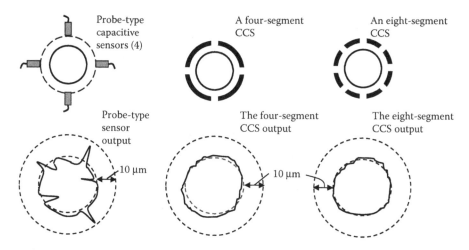

FIGURE 3.46 Comparison of measured orbits of the rotor with various sensor types.

design, can overcome the odd harmonic component in the sensor output signal and minimize the measurement error caused by the rotor's geometric errors. Thus the sensor output is purely based on the displacement of the spindle.

This CCS is superior to a probe-type displacement sensor used in AMB systems, because the latter is very sensitive to the surface quality of a rotor and requires additional and more complex algorithms to identify and compensate for the measurement error introduced by the geometric errors of a rotor. A comparison of the measurement results of this CCS with those of probe-type sensors are shown in Figure 3.46. Note that the conventional four-segment CCS is still sensitive to the odd harmonic components of the geometric error, but the eight-segment CCS, especially with random arc length, is less sensitive to the nonuniformity of the rotor geometry.

3.5.1.2.4 Conductive Liquid Level Measurement

Figure 3.47 shows a liquid level measurement setup. The probe acts as one plate of the cylindrical capacitor, while the high-conductivity liquid acts as a moving

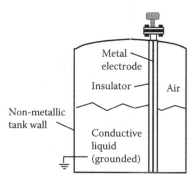

FIGURE 3.47 A capacitive liquid level sensor.

electrode (it is grounded). The surface of the probe electrode is coated with a thin insulating layer (e.g., Teflon or Kynar) to prevent an electric short circuit through the conductive liquid. It also acts as the dielectric medium of this cylindrical capacitor. As the liquid level varies, the capacitance changes. Note that the tank wall is non-metallic in this design.

3.5.2 DIELECTRIC-MEDIA-MOVEMENT-BASED SENSORS

3.5.2.1 Sensing Principle and Characteristics

In this type of CCSs, the level of the dielectric media changes, causing the total capacitance C to change (see Figure 3.48). The two capacitors C_1 and C_2 in Figure 3.48 are in a parallel connection. Thus, the total capacitance C is

$$C = C_1 + C_2 = \frac{2\pi\varepsilon_0(h_{max} - h)}{\ln(r_2/r_1)} + \frac{2\pi\varepsilon_0\varepsilon_r h}{\ln(r_2/r_1)} = \frac{2\pi\varepsilon_0}{\ln(r_2/r_1)}[h_{max} + (\varepsilon_r - 1)h] \quad (3.27)$$

or

$$C = C_0 + 2\pi\varepsilon_0(\varepsilon_r - 1)\frac{h}{\ln(r_2/r_1)} \quad (3.28)$$

where $C_0 = 2\pi\varepsilon_0 h_{max}/\ln(r_2/r_1)$ when the tank is empty; $C - C_0 = \Delta C = 2\pi\varepsilon_0(\varepsilon_r - 1)h/\ln(r_2/r_1)$.

The level h of the dielectric material can be found by

$$h = \frac{\Delta C \ln(r_2/r_1)}{2\pi\varepsilon_0(\varepsilon_r - 1)} \quad (3.29)$$

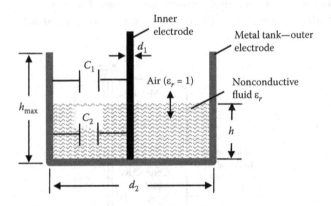

FIGURE 3.48 Schematic of a cylindrical capacitive liquid level sensor based on dielectric media movement.

In reality, the permittivity value is subject to variations and might cause systematic errors. These errors can be compensated by using an additional capacitive sensor that measures the permittivity.

EXAMPLE 3.16

A capacitive level sensor shown in Figure 3.48 is used to measure a liquid level. If h_{max} is 125 mm, ε_r is 80, and the measured capacitance change is 279.3 pF, what is the liquid level? (Given that the sensor's inner plate radius r_1 is 15 mm, the outer plate's diameter r_2 is 37.5 mm.)

SOLUTION

Apply Equation 3.29:

$$h = \frac{\Delta C \ln(r_2/r_1)}{2\pi\varepsilon_0(\varepsilon_r - 1)} = \frac{(279.3 \times 10^{-12} \, F)\ln(37.5/15)}{2(3.14)(8.85 \times 10^{-12} \, F \cdot m^{-1})(80 - 1)} = 58.29 \, mm$$

3.5.2.2 Design and Applications

Figure 3.49 shows a fluid level sensor developed by *OMEGA Engineering, Inc.*, Stamford, Connecticut, USA. The bare rod in the tank and the metallic vessel wall form a CCS with the vessel wall grounded. The capacitance change is based on the change in the amount of the dielectric material, the nonconductive fluid filled in between two electrodes. The capacitance changes can be measured by a bridge circuit. A potentiometer is usually built in to measure densities and dielectric constants of the fluid.

3.5.3 DIELECTRIC-CONSTANT-VARIATION-BASED SENSORS

3.5.3.1 Temperature Sensor

Figure 3.50 shows a temperature sensor with a cylindrical configuration [36]. Its inner axial electrode *A* and the outer cylindrical electrode *B* form a coaxial capacitor with the temperature-sensitive dielectric material between the electrodes. As

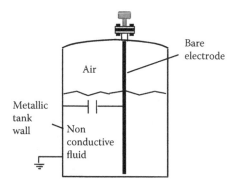

FIGURE 3.49 A capacitive liquid level sensor.

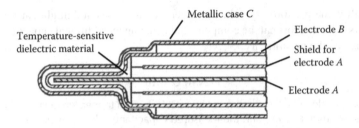

FIGURE 3.50   A capacitive temperature sensor with a coaxial configuration.

temperature changes, the dielectric constant of the material changes, causing capacitance change that is measured with a bridge. A metallic case C surrounds and shields the first two electrodes to eliminate the influence of the alternating electromagnetic fields around. All three electrodes (A, B, and C) are rigidly held in place by insulating supports. This three-terminal sensor nullifies the stray capacitance and provides larger capacitance without increasing the sensor's size.

The key advantages of dielectric-constant-variation-based temperature sensors are as follows:

- Broader dielectric materials whose temperature coefficients are sufficiently large can be used in the temperature sensors.
- A large temperature measurement range (from −273°C to above 2000°C) can be achieved by choosing proper materials that have different dielectrics. The same is true for the sensitivity—the sensor can be made extremely sensitive (to 10^{-6}°C or better) over a limited range of a few degrees.
- Physical sizes of the sensors can be made very small for a small capacitance value (e.g., 1 pF). This also results in a fast response time.

3.5.3.2   Oil Analyzer

Euro Gulf Group Company, Safat, Kuwait, has developed an EASZ-1 oil analyzer that uses the capacitance principle to measure the moisture content in oil. The inner cylindrical probe and outer barrel are fixed in size and form a cylindrical (coaxial) capacitor. The oil sample flows between the "plates" as a dielectric media. If the dielectric constant of the fluid changes, the capacitance of the assembly changes proportionally. The measured capacitance is then converted to a water content output by a microprocessor with the associated circuit to ensure stable and accurate readings. There is also a built-in temperature sensor for temperature compensation.

3.5.3.3   Water Quality Detector

A cylindrical capacitive water quality detector in Figure 3.51 consists of three coaxial capacitors [37]: the middle capacitor is the main sensing probe; the top and the bottom capacitors act as guard rings to reduce stray capacitance. Different types of water, such as distilled, tap, boiled, or salt water, have different dielectric constants that alter the capacitance. Thus, the system can examine the quality of water based on the capacitance values. The sensor's operating procedures include (1) measuring

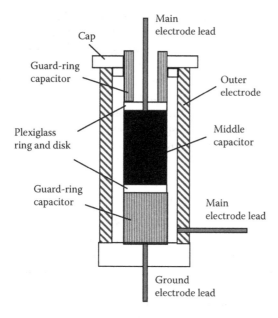

FIGURE 3.51   Structure of a water quality detector.

the air gap capacitance (dry signal), (2) filling the gap with water and measuring the capacitance (wet signal). The difference in measured capacitance can reach as high as 38.5 μF between salt and distilled water, which indicates high sensitivity and the capability to differentiate different water. The overall length of the sensor is about 12 cm and the sensing probe has a length of about 4 cm. The capacitor tube electrodes are made of aluminum.

3.6   SPHERICAL CAPACITIVE SENSORS

Spherical capacitive sensors are not seen nearly as often as the parallel or cylindrical configurations. This is largely due to the spherical design's complexity and higher manufacturing cost. However, the spherical geometry does provide several unique features, which neither parallel nor cylindrical capacitors have, such as higher capacitance within a limited or compact space, a shape that is more readily adaptable to measure irregular surfaces, spherical equipotentials, and wider bandwidth. The following are some design examples of spherical capacitive sensors.

3.6.1   GEOPHYSICAL FLUID FLOW CELL

A spherical capacitor provides the ideal shape for generating a nonlinear electric field gradient between its center electrode and the inner surface of the outer electrode. This unique feature was utilized by scientists at NASA (National Aeronautics and Space Administration) to create the *geophysical fluid flow cell* (GFFC) [38]. This cell uses spherical capacitors to simulate gravitational field conditions to study the behavior of fluids. By applying an electric field across a spherical capacitor filled with a dielectric liquid, a force analogous to gravity is generated around the fluid. The value of this artificial gravity is proportional to the square of the voltage applied across the sphere and can thus be imposed as desired. This force is also acts as a buoyant force and its magnitude is proportional to the local temperature of the fluid and in a radial direction perpendicular to the spherical surface. Thus, cooler fluid sinks toward the surface of the inner sphere, while warmer fluid rises toward the outer sphere. The significance of using this apparatus is that it simulates atmospheric flows around stars and planets, that is, the "artificial gravity" is directed toward the center of the sphere much like a self-gravitating body.

3.6.2 COATING THICKNESS DETECTOR

Researchers at the *University of Shanghai for Science and Technology* in China utilized the unique shape of the spherical capacitor to design a probe for measuring the thickness of coatings on metals [39]. This spherical capacitive probe is more accurate in measuring the thickness of nonconducting coatings on metals than common planar probes. Also, because it is a capacitive sensor, it is not subject to the material limitations like the magnetic induction method and the eddy current method—both are strongly influenced by the electrical conductivity and magnetic conductivity of the object.

3.6.3 ULTRA-PRECISION SPHERICAL PROBE

A 3D probing system using a spherical shape plate was developed [40]. It can make noncontact ultra-precision profile measurements of small structures with nanometer resolutions. The spherical capacitive sensor provides the identical sensing characteristic in any arbitrary spatial direction and converts the microgap between the sensor and the target being measured into a capacitance. This structure also allows most electric flux to concentrate within a small region between the sensor and the target, providing ideal properties for 3D noncontact probing, isotropy characteristics, near point sensing, and measurability of small structures with large aspect ratios. Experimental results indicate that with a 3 mm diameter spherical probing head, a resolution better than 5 nm can be achieved. The system's main components and its operating principles are shown in Figure 3.52, where the microgap δ between the probe and the target is directly related to the capacitance C.

3.7 CAPACITIVE SENSOR ARRAYS

Capacitive sensors can also be arranged in arrays to perform more sophisticated tasks. Recent advances in integrated and printed circuits, materials, MEMS, and nanotechnologies have further miniaturized capacitive sensors. This has helped to expand their use in severe environments (high temperature, strong magnetic fields, and massive radiation) as well as in noncontact and nonintrusive applications. Aggressive research and development programs around the world continue to drive the design and implementation of capacitive sensors. The following are some examples of capacitive sensor arrays.

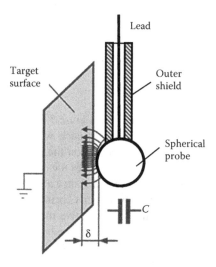

FIGURE 3.52 Components and operating principle of the spherical sensing plate.

3.7.1 SPHERICALLY FOLDED PRESSURE SENSOR ARRAY

Figure 3.53 shows a spherically folded capacitive pressure sensor array (1 mm thickness) for 3D measurements of pressure distribution in artificial joints [41]. The sensor array consists of 192 sensor elements that are arranged in a 16×16 matrix (Figure 3.53a), then folded spherically (Figure 3.53b) and placed in a 60 mm

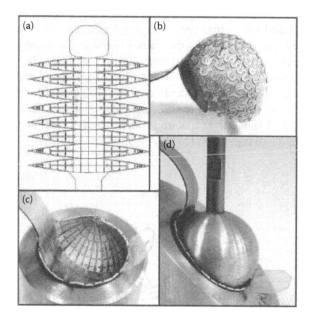

FIGURE 3.53 (a) Unfolded sensor array; (b) spherically folded sensor array; (c) sensor array placed in a cavity; (d) sensor array between ball and cavity.

diameter cavity (Figure 3.53c), followed by a 50 mm diameter ball joint (Figure 3.53d). Such a sensor allows for measuring pressure distributions along curved surfaces such as joints.

3.7.2 CAPACITIVE FINGER PRINT DETECTOR

The *Fingerprint Cards'* sensor [42] shown in Figure 3.54 contains tens of thousands of tiny capacitive plates (function as pixels), each with its own electrical circuit embedded in the chip. When a finger is placed on the sensor, electrical charges are created. Using these charges, the sensor measures the capacitance pattern across the surface. Where there is a ridge or valley, the distance above the plates varies, as does the capacitance, building a pattern of the finger's "print."

This capacitive sensing technique is an effective method to acquire fingerprints. Since an electrical field is measured and the distance between the skin and the pixels is very small, to achieve enough sensitivity, the protective coating must be very thin (a few microns). The drawback of this sensor is its vulnerability to strong external electrical fields (the most troublesome is electrostatic discharge—ESD).

3.7.3 CAPACITIVE TOUCHSCREEN

A capacitive touchscreen uses a surface (coated with conductive material, such as ITO) that can store a charge. The material conducts an electrical current across the panel along the *x*- and *y*-axes. When touched by something conductive, such as a finger, the electric field is altered and the point of contact in the *x–y* coordinate can be determined. In a situation using button-key touch, a discrete capacitive sensor is placed under that particular key location, and when the sensor's field is disturbed, the system notes the point of contact (in resistive touchscreen technology, a pen or a stylus can activate the screen) [43]. The advantages of capacitive touchscreen technologies include high resolution, high image clarity, and resistance to dirt, grease, and moisture. The primary disadvantage of this technology is its susceptivity to ESD and electromagnetic interference when touched. Figure 3.55 shows the capacitive touchscreen's principle (a) and the *x–y* trace array underneath the screen (b) [44,45].

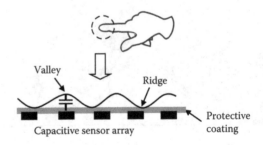

FIGURE 3.54 A capacitive fingerprint sensor.

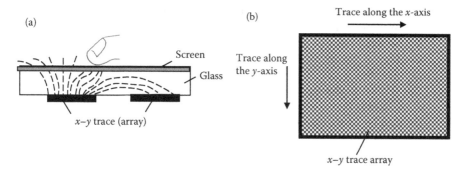

FIGURE 3.55 (a) Disturbance of electric field when screen is touched; (b) x–y trace array located under the touchscreen.

EXERCISES

Part I: Concept Problems

1. Capacitive sensors are designed based on the following variation(s) *except*
   A. Space between the plates
   B. Dielectric material property
   C. Overlap area between plates
   D. Excitation voltage

2. Which of the configurations in Figure 3.56 can hold most charges (if $D = d = 3r$)?

3. A Geiger counter has which of the following configurations?
   A. Parallel
   B. Cylindrical
   C. Spherical
   D. Array

4. Which of the following statements about a capacitor is true?
   A. A capacitor is an energy storage element.
   B. All capacitors are polarized.
   C. Capacitor is an active component.
   D. A capacitor works properly only under an AC source.

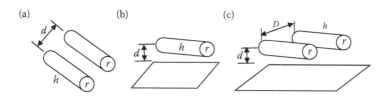

FIGURE 3.56 (a) Two parallel cylinders; (b) one cylinder and one plate; (c) two cylinders and one plate.

5. Which of the following materials has the highest RDP?
   A. Water (fresh)
   B. Alcohol
   C. Ice
   D. Sea water

6. One of the drawbacks of a capacitive proximity sensor is that it
   A. Can only detect a conductive target
   B. Can be easily affected by the media materials between the probe and the target
   C. Is less accurate even though within its measurement range
   D. Has a slow response

7. Capacitive proximity sensors can sense _____
   (conductive/nonconductive/both) materials.

8. The _____ (larger/smaller) the dielectric constant of a target material, the easier it is for a capacitive proximity sensor to detect it.

9. It is easier for a capacitive proximity sensor to detect ____ than porcelain.
   A. Teflon
   B. Marble
   C. Wood
   D. Paper

10. In the cell adhesion strength measurement shown in Figure 3.38, the resultant capacitance C is equal to
    A. $C_{cell} + C_0$
    B. $C_{cell} - C_0$
    C. $\dfrac{C_{cell}C_0}{C_{cell} + C_0}$
    D. $\dfrac{C_{cell} + C_0}{C_{cell}C_0}$

11. State the differences between the three capacitive level sensors shown in Figures 3.34, 3.47, and 3.48.

Part II: Calculation Problems

12. Find the equivalent capacitance between A and B in Figure 3.57.

13. A parallel capacitor is placed in vacuum and has a gap of 3.28 mm. Each plate carries a charge of 4.35×10^{-8} C, and the area of each plate is 12.2 cm^2.

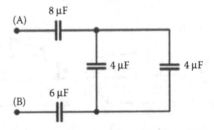

FIGURE 3.57   A capacitor circuit.

Find (1) the capacitance of the capacitor, (2) the voltage potential between the plates, and (3) the magnitude of the electric field between the plates.

14. A 10 μF capacitor has circular-shaped flat plates and is connected to a 12 V battery. Find (1) the charge on each plate, (2) the charge on each plate when the gap d between the plates is doubled, and (3) the charge on each plate if the radius of each plate r is doubled.

15. Compare the capacitance in unit area between a flat-plate capacitor, cylindrical capacitor, and a spherical capacitor. If for all capacitors, $r_1 = 5$ mm, $r_2 = 5.5$ mm, $d = r_2 - r_1$, $\varepsilon_r = 1$, (1) which capacitor has the biggest capacitance per unit area? (2) which capacitor has the smallest capacitance per unit area?

16. Using formulas in Table 3.2, calculate the capacitance of (1) the two 1-m-long parallel wires, $d = 2$ mm, $r = 0.5$ mm; (2) a 1-m-high cylindrical capacitor, $r_1 = 0.5$ mm, $r_2 = 2$ mm. Both are surrounded in a plastic media ($\varepsilon_r = 3$).

17. A 1.75 cm × 1.75 cm square quartz experiences a 153 N force. Assume that the force is evenly distributed over the quartz. If its piezoelectric coefficient is 2.38×10^{-12} C · N^{-1} at 25°C, find (1) the stress of the quartz σ, (2) the electric flux density D, and (3) the polarization charge Q.

18. A piezoelectric pressure sensor is made by fitting a 3 cm × 1.2 cm rectangular piezoelectric strip between two metal plates that transmit the dynamic pressure to the strip and also serve as electrodes to collect the electric charge yielded as output. If the piezoelectric coefficient of the strip at 25°C is 2.33×10^{-12} C · N^{-1} and a 100 kPa pressure is applied, find the polarization charge Q and the voltage output if the capacitance C is 1 nF.

19. A capacitive sensor can be made by inserting a piezoelectric sheet between two plate electrodes. Its capacitance can be described by $C = \varepsilon l b / t_h$, where l, b, and t_h are length, width, and thickness of the sensor, respectively. If $l = 700$ μm, $b = 20$ μm, $t_h = 0.8$ μm, and $\varepsilon = 70 \times 10^{-12}$ F · m^{-1}, find its capacitance.

20. A capacitive proximity sensor is chosen to detect (1) a slab (length: 30 mm, width: 20 mm, thickness: 5 mm; see Figure 3.58a) and (2) a cylinder (length: 25 mm, diameter: 30 mm; see Figure 3.58b). Find the maximum sensor size in diameter allowed to keep adequate spatial resolution for each case.

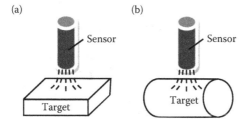

FIGURE 3.58 A capacitive sensor for detecting (a) a slab; (b) a cylinder.

21. A capacitive accelerometer has a similar structure as the one shown in Figure 3.14 except that C_1 and C_2 are connected in parallel. If $C_1 = 0.001$ pF and $C_2 = 0.002$ pF under a maximum acceleration, find the minimum number of silicon finger cells required to produce a measurable capacitance of 0.1 pF.

22. In Example 3.11, if everything remains the same except that the circuit connection is changed so that the two formed capacitors C_1 and C_2 are in parallel, instead of being in series, express the resultant capacitance C of the three-plate capacitor in terms of capacitance C_0 when the metal slab is not present.

23. The capacitance sensor shown in Figure 3.34 is used to determine the relative permittivity ε_r of an unknown liquid. Knowing that the plate width b is 15 mm, the gap d is 1 mm, h_{max} is 100 mm, the liquid level h is 65 mm, and the capacitance reading is 247.5 pF, find ε_r. What is the liquid most likely?

24. The capacitive level sensor in Figure 3.34 is built to monitor alcohol level ($\varepsilon_r = 16$). If the distance between plates d is 5 mm, the capacitance change ΔC is 10 pF, and the maximum alcohol level h_{max} is 20 mm, find the minimum plate width b required.

25. An area-variation-based capacitive accelerometer, as shown in Figure 3.24, has the following parameters: finger length $l_f = 300$ μm, mass length $l_m = 280$ μm, number of fingers $n = 100$, air gap $d_f = 1$ μm, and the relative permittivity $\varepsilon_r = 7$. If the measured capacitance change ΔC is 50.34 pF, find the angel θ in degrees (°).

26. A multiplate capacitive sensor has a circuit diagram shown in Figure 3.59. If the object moves right, which of the following best describes the resultant capacitance C_{S1} (where C_0 is the capacitance of C_{S1} when the object is in the middle position)?
    A. $C_{S1} = C_0 + \Delta C$ and $\Delta C > 0$
    B. $C_{S1} = C_0 - \Delta C$ and $\Delta C > 0$
    C. $C_{S1} = C_0 + \Delta C$ and $\Delta C < 0$
    D. $C_{S1} = C_0 - \Delta C$ and $\Delta C < 0$

27. The quantity of liquid available in its storage tank is often monitored by a cylindrical capacitive level sensor. By connecting the sensor to a capacitance-measuring instrument, the fraction F_{liquid} of the tank filled by liquid in terms

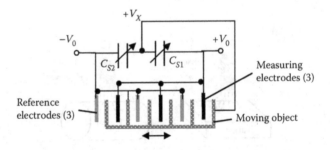

FIGURE 3.59 A multiplate capacitive sensor.

of the sensor's capacitance C can be determined. The sensor has inner and outer conductor radii r_1 and r_2, and height h_{max} that spans the entire height of the tank. The dielectric in the lower and upper region between the cylindrical conductors is the liquid ε_{liq} and its vapor ε_{vap}, respectively. The nonconducting liquid fills the tank to a height h (less than or equal to h_{max}) from the tank's bottom. (1) Derive a formula for F_{liquid}. (2) If $r_1 = 4.5$ mm, $r_2 = 5.0$ mm, $h_{max} = 2.0$ m, $\varepsilon_{liq} = 1.4$, and $\varepsilon_{vap} = 1.0$, what values of C will correspond to the tank being completely full and completely empty?

REFERENCES

1. *The New Encyclopedia Britannica*, Encyclopedia Britannica, Inc., Chicago, Illinois, USA, 1980.
2. The Leyden Jar Discovered, World Wide School® Library, www.worldwideschool.org/library/books/sci/history/AHistoryofScienceVolumeII/chap49.html
3. Fulmek, P.L. et al., Capacitive sensor for relative angle measurement, *IEEE Trans. Instrum. Meas.*, 51, 1145, 2002.
4. Giancoli, D.C., *Physics for Scientists & Engineers with Modern Physics*, 4th ed., Prentice Hall, Upper Saddle River, New Jersey, USA, 2008.
5. Thomson, J.J., Piezoelectricity and its application (quartz and tourmaline for explosion pressures), *Engineering*, 107, 543, 1919.
6. Capacitive sensors for noncontact position and displacement measurements and nano-positioning, ChenYang Technologies GmbH & Co. KG, Germany.
7. Piezo film sensors: technical manual, Measurement Specialties, Inc., Norristown, Pennsylvania, USA, 1999.
8. Jones, R.V. and Richards, J.C.S., The design and some applications of sensitive capacitance micrometers, *J. Phys. E*, 6, 589, 1973.
9. Jong, G.W. and Meijer, G.C.M., An efficient retrieving algorithm for accurate capacitive position sensors, *Sens. Actuators A*, 58, 75, 1997.
10. Capacitive sensors—an overview, Lion Precision, St. Paul, Minnesota, USA, 2010. www.capacitive-sensing.com.
11. Capacitive position sensors/controllers for nanometrology, Physik Instrumente, GmbH & Co. KG, Germany. www.capacitance-sensors.com/index.htm.
12. Jackson, R.G., *Novel Sensors and Sensing*, 1st ed., Institute of Physics Publishing, Bristol and Philadelphia, 53, 2004.
13. Capacitive accelerometers: series 3701, 3702, 3703, and 3801, PCB Piezotronics, Inc., Depew, New York, USA, 2004.
14. Baxter, L.K., *Capacitive Sensors: Design and Applications*, 1st ed., IEEE Press, New York, 1997.
15. Dutta, S. et al., Design and analysis of wet etching based comb type capacitive accelerometer, *Sens. Transduce. J.* 91, 31, 2008.
16. Selvakumar, A. and Najafi, K., A high-sensitivity-axis capacitive silicon microaccelerometer with a torsional suspension, *J. Microelectromech. S.*, 7, 192, 1998.
17. Kuijpers, A.A. et al., 2D-finite-element simulations for long-range capacitive position sensor, *J. Micromech. Microeng.*, 13, 183, 2003.
18. Capacitive proximity sensors: theory of operation, Siemens online course. www.automationmedia.com/Port1050%5CSiemensFreeCourses%5CSnrs_3.pdf.
19. Nibbelink, B., Capacitive humidity sensor presentation, MECH207: Mechatronics I, Oct. 23, 2009.
20. HS1100/HS1101 relative humidity sensor technical data, *Measurement Specialties*, Hampton, Virginia, USA, 2002.

21. Plum, T.J., Saxena, V., and Jessing, R.J., Design of a MEMS capacitive chemical sensor based on polymer swelling, *IEEE Workshop Microelectron. Electron Devices*, 2, 2006.

22. Ernst, A. et al., A capacitive sensor for non-contact nanoliter droplet detection, *Sens. Actuator A*, 153, 57, 2009.

23. Winn, W.P., An electrostatic theory for instruments which measure the radii of water drops by detecting a change in capacity due to the presence of a drop, *J. Appl. Meteorol.* 7(5), 929, 1968.

24. Puers, R., Capacitive sensors: when and how to use them, *Sens. Actuator A*, 37–38, 93, 1993.

25. Fabian, T. and Brasseur, G., A robust capacitive angular speed sensor, *IEEE Trans. Instrum. Meas.*, 47, 280, 1998.

26. Fulmek, P.L. et al., Capacitive sensor for relative angle measurement, *IEEE Trans. Instrum. Meas.*, 51, 1145, 2002.

27. Li, X., Meijer, G.C.M., and de Jong, G.W., A microcontroller-based self-calibration technique, *IEEE Trans. Instrum. Meas.*, 46, 888, 1997.

28. Tandeske, D., *Pressure sensors: selection and application*, Marcel Dekker, Inc., New York, USA, 1991.

29. Prakash, S.B. et al., A CMOS capacitance sensor for cell adhesion characterization, in *Proc. IEEE Int. Symp. Circuits Sys. (ISCAS)*, Kobe, Japan, 3495, 2005.

30. Engel, J.M. et al., Multi-walled carbon nanotube filled conductive elastomers: materials & application to micro transducers, *19th IEEE Int. Conf. Micro Electro Mech. Sys.*, Istanbul, Turkey, 246, 2006.

31. Guiducci, C. et al., Microelectrodes on a silicon chip for label-free capacitive DNA sensing, *Sens. J.*, 6, 1084, 2006.

32. Gong, J.L. et al., Capacitive chemical sensor for fenvalerate assay based on electropolymerized molecularly imprinted polymer as the sensitive layer, *Anal. Bioanal. Chem.*, 379, 302, 2004.

33. Chapman, P.D., A capacitive based ultraprecision spindle error analyzer, *J. Precision Eng.*, 7, 129, 1985.

34. Thomas, D.K. et al., Capacitive displacement sensor, US Patent 6,683,780, January 27, 2004.

35. Jeon, S. et al., New design of cylindrical capacitive sensor for on-line precision control of AMB spindle, *IEEE Trans. Instrum. Meas.*, 50, 757, 2001.

36. Robinson, M.C., Capacitance thermometer, US Patent 3,759,104, September 18, 1973.

37. Golnab, H. and Azim, P., Simultaneous measurements of the resistance and capacitance using a cylindrical sensor system, *Mod. Phys. Lett. B*, 22(8), 595, 2008.

38. Hart, J.E. et al., The geophysical fluid flow cell experiment, NASA Technical Report, September, 1999.

39. Zhang, R., Dai, S., and Mu, P., A spherical capacitive probe for measuring the thickness of coatings on metals, *Meas. Sci. Tech.*, 8, 1028, 1997.

40. Tan, J.B. and Cui, J.N., Ultraprecision 3D probing system based on spherical capacitive plate, *Sens. Actuator A*, 159, 1, 2010.

41. Muller, O. et al., Three-dimensional measurements of the pressure distribution in artificial joints with a capacitive sensor array, *J. Biomech.*, 37, 1623, 2004.

42. Fingerprint sensing techniques, retrieved on Sept. 20, 2007 at: http://perso.orange.fr/fingerchip/biometrics/types/fingerprint_sensors_physics.htm.

43. Adamas, L., Taking charge, *Appliance Design*, 16, 2008.

44. Philipp, H., Please touch! Explore the evolving world of touchscreen technology, *Electronic Design*, April 24, 2008.

45. Touchscreen technology for displays & monitors, Planar Systems, Inc., Beaverton, Oregon, USA, www.planarembedded.com/technology/touch/. Retrieved on July 21, 2010.

4 Inductive Sensors

4.1 INTRODUCTION

Inductive sensors are designed based on the operating principle and characteristics of an inductor. They are primarily used to measure electric and magnetic fields, or other physical quantities (e.g., displacement and pressure) that can be transformed into an electric or magnetic response. Inductive sensors do not require physical contact; thus, they are noncontact sensors and are particularly useful for applications which access presents challenges. Inductive sensors can only detect metals. They react differently to different metallic materials such as steel, copper, and aluminum. Thus, they can be used to identify different metals. Unlike capacitive sensors, inductive sensors are not affected by nonmetallic media materials between the probe and the target, so they are well adapted to harsh environments where oil, dust, dirt, or other substances are present. The key advantages of inductive sensors include

- Nearly infinite resolution
- Fast response
- Large operating temperature range
- High reliability
- Robustness
- Easy handling

There are historical examples of various inductive sensors. The Chattock–Rogowski coil was first described in 1887 [1,2]. Today, this sensor has been redesigned as an excellent current transducer and is used in measuring magnetic properties of soft magnetic materials. An Austrian patent in 1957 described the use of a needle sensor for investigating local flux density in steel [3].

An inductor is an electromagnetic component that relates the interaction between electrical and magnetic fields. Many electromagnetic principles can be employed to design inductive, magnetic, or electromagnetic sensors. Although there is no clear definition to distinguish inductive, magnetic, and electromagnetic sensors, the sensors that are primarily made of simple inductive coils, such as air coils, eddy-current sensors, and variable reluctance sensors are considered as inductive sensors and will be discussed in this chapter, while the rest of the magnetic or electromagnetic sensors will be presented in Chapter 5.

4.2 INDUCTORS, INDUCTANCE, AND MAGNETIC FIELD

4.2.1 INDUCTORS

An inductor is a passive electrical or electronic component that resists changes in current. The voltage–current relationship of an inductor is governed by

$$V = L\frac{dI}{dt} \tag{4.1}$$

where V is the voltage (in volts, V) across the inductor, L is the inductance (in henrys, H), I is the current (in amperes, A) flowing through the inductor, and t is the time (in seconds, s). The typical range of L is 1 µH–0.1 H.

An inductor is also an energy storage device. Energy is stored as a magnetic field in or around the inductor. The energy (in joules, J) stored in an inductor is equal to

$$U = \frac{1}{2}LI^2 \tag{4.2}$$

Inductors in series:

$$L_{eq} = \sum L_i \tag{4.3}$$

Inductors in parallel:

$$\frac{1}{L_{eq}} = \sum \frac{1}{L_i} \tag{4.4}$$

where L_{eq} is the resultant (or equivalent) inductance.

Inductors can be classified as wire-wound (air) inductors, radial inductors, chip inductors, and power inductors as shown in Figure 4.1. Their primary applications include magnetic energy storage, electric motors, transformers, and sensors.

Inductors can also be classified as the fixed, variable, or adjustable type. In fixed inductors, the turns of coils remain fixed. In variable inductors, the effective

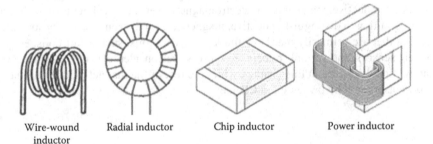

Wire-wound   Radial inductor   Chip inductor   Power inductor
inductor

FIGURE 4.1 Types of inductors.

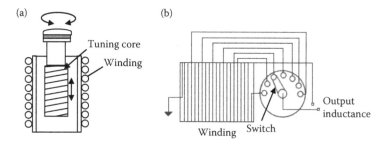

FIGURE 4.2   (a) An adjustable inductor; (b) a variable inductor.

inductance can be changed by a moving iron or ferrite core inside the coil form (see Figure 4.2a). The inductance increases or decreases based on how much the core has moved into the coil form. These coils are called *slug-tuned inductors*. Adjustable inductors either have taps for changing the number of desired turns, or consist of several fixed inductors that can be switched into various series or parallel combinations (see Figure 4.2b). The inductor-related circuit symbols are shown in Figure 4.3.

4.2.2   INDUCTANCE AND MAGNETIC FIELD

When current flows in a conductor, it creates a magnetic field and hence magnetic flux around the circuit. The inductance defines the ratio of the magnetic flux Φ (in webers, Wb, or volt-seconds, V · s) to the current I:

$$L = \Phi/I \tag{4.5}$$

The magnetic field can be described by magnetic flux density B (in teslas, T; or Wb · m^{-2}), and magnetic field strength H (in amperes per meter, A · m^{-1}). Their relationship can be expressed by

$$B = \mu H = \mu_0 \mu_r H \tag{4.6}$$

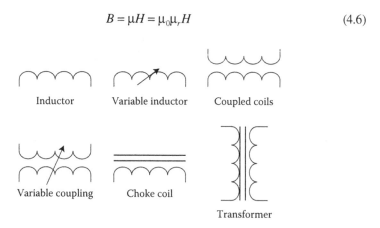

FIGURE 4.3   Circuit symbols for various inductors.

TABLE 4.1

Relative Permeability of Common Materials

Material	μ_r	Material	μ_r
Superconductors	0	Steel	100
Water, copper	0.99999	Nickel	100–600
Air, vacuum, plastic, teflon, wood	1	Ferrite (nickel zinc)	16–640
Aluminum	1.00002	Ferrite (manganese zinc)	≥640
Platinum	1.00027	Permalloy	2500–25,000
Cobalt	70–250	Mu-metal	20,000–100,000

where μ is the permeability of a material. It describes the ability of the material to support the formation of a magnetic field within itself. μ_0 ($4\pi \times 10^{-7}$ H · m^{-1}) is the permeability in vacuum (free space), and μ_r is the relative permeability ($\mu_r = 1$ for free space). The relative permeability for common materials is listed in Table 4.1.

B is often given in the non-SI unit of gauss: 1 T = 10^4 gauss. The tesla is a very large unit. The earth's magnetic field is about 0.5 gauss, or 0.05 mT. For comparison, a small bar magnet has $B \approx 10^{-2}$ T; magnetic resonance imaging (MRI) body scanner magnet $B \approx 2$T; a hair dryer $B \approx 10^{-7}$–10^{-3} T; sunlight $B \approx 3 \times 10^{-6}$ T; and the field at the surface of a neutron star is thought to be about 10^8 T.

EXAMPLE 4.1

The magnetic field of the earth is $B = 5 \times 10^{-5}$ T. What is the earth's magnetic field strength H in ampere per meter? Use $\mu_r = 1$.

SOLUTION

Apply Equation 4.6:

$$H = \frac{B}{\mu_0 \mu_r} = \frac{5 \times 10^{-5} \text{ T}}{(4\pi \times 10^{-7} \text{H} \cdot \text{m}^{-1})(1)} = 39.81 \text{ A} \cdot \text{m}^{-1}$$

The magnetic field depends on the trajectory of the moving charge or the shape of the conductor carrying the current. Table 4.2 lists the inductance L (in H), magnetic field H (in A · m^{-1}), and current I (in A) for various configurations.

EXAMPLE 4.2

Find the magnetic flux density at a distance $R = 2$ mm in free space from a straight long wire conductor carrying a current of 20 A.

SOLUTION

In free space $\mu_r = 1$, from Equation 4.9

$$B = \frac{\mu_0 \mu_r I}{2\pi R} = \frac{(4\pi \times 10^{-7} \text{V} \cdot \text{A}^{-1} \cdot \text{m}^{-1})(1)(20 \text{ A})}{(2\pi)(0.002 \text{ m})} = 0.002 \text{ (V} \cdot \text{m}^{-2})$$

TABLE 4.2

Inductance and Magnetic Field for Various Configurations

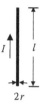

2r

A straight wire with length l:

$$L = 200l \left(\ln \frac{2l}{r} - 1 \right) \times 10^{-19} \qquad (4.7)$$

where r is the radius of the wire (in m) and I is the current (in A).

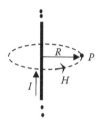

Magnetic field at point P around a straight very long wire:

$$H = \frac{I}{2\pi R} \qquad (4.8)$$

$$B = \frac{\mu_0 \mu_r I}{2\pi R} \qquad (4.9)$$

where R is the distance from point P to the wire (in m).
Infinite cylindrical coil helix:

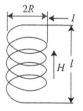

$$L = \mu N^2 I A_{coil} \qquad (4.10)$$

$$H = NI \text{ (for } l \gg R) \qquad (4.11)$$

where N is the number of turns of the coil and A_{coil} is the area of the coil (m²).

continued

TABLE 4.2   (continued)

Inductance and Magnetic Field for Various Configurations

Torus (toroidal coil):

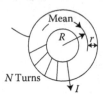

$$L = \frac{\mu N^2 r^2}{2R} \tag{4.12}$$

$$H = \frac{NI}{2\pi R}\left(\text{for } r \ll R\right) \tag{4.13}$$

Coaxial cable (high frequencies):

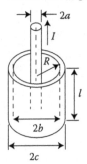

$$L = \frac{\mu I}{2\pi}\ln\left(\frac{b}{a}\right) \tag{4.14}$$

$$H = \frac{I}{2\pi R}\quad\left(a < R < b\right) \tag{4.15}$$

Parallel transmission lines (high frequencies):

$$L = \frac{\mu I}{\pi}\ln\left(\frac{d-r}{r}\right) \tag{4.16}$$

$$H = \frac{2I}{\pi d} \tag{4.17}$$

TABLE 4.2 (continued)

Inductance and Magnetic Field for Various Configurations

A long and thin coil (length of the coil $l \gg R$):

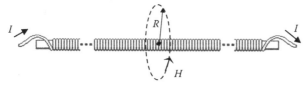

The magnetic flux density B within the coil:

$$B = \mu_0\mu_r NI/l \tag{4.18}$$

A circular conductive loop (single turn coil):

$$L = \mu_0\mu_r R[\ln(8R/r) - 2 + Y] \tag{4.19}$$

where R is the radius of the loop, r is the radius of the conductor, and Y is a constant ($Y = 1/4$ when the current is homogeneous across the wire; $Y = 0$ when the current flows on the surface of the wire).

An N-turn circular conductive loop (r—radius of the wire):

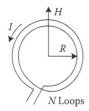

$$L = \frac{R^2 N^2}{(2R + 5.6r) \times 10^5} \tag{4.20}$$

$$H = \frac{NI}{2R} \quad \text{(at the center of the coil)} \tag{4.21}$$

Source: Linberger M., *FE Review Manual*, 3rd ed., Professional Publications, Inc. (PPI), Belmont, California, USA, 2011.

4.3 PHYSICAL LAWS AND EFFECTS GOVERNING INDUCTIVE SENSORS

4.3.1 LORENTZ FORCE

When an electron moves (or electrical current I flows) through a magnetic field B, it experiences a force F (see Figure 4.4a), called *Lorentz force* (in newtons, N), defined by the vector equation

$$\vec{F} = q\vec{v} \times \vec{B} \tag{4.22}$$

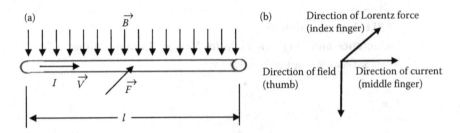

FIGURE 4.4 (a) Graphical indication of Lorentz force; (b) right-hand rule to determine Lorentz force direction.

where q is the charge of an electron (1.602×10^{-19} C), $\vec{v}$ is the velocity of the electron (in m · s⁻¹), and $\vec{B}$ is the magnetic field flux density (in T). Lorentz force is always perpendicular to $\vec{B}$ and $\vec{v}$, causing the moving electron to be deflected into a circular path.

Lorentz force is caused by the interaction (attraction or repulsion) between the magnetic field created by the moving charge (e.g., an electron) and the external magnetic field $\vec{B}$. The direction of the Lorentz force for a given current and magnetic field can be determined by the *right-hand rule* (see Figure 4.4b).

EXAMPLE 4.3

A current I flows through a conductor (with a length l) located in a magnetic field $\vec{B}$. If the conductor moves right with a velocity $\vec{v}$, as shown in Figure 4.5a, find the Lorentz force acting on the conductor, and also indicate its direction in the figure.

SOLUTIONS

From Equation 4.22

$$|\vec{F}| = |q\vec{v} \times \vec{B}| = |(q)(v)(B)\sin 90°| = qvB$$

The Lorentz force will point to the left (see Figure 4.5b).

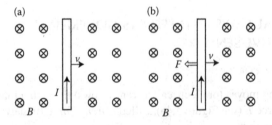

FIGURE 4.5 (a) A current-carrying conductor moving in a magnetic field; (b) the Lorentz force $\vec{F}$ acting on the moving conductor.

4.3.2 FARADAY'S LAW OF ELECTROMAGNETIC INDUCTION

Faraday's law of electromagnetic induction states that when the magnetic field that passes through a conductor changes, a voltage or current (if it is a closed loop) will be induced in the conductor. The induced voltage is called the *electromotive force* (*EMF*) or *transformer EMF*.

For a straight wire or a one-loop wire, the induced voltage V_L is [5]

$$V_L = -\frac{d\Phi}{dt} \tag{4.23}$$

where Φ is the magnetic flux (in Wb) and $d\Phi/dt$ is the rate of change in the magnetic flux (in Wb $\cdot$ s^{-1}). The negative sign indicates that the induced EMF produces current that opposes the original flux change.

For an *N*-loop coil, the above induced voltage V_L becomes

$$V_L = -N\frac{d\Phi}{dt} = -NA\frac{dB}{dt} = -\mu NA\frac{dH}{dt} \tag{4.24}$$

where A is the area of the coil and N is the number of turns.

The EMF generated by moving a wire in a magnetic field is called *motional EMF* described by

$$V_L = -\frac{d\Phi}{dt} = -\frac{BdA}{dt} = -\frac{Blvdt}{dt} = -Blv \tag{4.25}$$

where l is the wire length; v is wire's moving velocity; $lvdt$ is the area swept out by the wire in time interval dt.

Faraday's law is widely applied in design of inductive sensors, transformers, electric generators, sound systems, and computer memory.

EXAMPLE 4.4

Derive the open circuit voltage between the brushes on a Faraday's disk as shown in Figure 4.6.

SOLUTION

From Equation 4.25

$$V_L = -(BdA/dt) \Rightarrow \int V_L dt = -\int BdA$$

Choose a small line segment of length dr at position r from the center of the disk between the brushes. The induced EMF in this elemental length is then ($v = r\omega$)

$$\int V_L dt = -\int BdA = -\int_{r_1}^{r_2} Bvdr = -\int_{r_1}^{r_2} B(\omega r)dr = -\frac{\omega B}{2}(r_2^2 - r_1^2)$$

A Faraday's disk is also known as a *homopolar generator*.

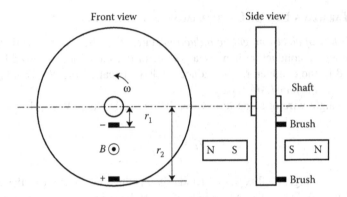

FIGURE 4.6 A Faraday's disk.

4.3.3 BIOT–SAVART LAW

Biot–Savart law relates a *steady* current I flowing through a closed loop and the magnetic field $\vec{B}$ generated:

$$\vec{B} = \frac{\mu I}{4\pi} \oint \frac{d\vec{l} \times \vec{r}}{r^2} \qquad (4.26)$$

where $d\vec{l}$ is the integration element of length l along the current path and $\vec{r}$ is the position vector pointing from the element $d\vec{l}$ to the field point at which the magnetic field $\vec{B}$ is to be calculated.

4.3.4 AMPERE'S LAW

Ampere's law is one of the four Maxwell's equations. It relates a magnetic field $\vec{B}$ to an electric current I that produces:

$$\oint \vec{B} \cdot d\vec{l} = \mu I \qquad (4.27)$$

where $\oint$ denotes the integration along a closed path, $d\vec{l}$ is an infinitesimal element of length along the integration path, and I is the applied current. Using Ampere's law, one can determine the magnetic field associated with a given current, or the current associated with a given magnetic field.

EXAMPLE 4.5

A solenoid is a tightly wound cylindrical helix of current-carrying wire, widely used as an electromagnetic valve (at least several in every car). Find the magnetic field inside a solenoid using Ampere's law.

SOLUTION

Refer to Figure 4.7 [6], an imaginary rectangular path is used for integration. The only nonzero contribution to the integral is along the length l inside the solenoid (since

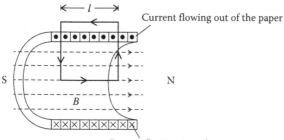

Current flowing out of the paper

Current flowing into the paper

FIGURE 4.7 The integration of magnetic field strength H along a closed path to find the MMF for a long solenoid. (From Morgan, K., *Ampere's Law*, HandiComputing, Charlotte, Michigan, USA, 2000.)

$B = 0$ outside the solenoid, and the other two edges canceled out after integration due to their opposite signs). The total net current through the length l is nIl, where n is the number of turns per unit length (nl, thus, is the number of turns in length l) and I is the current in the wire. Plugging $\oint \vec{B} \cdot \vec{dl} = Bl$ and nIl into Equation 4.27 yields

$$Bl = \mu nIl$$
$$\Rightarrow B = \mu nI \tag{4.28}$$

EXAMPLE 4.6

A toroid is a solenoid that has been bent into a circle (the space-saving shape, see Figure 4.8a). Find the magnetic field inside an N-turn toroid carrying a current I using Ampere's law.

SOLUTION

Apply Ampere's law through a circular integration path (see Figure 4.8b):

$$\oint \vec{B} \cdot \vec{dl} = 2\pi R B = \mu_0 NI$$
$$\Rightarrow B = \frac{\mu_0 NI}{2\pi R} \tag{4.29}$$

The magnetic field is tangential to the circular integration path. Unlike the solenoid, the magnetic field inside the toroid is not constant over the cross section of the coil but varies inversely as the distance R. For points outside a toroid, the field is essentially zero if the turns of the wire are very close together.

4.3.5 Magnetomotive Force

Any physical cause that produces magnetic flux is called *magnetomotive force* (MMF).

The total MMF $\Im$ along a closed path encircling a current-carrying coil is proportional to the ampere-turns NI in its winding:

$$\Im = \oint H dl = NI \tag{4.30}$$

(a)   (b)

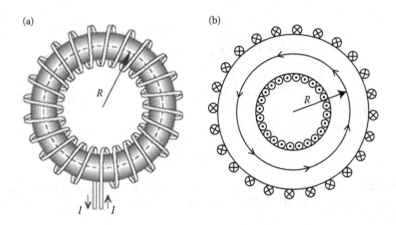

FIGURE 4.8 (a) An N-turn toroid carrying a current I; (b) the integration path for the toroid. (From Morgan, K., *Ampere's Law, HandiComputing*, Charlotte, Michigan, USA, 2000.)

where the magnetic field strength H is expressed in ampere-turns. It is analogous to EMF in electricity. The MMF in an N-turn coil is given by

$$\Im = NI = \Phi \Re \tag{4.31}$$

where $\Re$ (in $A \cdot Wb^{-1}$) is the reluctance of the magnetic circuit. $\Re$ is analogous to electrical resistance and it depends on the path area A, length l, and permeability μ:

$$\Re = \frac{l}{\mu A} \tag{4.32}$$

4.3.6 Eddy Current

An *eddy* or *Foucault current* is an electromagnetic phenomenon discovered by the French physicist Leon Foucault in 1851 [7]. When a conductor is exposed to a changing magnetic field (e.g., created by an AC coil, as shown in Figure 4.9), a circulating flow of electrons (eddy current) is induced in the conductor.

    The generation of these eddy currents takes energy from the coil and appears as an increase in the electrical resistance of the coil. The eddy currents also generate their own magnetic fields that oppose the magnetic field of the coil, and thus change the inductive reactance of the coil. Resistance and inductive reactance vectors add up to the total impedance of the coil. This impedance increase can be measured with an eddy-current probe, from which much information about the test material can be obtained. Typical eddy-current testing devices are designed to measure these energy losses. The eddy-current loss, P_E (in watts), is measured by

$$P_E = K_E B_{max}^2 f^2 t_h^2 V_{vol} \tag{4.33}$$

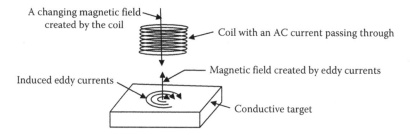

A changing magnetic field — created by the coil

Coil with an AC current passing through

Magnetic field created by eddy currents

Induced eddy currents

Conductive target

FIGURE 4.9 Eddy-current principle.

where K_E is a constant whose value depends on the electrical resistance of the conductor and the systems used, B_{max} is the maximum flux density in Wb $\cdot$ m^{-2}, f is the frequency (in Hz) of magnetic reversals, t_h is the thickness of laminations (in m), and V_{vol} is the volume of the conductor (in m^3). Winding loss at high frequencies is caused by eddy-current effects. It is usually calculated by the *finite element analysis* (FEA) method in order to get accurate results.

4.3.7 Skin Effect

If a conductor is carrying a high alternating current, the distribution of the current is not evenly dispersed throughout the cross section of the conductor. This is due to two independent effects known as *skin effect* and *proximity effect*.

Skin effect often occurs in a single conductor (i.e., no other conductors are nearby), in which the current density near the surface of the conductor is largest and the current density decreases as the depth increases. For an eddy current case the skin effect suggests that the induced eddy current flows along the "skin" of the conductor. This results in the current density near the surface of the conductor being greater than that at its core. The average depth of current flow is called *skin depth*, defined as the distance over which the current falls to 1/e or 37% of its original value ($e = 2.71828$). The *skin depth*, Δ_S, can be calculated as follows:

$$\Delta_S = \sqrt{\frac{2\rho}{\omega\mu}} \quad \text{or} \quad \Delta_S = \sqrt{\frac{\rho}{\pi f \mu_0 \mu_r}} \tag{4.34}$$

where ρ is the resistivity of the conductor, ω is the frequency (in radians) of the current ($\omega = 2\pi f$, f is the AC excitation frequency in Hz), and μ is the absolute magnetic permeability ($\mu = \mu_0 \mu_r$). In general, the minimum thickness of a target should be three times of the skin depth Δ_S.

The *skin factor*, F_S, can be calculated by [8]

$$F_S = x_S^4/(192 + x_S^4) \tag{4.35}$$

where $x_S^2 = 2\mu_0 f k_S/R_{DC_T}$, k_S is a design/construction dependant coefficient of the conductor ($k_S = 1$ for circular, compacted, and sectored), and R_{DC_T} is the DC resistance at temperature T.

TABLE 4.3

Minimum Thickness (Three Skin Depths) of Common Ferrous and Magnetic Materials

Material	ρ ($\Omega \cdot m$)	μ_r	f (MHz)	Minimum Thickness (mm)
416 SST	57×10^{-8}	700	1	0.04
Iron	11×10^{-8}	100	1	0.05

Skin effect causes the effective resistance of the conductor to increase at higher driving frequencies where the skin depth is smaller. Tables 4.3 and 4.4 list some minimum thicknesses for common materials with a typical 1 MHz driving frequency [7,9]. Table 4.5 gives the resistivity and temperature coefficients for common materials.

EXAMPLE 4.7

A brass target is detected by an inductive probe. If the probe is driven by a 500 kHz AC current, what is the brass target's minimum thickness?

SOLUTION

From Table 4.4, the resistivity of brass is 6.4×10^{-8} $\Omega \cdot m$, and the skin depth (often expressed in mm) for brass with $f = 500$ kHz is

$$\Delta_S = \sqrt{\frac{\rho}{\pi f \mu_0 \mu_r}} = \sqrt{\frac{6.4 \times 10^{-8}\ \Omega \cdot m}{(3.14)(5 \times 10^5\ Hz)(4 \times 3.14 \times 10^{-7}\ H \cdot m^{-1})(1)}} = 0.18(mm)$$

The minimum thickness is therefore

$$\text{Minimum thickness} = 3\Delta_S = 3 \times (0.18\ mm) = 0.54\ mm$$

4.3.8 PROXIMITY EFFECT

Proximity effect occurs where two or more conductors are close to each other. Refer to Figure 4.10, if each conductor carries a current flowing in the same direction, the

TABLE 4.4

Minimum Thickness (Three Skin Depths) of Common Nonferrous Materials

Material	ρ ($\Omega \cdot m$)	μ_r	f (MHz)	Minimum Thickness (mm)
Silver	1.59×10^{-8}	1	1	0.19
Copper	1.72×10^{-8}	1	1	0.2
Gold	2.21×10^{-8}	1	1	0.22
Aluminum	2.65×10^{-8}	1	1	0.25
Zinc	5.97×10^{-8}	1	1	0.37
Brass	6.4×10^{-8}	1	1	0.38
Tin	11.5×10^{-8}	1	1	0.51
Lead	20.8×10^{-8}	1	1	0.69
Titanium	47×10^{-8}	1	1	1.03

TABLE 4.5
Electrical Resistivity and Temperature Coefficients for Various Materials

Material	Resistivity ($\Omega \cdot m$) at 20°C	Temperature Coefficient (K^{-1})
Silver	1.59×10^{-8}	0.0038
Copper	1.72×10^{-8}	0.0039
Gold	2.44×10^{-8}	0.0034
Aluminum	2.82×10^{-8}	0.0039
Tungsten	5.60×10^{-8}	0.0045
Zinc	5.90×10^{-8}	0.0037
Iron	1.0×10^{-7}	0.005
Tin	1.09×10^{-7}	0.0045
Platinum	1.06×10^{-7}	0.00392
Lead	2.2×10^{-7}	0.0039
Manganin	4.82×10^{-7}	0.000002
Constantan	4.9×10^{-7}	0.000008
Mercury	9.8×10^{-7}	0.0009
Nichrome	1.10×10^{-6}	0.0004
Carbon	3.5×10^{-5}	−0.0005
Germanium	4.6×10^{-1}	−0.048
Silicon	6.40×10^{2}	−0.075

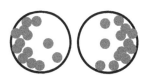
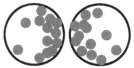

Two wires with the same current direction

Two wires with the opposite current direction

FIGURE 4.10 Proximity effect for two wires with same or different current direction.

areas of the conductors in close proximity experience more magnetic flux than the remote areas. Consequently, current distribution is not even throughout the cross section, and a greater proportion of current is carried by the remote areas. This phenomenon can also be explained by Lorentz force that pushes the charge carries (e.g., electrons or holes) and causes the higher charge densities in the remote areas. If the currents are in opposite directions, the areas in close proximity will carry the greater density of current. This effect is known as *proximity effect*. For two cable conductors, the proximity effect factor F_p can be found by [8]

$$F_P = x_P^4/(192 + 0.8x_P^4) * (2r/d)^2 * 2.9 \tag{4.36}$$

where $x_P^2 = 2\mu f k_P / R_{DC_T}$, k_P is a constant determined by conductor construction ($k_P = 1$ for circular, stranded, compacted, and sectored, $k_P = 0.8$ if the above conductors are dry and impregnated), r is the radius of the conductor (in m), R_{DC_T} is DC resistance at temperature T, and d is the distance between the conductor centers (in m).

Proximity effect increases the wire loss and effective resistance. As compared to the skin effect, at high frequencies, proximity effect can increase resistance values by factors of 10–1000 above the skin effect.

4.3.9 Electric and Magnetic Field Analogies

The structure of magnetic laws suggests an electric circuit analogy shown as follows:

$$V \Leftrightarrow NI, \quad I \Leftrightarrow \Phi, \quad R \Leftrightarrow \Re$$

that is, voltage V is analogous to NI, current I is analogous to flux Φ, and resistance R is analogous to reluctance $\Re$. The analogy between resistance and reluctance is

$$R = \frac{l}{\sigma_p A} \Leftrightarrow \Re = \frac{l}{\mu A}$$

4.4   CHARACTERISTICS AND MATERIALS OF INDUCTIVE SENSORS

4.4.1 Terminologies

Terminologies involved in inductive sensors are

Nominal sensing distance S_n: the maximum distance from a sensor to a standard 1-mm-thick square plate that can trigger a change in the output of the sensor. The standard square plate is made of carbon steel FE 360 as defined in ISO 630:1980. Its side length is equal to the diameter of the active area of a shielded inductive sensor, and 1.5 times of the diameter of a nonshielded inductive sensor. The tests are performed at 20°C with a constant voltage supply.

Maximum switching current: the maximum amount of continuous current allowed to flow through a sensor without causing damage to the sensor.

Minimum switching current: the minimum value of current that should flow through a sensor in order to guarantee its operation.

Maximum peak current: the maximum current value that a sensor can bear in a limited period of time.

Residual current: the current that flows through a sensor when it is in the open state.

Power drain: the amount of current required to operate a sensor.

Voltage drop: the voltage drop across a sensor when driving the maximum load.

Operating frequency: the maximum number of on/off cycles that a sensor device can have in one second.

Operating sensing distance: within the operating sensing distance, a sensor can operate reliably taking into account all the possible situations.

Repeatability (in percentage of S_n): the variation measured at any distance within the operating range in an 8-hour period at a temperature between 15°C and 30°C and a supply voltage with a ≤5% deviation.

Switching hysteresis (in percentage of S_n): the distance between the switching-on point (approach) and the switching-off point (retreat or moving away).

Protection degree: enclosure degree of protection according to the International Electrotechnical Commission (IEC). IP 65: Dust tight. Protection against water jets; IP 67: Dust tight. Protection against the effects of immersion.

Offset: the distance between the sensor face and the beginning of the measuring range. A target and a sensor should never be closer to each other than the specified offset. An offset is necessary to keep a sensor as a noncontact device, and to put the sensor's inductive characteristics in more linear portion of its range. Standard offset is optimized for the best performance.

Power up time delay: the time that a sensor needs to be ready for operation after connecting the power supply. It is in the millisecond range.

Nonparallelism: a target and a sensor should be parallel to each other for ideal performance. The error associated with nonparallelism is called *cosine error*, which results in increased nonlinearity. A nonparallelism of up to 3° will increase nonlinearity less than 0.5% of the full scale. A nonparallelism of 10° will increase nonlinearity approximately 4% of the full scale.

4.4.2 COIL MATERIALS

The key component of an inductive sensor, an inductor, is usually made of a copper (Cu) coil with or without winding around a magnetic core. Parameters commonly used in the inductive sensor design are listed in Table 4.6.

4.4.3 CORE MATERIALS

Characteristics of an inductive sensor mainly depend on coil and core materials, turns of coil, shape of coil, and excitation frequency. A magnetic core is used to concentrate the magnetic field in the region around a coil and to achieve higher inductance values in a smaller size.

Core materials used in inductive sensors are either (1) ferromagnetic materials, including crystalline metals and alloys (iron, nickel, cobalt, manganese, or their compounds), amorphous metals and alloys, and ferrites (ferromagnetic oxides) [10], or (2) nonmagnetic materials such as air (no core), wood, or plastic. A nonmagnetic core is used to increase the stiffness of the coil and provide a support for the metal wires to be wound on. The main function of a magnetic core is to provide an easy path for flux to facilitate flux linkage, or coupling between two or more magnetic elements. Ideal magnetic material characteristics should have a square-shaped *B–H* curve, extremely high permeability (e.g., 60,000 H · m^{-1}), high saturation flux density (e.g., 0.9 T), high resistivity (to reduce eddy current and its losses), and insignificant energy storage.

TABLE 4.6

Parameters Commonly Used in Inductive Sensor Design

Term	SI	CGS	CGS to SI
Permeability μ_0	$4\pi \times 10^{-7}$ H · m^{-1}	1	$4\pi \times 10^{-7}$
Relative permeability μ_r	–	–	–
Magnetic flux density B	Tesla, T	Gauss, Gs	10^{-4}
Magnetic field strength H	A · m^{-1}	Oersted, Oe	$1000/4\pi$
Area A	m^2	cm^2	10^{-4}
Length l	m	cm	10^{-2}
Total flux $\Phi = \oint BdA$	Weber, Wb	Maxwell, Mx	10^{-8}
Total field MMF, $\Im = \oint Hdl$	A · T	Gilbert, Gi	$10/4\pi$
Reluctance $\Re = \Im/\Phi$	–	–	$10^9/4\pi$
Permeance $P = 1/\Re = L/N^2$	–	–	$4\pi \times 10^{-9}$
Inductance $L = P\,N^2$	Henry, H	Henry, H	1

Metal alloy cores (usually permalloy) are often made up with very thin tape-wound laminations to minimize losses due to induced eddy currents. Tape thickness can be made extremely thin (12.5 μm). Tape-wound cores are primarily used at 50, 60, and 400 Hz line frequencies. *Powdered metal cores* composite powdered metals (e.g., permalloy powders) that can store considerable energy and are therefore used in inductors and flyback transformers. Note that the energy is not stored in the very high permeability magnetic metal portions of the core but in the nonmagnetic gaps (distributed throughout the entire core between the magnetic particles in the binder that holds the cores together). *Ferrite cores* have permeability in the range of 1500–3000 H · m^{-1}. Ferrites are ceramic materials made by sintering a mixture of iron oxide with oxides or carbonates of either manganese and zinc or nickel and zinc. MnZn ferrites are used in applications up to 1 or 2 MHz; NiZn ferrites have lower permeability and much higher resistivity, and hence lower losses. They are used from 1 MHz to several hundred MHz.

4.4.3.1   B–H Characteristics of Magnetic Core Materials

A *B–H* curve is a plot of flux density *B* versus field strength *H*. It reveals important characteristics of a magnetic core material. Figure 4.11 shows a set of *B–H* curves for air, iron, and steel core materials [10].

The slope of the *B–H* curve is the permeability (i.e., $\mu = B/H$, as defined in Equation 4.6). Notice that the *B–H* curve for air-cored or any nonmagnetic medium-cored (such as wood or plastic) coil is linear (a straight line) and their permeability is constant ($\mu_r = 1$). But for a ferromagnetic-cored (iron or steel) coil, the *B–H* curve is nonlinear, meaning that the permeability μ is not constant. The flux density *B* increases in proportion to the field strength *H* until the maximum flux density point is reached—the *magnetic saturation* or *saturation of the core*. This is because there is a limit to the amount of flux density that can be generated by a core, in which all

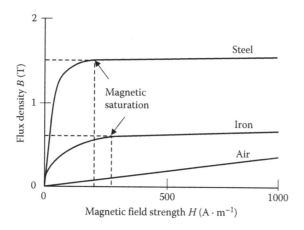

FIGURE 4.11   *B–H* curves for air, iron, and steel core materials.

the magnetic domains in the core material are perfectly aligned. A *B–H* curve is also known as a *magnetization curve* or a *magnetic hysteresis curve*.

4.4.3.2 Magnetic Hysteresis Loop

If the current flowing through the coil is removed, the magnetic flux does not disappear completely. As a result, a magnetic hysteresis loop is formed (see Figure 4.12a). This is because the magnetic core material still retains some magnetism due to some of the tiny molecular magnets still aligned in the direction of the previous magnetizing field. The ability to retain some magnetism in the core is called *retentivity* or *remanence*, while the amount of flux density still present in the core is called *residual magnetism*. Some ferromagnetic materials have a high retentivity (*magnetically hard*, see Figure 4.12b), which are excellent for making permanent magnets, while other ferromagnetic materials (e.g., iron or silicon steel) have low retentivity

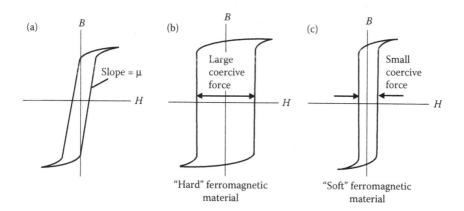

FIGURE 4.12   (a) A typical magnetic hysteresis loop; (b) a magnetic hysteresis loop for a "hard" ferromagnetic material; (c) a magnetic hysteresis loop for a "soft" ferromagnetic material.

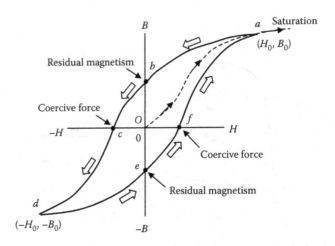

FIGURE 4.13   A magnetic hysteresis loop.

(*magnetically soft*, see Figure 4.12c), making them ideal for electromagnets, solenoids, relays, or transformers as they can be easily magnetized and demagnetized.

A complete hysteresis loop can be examined through Figure 4.13. Starting with an unmagnetized core when both B and H are zero at origin O, as the current in the coil I increases, the magnetic field strength around the core, H, increases proportionally and the flux density B also increases until the saturation point a is reached (path $O \to a$). If I is reduced to zero, H reduces to zero but B does not reach zero due to the residual magnetism remaining within the core ($a \to b$). To reduce B to zero, a reverse current needs to flow through the coil and produces a negative H (called *coercive force*, H_c) to realign molecular magnets. When the reverse H reaches c, the core becomes unmagnetized. Further increasing the reverse current causes the core to be magnetized in the opposite direction until the core reaches the opposite saturation point d ($c \to d$). If I is reduced to zero again, the residual magnetism present in the core will be equal to the previous value but in reverse at point e. By constantly changing the direction of the magnetizing current through the coil from a positive direction to a negative direction, as would be the case with an AC supply, the B–H curve forms a loop, following the path of a–b–c–d–e–f–a.

4.4.4   Housing, Cable, and Target Materials

Inductive sensors are mounted in plastic or metal housings and encapsulated with epoxy casting resin. The plastics used for the housings are

- PVC (polyvinylchloride)
- PA (polyamide) 6.6 glass-fiber reinforced
- PC (polycarbonate)
- PTFE (polytetrafluorethylene)
- PEEK (polyetheretherketone)

TABLE 4.7
Sensing Target versus Correction Coefficients

Target Material	Correction Coefficient
Iron	1
Stainless steel (SUS304)	0.76
Brass	0.50
Aluminum	0.48
Copper	0.3

The metal housing materials are

- Brass/chrome- or nickel-plated
- VA stainless steel, material No. 1.4301 or No. 1.4305
- Aluminum die-cast

Commonly used cable materials are PVC or PUR (polyurethane). PVC is a general purpose thermoplastic jacketing material. It has good mechanical strength, and is resistant to chemicals. PUR is broad class of polymers known for its resistance to abrasion and tolerance to solvents. PUR/PVC contains an inner wall of PVC and an outer wall of PUR material. PVC is not suitable for use in applications with oil-based liquids or with (ultraviolet) radiation. PUR is not suitable for continuous contact with water. For special applications, silicone or PTFE cables can be used. One has to take into consideration that cables should not be moved with ambient temperatures below −5°C.

Two basic types of conductive target materials are ferrous and nonferrous metals. Ferrous materials contain iron (e.g., cast iron, carbon steel, and stainless steel). Nonferrous metals do not contain any iron (e.g., aluminum, beryllium, copper, lead, magnesium, nickel, tin, titanium, and zinc). Some inductive sensors can work with both materials, while others can only work with one type of materials. Sensors must be calibrated first prior to detecting a different target material [11]. This is because each target material has different influence over the inductive effect and is usually less detectable than iron. Table 4.7 lists the correction coefficients used in an inductive coil sensor design for detecting different targets. Using these coefficients, the sensing distance corrections for typical inductive proximity sensors are follows:

Stainless steel: standard sensing distance × 0.76
Brass: standard sensing distance × 0.5
Aluminum: standard sensing distance × 0.48
Copper: standard sensing distance × 0.3

4.4.5 Power Losses in Inductive Sensors

Power losses in inductive sensors are through several different mechanisms:

1. Resistance of the windings (*copper loss* or *winding loss*)
2. Magnetic friction in the core (*hysteresis loss*)
3. Electric currents induced in the core (*eddy-current loss*)

4. Physical vibration and noise of the core and windings
5. *Dielectric loss* in materials used to insulate the core and windings

Copper loss or *winding loss* describes the energy dissipated by resistance in the winding wire. Most winding wire is made of copper, whose resistivity at 20°C is about $1.72 \times 10^{-8} \, \Omega \cdot \text{m}$. Thus, the resistance of a wire is $R = \rho l/A \approx 0.022/d_W^2$ ohms per meter (d_W is the diameter of the wire). The copper loss is then

$$P_W = I^2 R = 0.022 \, l_W (I/d_W)^2 = 0.022 \, (N\pi D_{avg})(I/d_W)^2 \qquad (4.37)$$

where I is the current flowing through the wire, l_W is the length of the wire, D_{avg} is the average diameter of the winding, and N is the number of turns. To find the resistivity above 20°C, one can multiply a factor, which is 1.079 for 40°C, 1.157 for 60°C, 1.236 for 80°C, and 1.314 for 100°C.

Hysteresis loss is equal to the area enclosed by the hysteresis loop and it is unrecoverable energy loss. Because the loss is related to an *area*, the hysteresis loss is roughly proportional to the square of the working flux density B^2. The nonlinearity of an iron core will reduce this loss to about $B^{1.6}$. A hysteresis loss incurs each time the core cycles from positive to negative values of B; thus, its loss rate P_H (in watts per unit mass) is directly proportional to the operation frequency f as

$$P_H = K_H f B^n \qquad (4.38)$$

where K_H is a core-material-dependent constant and n is the *Steinmetz exponent* (for iron, it is 1.6; for ferrite grade 3C8, it is 2.5). The above equation indicates that the hysteresis loss in a transformer is independent of load current, but depends on the flux density B and hence the voltage. That is why it is not advisable to subject a transformer to a voltage (hence flux) overload. It will become very hot.

Eddy-current loss occurs whenever the core material is electrically conductive. It is the heat dissipation caused by the eddy current I_e flowing through the core resistance R (can be calculated by $I_e^2 R$). Since power is proportional to the square of the applied voltage in a circuit, i.e., $P = V^2/R$ while the induced voltage is proportional to fB, the eddy-current losses are proportional to $f^2 B^2$, while B itself is related to the size of the loop. To reduce the eddy-current loss, a laminated core (consists of a stack of thin slices instead of a solid core) is used. The eddy-current loss is inversely proportional to the square of the number of laminations. To further reduce the eddy-current loss, an iron dust core that divides up the iron into fine particles (to prohibit the eddy-current flow) is invented, where the iron is ground into a powder, mixed with some insulating binder or matrix material and then fired to produce a core shape. Such a core can operate at several MHz, but its permeability is lower than a solid iron.

Hysteresis and eddy-current losses are collectively known as *core loss*. Core loss is the most important core limitation in many applications. At low frequencies, core loss is almost entirely hysteresis loss. Eddy-current loss overtakes hysteresis loss at 200–300 kHz. In metal alloy cores, eddy-current loss becomes dominant when the frequency is above a few hundred hertz.

4.4.6 QUALITY FACTOR Q

Q or *quality factor* of a coil is defined as the ratio of the energy stored in a circuit or component to the energy lost during one cycle of operation. It is important when used in a resonant circuit because it affects the "sharpness" of the response curve. All practical inductors exhibit losses due to the resistance of the wire or other causes mentioned earlier. Q factor of an inductor or a coil is the ratio of its reactance to its electrical resistance (frequency dependent, typical Q is 5–1000):

$$Q = \frac{\omega L}{R} \tag{4.39}$$

Cores with an air gap have the highest Q *factor* and temperature stability.

4.4.7 NUMBER OF TURNS N

From the fundamental Faraday's law of induction (Equation 4.24), a large coil sensitivity can be obtained by using a large number of turns N and a large active area A. A large N also increases the magnetomotive force $\mathfrak{I}$ (see Equation 4.30) and magnetic flux density B (see Equation 4.31). Increasing area A, however, has a practical limitation.

4.4.8 FREQUENCY RESPONSE OF INDUCTIVE SENSORS

Based on Equation 4.24, in order to obtain an output voltage signal from an inductive sensor, the magnetic flux must be varying with time (i.e., $d\Phi/dt \neq 0$). Therefore, inductive coil sensors are capable of measuring dynamic (varying or AC) magnetic fields. A typical frequency characteristic of a coil sensor is presented in Figure 4.14. Well-designed inductive sensors can achieve a bandwidth of several mHz to several MHz or even a GHz range [12]. The excitation current can be AC or pulse current. In case of the DC magnetic fields, the variation of the flux can be achieved by moving the coil. Through using a sensitive amplifier and a large coil, it is possible to measure

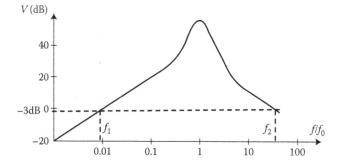

FIGURE 4.14 Typical frequency characteristic of an induction coil sensor.

low-frequency magnetic fields [13,14]. It is also possible to investigate quasi-static magnetic fields with fixed-coil (unmovable) sensors.

The frequency-response requirements for inductive sensors are application specific. In static target measurements, an inductive sensor can have very high resolutions. In dynamic/moving target measurements, the inductive sensors' ability to detect the motion depends on their frequency characteristics. The frequency response or performance of an inductive sensor can be specified by their cutoff frequencies at −3 dB point. In the case shown in Figure 4.14, the sensor performs well within the frequency range between 0.01 and 60 Hz (logarithmic scale). It is important to determine the speed of a moving target and comparing it to the frequency response of the sensor that to ensure there is fast enough frequency response to collect meaningful data about the movement.

4.4.9 STABILITY OF INDUCTIVE SENSORS

The stability of an inductive sensor system can be impacted by several factors, two of which are related to temperature and electronic components.

4.4.9.1 Thermal Stability

Temperature change is most likely the cause of measurement errors for an inductive sensor. Temperature impacts the stability of a sensor system in two ways: *thermal zero shift* and *thermal sensitivity shift*. The former is an offset shift and therefore affects all points in measuring range equally. It can be easily compensated by adding or subtracting a constant value. The latter is a shift in the slope of the sensor's transfer function. It takes place primarily in the upper 50% of the measuring range and becomes worse as the range increases. Thermal sensitivity shift is of most concern since it affects the system's performance—the output to input relationship. Other than controlling or stabilizing the temperature, the thermal sensitivity shift can be minimized by temperature compensation over the measurement range or by only utilizing the first 50% of the sensor range.

4.4.9.2 Long-Term Stability

Long-term stability of an inductive sensor is a measure of the sensor's performance over time and is commonly referred as a drift. It is typically specified as a percent of full scale output (FSO) over a period of time, e.g., 0.1% FSO per month. Typically, this drift is caused by degradation of the components in the sensor electronics. It can be sensitivity error that affects the slope of the transfer function if it is caused by sensor cabling.

4.5 TYPES AND OPERATING PRINCIPLES OF INDUCTIVE SENSORS

4.5.1 TYPES OF INDUCTIVE SENSORS

Inductive sensors can be classified in many ways. Based on measurants, inductive sensors can be classified as distance sensors, vibration sensors, metal detectors,

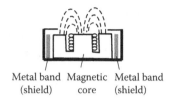

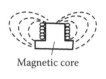

Metal band   Magnetic   Metal band
(shield)   core   (shield)

Magnetic core

In a shielded configuration, a metal band surrounds the magnetic core and coil arrangement. This will direct the electromagnetic field to the front of the sensor.

In a non-shielded configuration, no metal band surrounds the sensing coil. Therefore, the sensor can be side sensitive.

FIGURE 4.15   Shielded and non-shielded inductive sensors.

liquid level sensors, velocity sensors, magnetic field sensors, or current sensors. Based on applications, they can be categorized as transformers, flaw or crack detectors, weld seam detectors, proximity sensors, noncontact switches, and so on. Based on the shapes and sizes of the sensing coils, inductive sensors can be divided into cylindrical, rectangular, spherical, flat, pancake, miniature, and more. Most inductive sensors are of the standardized cylindrical threaded barrel type. Inductive sensors sometimes are also grouped based on shielding (shielded type and nonshielded type, see Figure 4.15), mounting (see Figure 4.16), structure (with or without a core; single coil or dual coils), number of turns, and output (AC or DC voltage, analog or digital output), and sensing directions (front sensing, top sensing).

When two or more sensing surfaces of inductive sensors are close to each other, a minimum spacing is required between adjacent or opposing surfaces in order to eliminate the interference of the electromagnetic fields between sensors. Refer to Figure 4.16, in *shielded mounting*, the sensing surfaces may be flush with the metal in which the sensors are mounted (can be put closer); in a *nonshielded mounting*, there is a free zone (i.e., no metal) between the sensing surfaces, while in an opposite mounting (the sensing surfaces are opposite to each other), there must be a minimum distance between them.

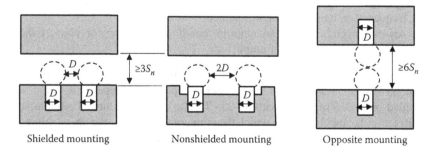

Shielded mounting   Nonshielded mounting   Opposite mounting

FIGURE 4.16   Mounting types.

In this chapter, inductive sensors will be grouped and discussed as follows:

- *Inductive air coil sensors:* In this type of sensor, there is no magnetic core—either no core at all (air coil sensors) or has a non-magnetic core (made of wood or plastics).
- *Inductive coil sensors with ferromagnetic cores:* This type of inductive sensor usually has one coil and one ferromagnetic core, with the coil wound around the core. The core may or may not move related to the coil. The sensors often have a probe form. Most inductive displacement sensors or eddy current sensors belong to this type.
- *Transformer-type inductive sensors:* This type of sensor has two (*primary* and *secondary*) or more coils and at least one core. *Linear variable differential transformers* (LVDT), *rotary variable differential transformers* (RVDT), *fluxgate, synchros, resolvers,* and *inductosyns* fall into this category. The coils and/or the core can be stationary or movable, depending on specific applications.

4.5.2 OPERATING PRINCIPLES

Although there are many inductive sensor designs, their operating principles are all similar: a coil (inductor) carrying an alternating current (AC) creates an alternating magnetic field. This varying magnetic field interacts with a second magnetic field, causing a change in the inductor's inductance or impedance in terms of amplitude and/or phase angle. The secondary magnetic field could be: (1) a permanent magnet attached to a moving object, or (2) created by the eddy currents in a nearby conductive target induced by the inductor; or (3) generated by the current flowing in the secondary coil that is induced by the inductor. Note that the induced secondary magnetic field is always opposed to the primary magnetic field. The change in inductance or impedance can be measured by an evaluating circuit.

Thus, for an inductive sensor to work, the following elements are necessary:

1. An inductor (a coil with or without a core)
2. An oscillator to create and emit a high-frequency alternating current
3. An electrically conductive or magnetically permeable object (could be a target or another coil)
4. A sensing or measurement circuit that can detect and measure the change in magnetic field in terms of inductance, reluctance, impedance, natural frequency, voltage, current, or magnetic field strength
5. An output circuit that can amplify, condition, interpret, or convert the detected signal to a proper output

An inductive sensor that contains all these components is called a *self-contained* sensor as shown in Figure 4.17. The normal operating temperature for a self-contained inductive sensor is within –25–70°C to prevent failure of silicon-based circuitry. For the applications in which the temperature exceeds the above range, the circuitry should be separated from the sensing coil. The detection circuitry can be located safely away in a remote environmentally controlled area. Such sensors can resist temperatures as high as 200°C.

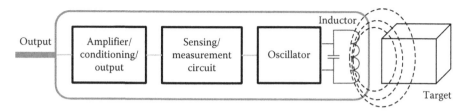

FIGURE 4.17   Basic structure of a self-contained inductive sensor.

Inductive sensors are typically rated by frequency, or on/off cycles per second. Because of magnetic field limitations, inductive sensors have a relatively narrow sensing range—from fractions of millimeters to 60 mm on average, although a longer range with specific design can be achieved. Because of the limited sensing range and the tight confines in practical applications, mounting and geometric styles are important. Improper mounting and the amount of conductive material in or near the mounting fixture will impact sensor performance. A sensor is "side-loaded" when its field interacts with conductive material other than the target. Shielded (flush) sensors, or keeping conductive material out of the sensing field, or *in situ* calibration can reduce this effect. "Tubular sensors are by far the most popular configuration and available with diameters of 3–40 mm. When it comes to miniaturization, a 6 mm × 6 mm × 18 mm rectangular proximity sensor with an extended sensing range of 1.6 mm has been developed by *Panasonic USA*, Newark, New Jersey, USA. Inductive sensor housing is typically nickel-plated brass, stainless steel, or PBT (polybutylene terephthalate) plastic.

4.6   INDUCTIVE AIR COIL SENSORS

An air coil sensor, also known as an *induction coil sensor, search coil, pickup coil, magnetic antenna*, and *magnetic loop sensor*, does not contain a magnetic core. It is the simplest and the oldest type of inductive sensors. Air coil sensors also include sensors with coils wound around a plastic or ceramic core in addition to those made of stiff, self-supporting wire. The coil is excited by a high-frequency AC.

4.6.1   TYPES OF AIR COILS

Figure 4.18 shows various types of air coil designs: (a) a single magnetic loop antenna for indoor shortwave detection; (b) a single coil for flux density and magnetic field detection; (c) three mutually perpendicular coils for three-axis magnetic field measurements; (d) a spherical coil for magnetic field measurement at the center of the spherical coil; and (e) a planar thin-film coil for eddy-current sensing [15]. Planar coils are often made in printed circuit board (PCB) using thin-film technology and connected to an on-chip complementary metal oxide semiconductors (CMOS) electronic circuit. PCB-fabricated planar coils are very flexible and can be made in various shapes, but have a limit on the number of turns that can be achieved. CMOS coils often have higher resistance.

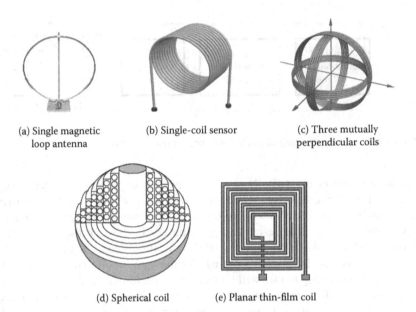

(a) Single magnetic loop antenna   (b) Single-coil sensor   (c) Three mutually perpendicular coils

(d) Spherical coil   (e) Planar thin-film coil

FIGURE 4.18 Various air coil configurations: (a) a single magnetic loop antenna; (b) a single coil sensor; (c) three mutually perpendicular coils; (d) a spherical coil; (e) a planar thin film coil. (From Tumanski, S., Induction Coil Sensors—A Review, *Measurement Science and Technology*, Institute of Physics Publishing, 18, pp. 31–46, 2007.)

Copper is a common material used for coils, but aluminum can be used where weight is critical. As shown in Table 4.8, the density of aluminum is three times lower than that of copper, although aluminum has a higher resistivity.

4.6.2 FEATURES OF AIR COIL SENSORS

Air coils are very stable, rugged, and easy to handle. Its *B–H* characteristic curve is linear as shown in Figure 4.11. Air coil sensors have limited sensitivity. They are only sensitive to the flux that is perpendicular to their main axis. Large coil sensitivity can be achieved by increasing the number of turns and coil diameter, and decreasing the wire length. To reach low demagnetization, however, they should be long and thin. Thus,

---

TABLE 4.8
Physical Properties of Wire Materials

Material	Density $(g \cdot cm^{-3})$	Specific Resistivity $(\Omega \cdot m)$	Temperature Coefficient of Resistance $(\% \cdot {}^\circ C^{-1})$	Coefficient of Linear Expansion $(ppm \cdot {}^\circ C^{-1})$
Copper	8.9	1.72×10^{-8}	0.39	16.6
Aluminum	2.7	2.65×10^{-8}	0.4	25

Note: ppm—parts per million.

---

a tradeoff must be made in design for optimal performance. Velocity and position air coil sensors provide high resolution, but they cannot be used at low or zero speeds due to little or no flux density change. Proper shielding and grounding of air coil sensors and their cables are important to prevent capacitive and/or inductive couplings.

Air coils have lower inductance than similarly sized ferromagnetic core coils, but have less energy losses (e.g., *core loss*) than magnetic cores. The absence of core loss permits a higher quality factor Q [16]. Therefore, air-core sensors are often used in high-frequency resonant circuits. They are also widely used as eddy current proximity sensors, where spatial resolution is more critical than sensitivity.

4.6.3   DESIGN CONSIDERATIONS OF AIR COIL SENSORS

4.6.3.1   Coil Diameter

Coil design involves determining coil size, shape, and number of turns; selecting coil material; considering the target size and material characteristics, as well as the application conditions. The smaller the sensor is, the larger the shock and vibration it can withstand. A coil sensor's measuring range is directly proportional to the coil's diameter. For example, an air coil sensor with 10,000 turns and diameters of 1 m can detect the magnetic flux density in the pT range for magnetocardiograms. The sensing range of a coil sensor, measured from the face of the sensor, is typically 30–50% of its coil diameter. In case of highly conductive targets (brass, aluminum, copper, or silver), the range can be extended to 65% of the coil diameter with a reduction in linearity and overall system stability. At distances beyond this range, the sensor's magnetic field strength decreases rapidly and the sensor often presents an unstable performance. For example, a sensor with a coil diameter of 3 mm has an effective measurement range of 1 mm; and a sensor with a coil diameter of 76.2 mm has an effective measurement range of 30.5 mm. Sensors with larger coil diameters provide greater measuring ranges but less resolution than the smaller coil sensors do. In addition, space constraints often limit a coil's diameter and mounting configuration. Using only a portion of a sensor's measuring range improves the overall performance.

4.6.3.2   Target Thickness and Temperature Stability of Target Material

For a non-magnetic conductive target, the resistivity increase of the target as temperature increases can be compensated by increasing the target's thickness. This is because the increased target thickness allows an increase in skin depth, resulting in the effective resistance along the eddy current path remaining constant. As a comparison, a ferromagnetic target is less stable thermally due to its non-linear behavior of permeability, which also varies in skin depth. Temperature variations may cause sensitivity shift, which affects gain and linearity.

4.6.3.3   EMI Interference

Inductive sensors are susceptible to EMI (electromagnetic interference), defined as any electromagnetic disturbance that interrupts, obstructs, or otherwise degrades or limits the performance of a sensor or an electronic device. It appears as an undesirable signal superimposed on the output of a sensor. For an inductive sensor, EMI generally comes from two main sources: (1) proximity between sensors and

(2) electromagnetic fields in the mounting environment. Shielding around a coil sensor can reduce interference. If a coil sensor must operate in an electromagnetic field, non-ferrous target materials are recommended, because EMI can cause ferrous target materials' permeability change, causing a measurement error [17]. Devices with strong local fields or noise sources in the lower frequency range (e.g., long, middle, or short-wave transmitters) should not be operated close to the inductive sensors.

4.6.3.4   Other Considerations

Velocity variation of a target, caused by "eddy current drag," occurs when the velocity of a target exceeds the recommended value. Extreme velocity produces an erroneous output that appears as an increase in displacement when there actually is no increase in displacement. To obtain less than 1% velocity error, the oscillator cycles or the coil driving frequency should be greater than 50 for absolute displacement measurement and greater than 10 for tachometer applications.

One problem that inductive sensors face is that there is a slow buildup of metal filings or "chips" on the sensing face over a period of time. Modern intelligent coil sensors have the ability to detect these "chips" and teach the inductive sensor to ignore their existence. This type of coil sensors is called a "chip immune" type. Another method to prevent or reduce chip buildup is the flat-pack sensor design, making the sensing face exposed vertically to be virtually unaffected by chip buildup on its slim horizontal component.

In extreme cases, inductive peak voltages can destroy the sensors despite the integrated protective circuit. Screened cable or twisted lines are recommended, especially for longer cables (>5 m). Direct control of electric light bulbs should be avoided, because during the switch-on moment, cold current is many times of the rated current and could destroy the output circuit of the sensor.

EXAMPLE 4.8

IEC 60947-5-2 defines the recommended target material and size for an inductive proximity sensor. The standard target material is carbon steel (Fe 360) with a rolled finish, and the standard target size is square in shape, 1 mm thick, and with length or width equal to or *greater* than: (1) the size of the sensor's active surface (e.g., 8, 12, 18, and 30 mm for standard tubular sensors); or (2) three times the rated operating distance (S_n) of the sensor. Calculate the target size for a standard range M18 sensor with a rated sensing distance (S_n) of 5 mm.

SOLUTION

A standard range M18 sensor's active surface is 18 mm^2. Based on Rule (1), the target length should be at least 18 mm. Based on Rule (2), the target should be at least three times of S_n: 3×5 mm = 15 mm.

Since 18 is larger, the target size should be at least 18 mm × 18 mm to achieve the sensor's rated sensing distance.

For a normal sensing range, Rule 1 will always win, while for an extended sensing range (two, three, or four times the normal sensing range), Rule 2 will typically win. Extended sensing range models become more popular today, which means Rule 2 will typically be the one to determine the target size in order to achieve the full rated sensing range.

EXAMPLE 4.9

Explain why target size affects an inductive sensor's detection range.

Answer

An inductive sensor generates a magnetic field from its active surface, which induces eddy currents on the conductive target. As the target gets closer, the eddy currents increase and lower the amplitude of the sensor's oscillator signal. If the conductive target is smaller than required, it will not generate as many eddy currents, thus the target must be placed closer to the sensor to have the same affect. Thus, targets smaller than the recommended size has a shorter sensing range.

4.6.4 Applications of Air Coil Sensors

In general, inductive coils are well suited for all types of environments—submerged in salt water, antifreeze, diesel fuel, grease, brake fluids, and oils; under a temperature from cryogenic temperature to 1100°F. There are numerous applications of inductive coil sensors, including geophysics when measuring micropulsations of the earth's magnetic field (1 Hz–1 MHz frequency range), audio frequency, and magnetic recording systems; measurement of secondary magnetic fields caused by the earth's currents (called *magnetotelluric exploration*); plasma experiments; space research; submarines and trains; cable location; and electromagnetic surveillance. Air coil sensor can also be used as limit switches in machines, systems, and vehicles; contactless position switches for monitoring and positioning; pulse generators for counting tasks, and distance and speed sensors.

In archeological surveys, coil sensors are used to detect buried artifacts, gravesites, and other cultural resources. They are also used for the location of ferrous or metallic objects such as underground storage tanks and so on.

Figure 4.19 shows the operating principle of an EX-500 series inductive displacement sensor developed by *Keyence Corporation*, Itasca, Illinois [18]. A high-frequency current passes through the sensor head coil and creates a high-frequency magnetic field. When a metal target is present in the changing magnetic field, eddy

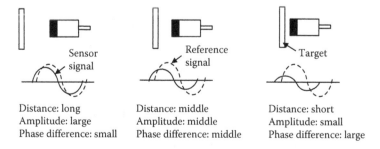

Distance: long   Distance: middle   Distance: short
Amplitude: large   Amplitude: middle   Amplitude: small
Phase difference: small Phase difference: middle Phase difference: large

FIGURE 4.19 Amplitude and phases change in the detecting coil as the distance changes. (From Inductive (eddy current) displacement, SensorCentral.Com. Available at: www.sensorcentral.com/displacement/laser03.php.) (Courtesy of *Keyence Corporation*.)

currents are induced in the target that are perpendicular to the magnetic flux. This changes the impedance of the sensor coil. Based on the distance between the sensor coil and the target, the oscillation amplitude and the phase difference in the coil output can be small or large.

4.7   INDUCTIVE SENSORS WITH FERROMAGNETIC CORES

A magnetic core is a piece of magnetic material with a high permeability used to confine, guide, and concentrate magnetic fluxes in the core material. It is made of ferromagnetic metal such as iron, or ferromagnetic compounds, such as ferrites. The magnetic field is created by a coil of wire around the core that carries a current. The use of a magnetic core can increase the magnetic field of a coil by a factor of several thousand over what it would be without the core and greatly increases a sensor's sensitivity.

Coil sensors with a ferromagnetic core provide higher sensitivity and high permeability in comparison with air coil sensors, but they are less stable, have more energy loss, and are nonlinear. These types of sensors are often used in applications requiring high sensitivity or those with size limitations.

One of the most important inductive sensors is an eddy-current sensor. An eddy-current sensor consists of an excitation coil with or without a magnetic core, a detection element (to identify the perturbation of the currents caused by cracks or other defects), and a driver (the electronics that drives the probe and generates an output voltage proportional to the measurand). The detection element can be another coil, a superconducting quantum interference detector, or a solid-state magnetic sensor (e.g., a Hall sensor, or a magnetoresistive sensor, or a spin-dependent-tunneling sensor). In some eddy-current sensors, the driver is physically integrated into the probe body.

4.7.1   CHARACTERISTICS OF INDUCTIVE SENSORS WITH MAGNETIC CORES

Like most inductive sensors, eddy-current sensors have the following features:

- Offer nondestructive detection (NDD)
- Be able to detect very small defects
- Have high and constant sensitivity over a wide range of frequencies
- Penetrate any nonmagnetic metals without loss of accuracy
- Good temperature stability
- High detecting speed
- Tolerant of dirty or harsh environments
- Not affected by material in the gap between the probe and target
- Less expensive than other NDD devices

Eddy-current sensors are not a good choice in the following conditions:

- Extremely high resolution required (capacitive sensors are ideal)
- Large gap between sensor and target is required (optical and laser sensors are better)
- Conductivity is sensitive to cracks and material inhomogeneities

TABLE 4.9
Technical Data of *WayCon*'s TX Series Eddy-Current Sensors

Full scale	0–10 mm
Linearity	±0.25%
Resolution	10 Hz: 0.007%; 1 kHz: 0.018%; 35 kHz: 0.1%
Filter corner frequency (at –3 dB)	10 Hz/100 Hz/1 kHz/10 kHz/35 kHz
Dynamics	Output rate 120 kHz (1-Kanal), 70 kHz (2-Kanal)
Output	0–10 V or 0–5 V or –5 V to +5 V or 0–20 mA or 4–20 mA
Operating temperature	–35°C to +185°C
Electronics temperature range	–10°C to +70°C
Power supply	9–36 VDC wide input
Maximum power consumption	190 mA (24 V); 300 mA (12 V); 390 mA (9 V)
Short circuit resistance	Yes
Resistance reverse polarity	Yes
Sensor cable	PTFE-Koax, length 3 m/6 m and customized length

Source: *Eddy Current Distance and Displacement Transducer, Product Handbook*, WayCon, Germany. p.3. Available at: www.waycon.de/fileadmin/pdf/Eddy_Current_Probe_TX.pdf.

Oscillating frequency greatly affects eddy-current sensors' responses—not only the strength of the response from flaws and the effective depth of penetration but also the phase relationship [19]. Table 4.9 shows the technical data of TX series eddy-current sensors made by *WayCon*'s.

4.7.2 Sensor Design

4.7.2.1 Ferromagnetic Core Design
Design of a magnetic core involves the following considerations:

- Geometry of the magnetic core
- Amount of air gap in the magnetic circuit
- Properties of the core material (especially permeability and hysteresis)
- Operating temperature of the core
- Lamination of the core

Typical core geometries are presented in Figure 4.20. A *cylindrical core* or *rod*, commonly made of ferrite, is especially used for some applications such as tuning an inductor. The core sits in the middle of the coil and slightly adjusting the core's position will fine tune the inductance (the core is threaded to allow adjustment with a screwdriver). Using a high permeability core material increases the inductance, but the field must still spread into the air at the ends of the rod. The path through the air ensures that the inductor remains linear. In this type of sensors, radiation occurs at the end of the rod and EMI may be a problem in some circumstances. *U-* or *C-shaped cores* are the simplest closed core shape. Windings may be put on one or

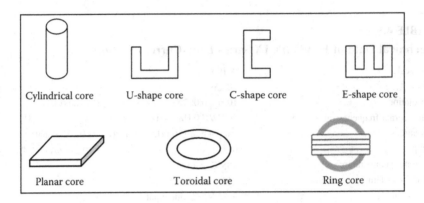

FIGURE 4.20 Typical core geometries used in inductive sensor design.

both legs of the core. *E-shaped core* are the more symmetric solutions for providing a closed magnetic system. Most often, a wire is wound around the center leg, whose section area is twice that of each individual outer leg. The shape of a *pot core* (is not shown in Figure 4.20) is round with an internal hollow that almost completely encloses the coil, providing a shielding effect that prevents radiation and reduces EMI. In a toroid core (the same shape as a doughnut), a coil is wound through the hole and distributed evenly around the circumference of the torus, creating circular magnetic loops inside the core. The lack of sharp bends constrains virtually all of the field to the core material. This not only makes a highly efficient transformer but also reduces the EMI radiated by the coil. A *planar core* is typically used with a flat coil. It is excellent for massive production and allows a high-power, small-volume transformer to be constructed at low cost. It is not as ideal as either a pot core or a toroidal core but costs less to produce. A *ring core* is basically identical in shape and performance to the toroid, except that windings usually pass only through the center of the core, without wrapping around the core multiple times [20].

In many applications, such as transformers, it is undesirable for the core to retain magnetization when the applied field is removed. This property, called *hysteresis*, can cause energy losses. Therefore, soft magnetic materials with low hysteresis, such as silicon steel, rather than the hard magnetic materials used for permanent magnets, are often used in cores.

To make the calculation of the inductance of a core easier, the *inductance factor* A_L (known as the A_L *value* in henry per square turns), is given in manufacturers' data sheet. The inductance L of a core is defined as

$$L = N^2 A_L \tag{4.40}$$

where N is the number of turns.

4.7.2.2 Probe Design

There are several designs for inductive sensing probes. Figure 4.21 shows the simplest design of inductive sensing probes. It contains a coil, a plastic tip, a stainless-steel

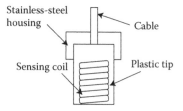

FIGURE 4.21 The internal structure of a simple probe.

housing, and a cable. Tips are usually made of dielectric materials (e.g., plastic or epoxy). The sensing coil is encapsulated in the tip and installed in a stainless-steel housing.

Some probes have two coils in one tip: one is the excitation coil and the other is a receiving coil (see Figure 4.22). The excitation coil creates an alternating magnetic field to induce eddy currents in a conductive target, while the receiving coil picks up the change of the coil magnetic field due to its interaction with the induced eddy-current magnetic fields. The receiving coil is placed in the very front of the probe where the strength of the induced field is higher. The coils may be fiberglass-insulated copper wires. The coil former can be made of machinable ceramic. The coils are encapsulated in a ceramic tube filled with ceramic sealant [21].

Probe bodies and coils are available in a large variety of shapes and sizes, and can be custom designed as well. Figure 4.23 shows the tangential and pancake types. Tangential coils cannot detect certain cracks effectively. For example, circumference cracks on a tubic target because the cracks and the flow of the induced eddy current have the same direction, while pancake coil, however, can ensure that the circumference cracks be interrupted by the eddy-current flow (see Figure 4.24).

Probes can also be classified into surface, outside diameter (OD), and inside diameter (ID) types, as shown in Figure 4.25. Surface-type probes are most commonly seen in eddy-current sensors. OD probes are encircling probes, in which the coil or coils encircle a test piece and inspect it from the outside in. ID probes are inserted

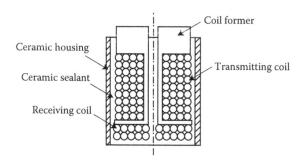

FIGURE 4.22 A two-coil probe design. (From Vetterlein, J. et al., Eddy current testing at high temperatures for controlling heat treatment processes, *International Symposium on Non-Destructive Testing in Civil Engineering*. 2003 (NDT-CE 2003), September 16–19, 2003, Berlin, Germany, p. 22.)

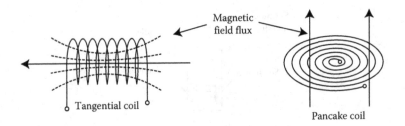

FIGURE 4.23   Tangential and pancake coil configurations.

into a test piece and inspect it from the inside out. These three configurations are used in most flaw detection applications.

Furthermore, based on the operating mode, a probe can have one of the four modes: *absolute, differential, reflection,* and *hybrid* [22].

An *absolute probe* has a single coil wound to a specific value (used to induce eddy currents and sense the field changes) and gives an "absolute" reading. Absolute probes can be used for flaw detection, conductivity measurement, and thickness determination. They can detect both sharp and gradual changes in impedance or magnetic fields.

A *differential probe* has two sensing coils usually wound in opposite directions to measure the difference between the two coil readings, providing a greater resolution for sharp discontinuity detection. Figure 4.26 shows the response of a differential probe when passing over a defect. As can be seen, when the two coils are over a flaw-free area of the target, there is no differential signal developed between the coils since they are both inspecting identical material. However, when one coil is over a defect and the other is over good material, a differential signal is produced.

A *reflection probe* (often referred to as *driver-pickup probe*) also has two coils, but one coil is used for excitation to induce eddy currents in the test material and the other

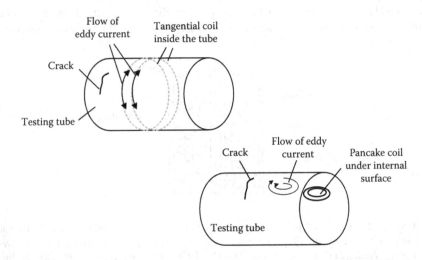

FIGURE 4.24   Tangential and pancake coil configurations.

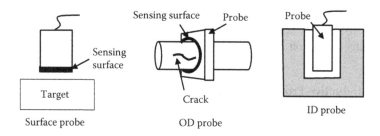

Surface probe   OD probe   ID probe

FIGURE 4.25   Three types of inductive sensing probes: surface, OD, and ID.

senses changes in the material. The advantage of a reflection probe over an absolute probe is that the driver and pickup coils can be separately optimized for their intended purpose (e.g., the driver coil can be made so as to produce a strong and uniform magnetic field, while the pickup coil can be made very small so that it is sensitive to very small defects).

A *hybrid probe* is a combination of the above different probe modes. For instance, the split-D probe is the combination of a differential probe and a reflection probe [22]. It has a driver coil that operates in the reflection mode, and a pickup coil formed by the two D-shaped sensing coils that operates in the differential mode. This type of probe is very sensitive to surface cracks. Hybrid probes are usually specially designed for a specific inspection application.

4.7.3 Applications of Inductive Sensors with Ferromagnetic Cores

Inductive cored coil sensors have many applications. In geophysics, they can measure micropulsations of the earth's magnetic field in 1 mHz to 1 Hz frequency range.

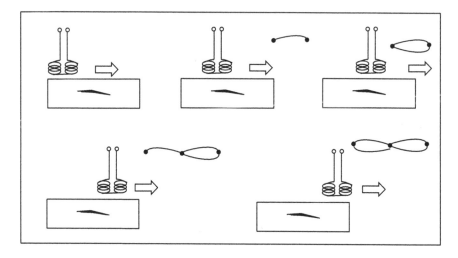

FIGURE 4.26   Response of a differential probe passing over a defect. (From Probes—mode of operation, NSF-NDT Resource Center. Available at: www.ndt-ed.org/EducationResources/ CommunityCollege/EddyCurrents/ProbesCoilDesign/ProbesModeOp.htm.)

They are also seen in magnetic recording, navigation and telecommunication, proximity and distance measurement, thickness and conductivity gauging, and crack and surface defect detection. Below are some application examples of ferromagnetic core sensors.

4.7.3.1 Inductive Proximity Sensor

One of the common inductive sensors is an *inductive proximity switch* (see Figure 4.27), primarily used to detect presence of a conductive object in front of it. This type of sensor has a coil wound around an iron core, creating an electromagnetic field when an oscillating current flowing through the coil. If a magnetic or conductive object, such as a metal plate, is placed within the magnetic field around the sensor, the inductance of the coil changes. The sensor's detection circuit detects this change and produces an output voltage to tigger the switch ON.

Inductive proximity sensors are widely used for traffic light control at intersections. Rectangular inductive loops of wire are buried into the road surface. When a vehicle passes over the loop, the metallic body of the vehicle changes the loop's inductance and triggers the circuit to change the traffic lights. One disadvantage of this type of sensors is that they are "omnidirectional," meaning that they will sense a metallic object above, below, or to the side of it. Also, they cannot detect nonmetallic objects (capacitive sensors can detect nonmetallic objects with a typical sensing range of 0.1–12 mm.

4.7.3.2 Inductive Displacement Sensor

When a permeable core is inserted into an inductor as shown in Figure 4.28, it increases the inductance of the coil. At each position, the core produces a different inductance. Therefore, it is called a variable inductor. If the movable core is attached to an object, it can measure the displacement of the object.

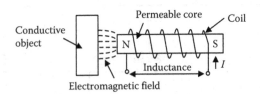

FIGURE 4.27   An inductive proximity switch.

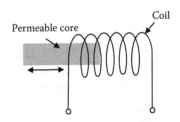

FIGURE 4.28   Variable inductance displacement sensor.

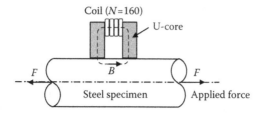

FIGURE 4.29 An eddy-current force sensor. (From Inductive eddy current sensors for force and stress measurements, *Chen Yang Technologies GMBH & Co. KG*, Finishing, Germany, 2014. Available at: www.chenyang-ism.com/EddyCurrent.htm. With permission.)

4.7.3.3 Eddy-Current Force Sensor

An eddy-current force sensor (see Figure 4.29) was developed by *Chen Yang Technologies GMBH & Co. KG*, Finsing, Germany [23]. It uses a U-core (made of a soft magnetic ferrite material) and a 160-turn coil. When a force is applied on the testing steel specimen, the sample's permeability changes, causing the impedance of the sensor to change, which is converted to a voltage change. The sensitivity, linearity, and hysteresis depend on sensor geometry and measurement conditions (e.g., exciting current and frequency). The exciting or testing frequency is normally selected between 100 Hz and 10 kHz to produce an optimal exciting current (25–50 mA) and a hysteresis error (less than 3%). This sensor can also be used in force and stress measurement for (metal) bridge monitoring.

4.7.3.4 Thread Detector

Keely NDT Technologies, Inc. (Pontiac, Michigan, USA) developed thread detectors that can detect presence or absence of threads in a tapped hole or verify if a thread is properly made [24]. In Figure 4.30a, a probe with a single-coil winding extracts the signature of a hole. Then the signature is compared to the signatures of the hole with or without threads. In Figure 4.30b, two coils are placed in a probe, and the difference between their signatures is used for detection. If the signature difference between the dual coils is within a threshold, the threads are considered as good threads. Otherwise, the part will be rejected.

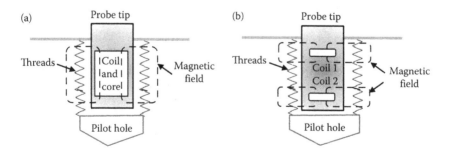

FIGURE 4.30 (a) Single-coil thread detector; (b) dual-coil thread detector.

4.8 TRANSFORMER-TYPE INDUCTIVE SENSORS

4.8.1 INTRODUCTION

A transformer is an inductive device that transfers energy by inductive coupling between its primary winding and secondary winding. In transformer-type inductive sensors, the primary coil is driven by an AC excitation current, which creates a varying magnetic field. This changing field induces an AC voltage or current in the secondary coil (usually two or three windings) that can be measured and sent to an output circuit. Based on the type of the motion mechanism and the coil/core configurations, transformer-type inductive sensors can be categorized as

- *Linear variable differential transformer*: Has a primary coil, two secondary coils, and a linearly movable core (see Figure 4.31a), and can measure linear displacement, force, acceleration, and pressure.
- *Rotary variable differential transformer*: Has a rotating core (Figure 4.31b) and can measure angular position.
- *Fluxgate sensor*: Uses a high-permeability ferromagnetic core (motionless) and operates at its saturation (gating) state (Figure 4.31c and d). The periodic saturation of the core due to the alternating current flowing through the excitation coil causes the permeability of the core to drop and a nearby DC magnetic field to decrease. The sensor output will be proportional to the intensity of the external magnetic field.
- *Synchro*: Has a single winding rotor that rotates inside a stator of three windings, much like an electric motor as shown in Figure 4.31e. The primary winding wound around the rotor is excited by an alternating current, which induces currents to flow in three Y-connected secondary windings (oriented 120° apart). The relative magnitudes of secondary currents are measured to determine the angle of the rotor relative to the stator, or the currents can be used to directly drive a receiver synchro that will rotate in unison with the synchro transmitter. In the latter case, the whole system is often called a *selsyn* (a portmanteau of *self* and *synchronizing*). A synchro provides accurate angular and rotational information.
- *Resolver*: Has a single-winding rotor that rotates inside a stator of two windings (oriented 90° apart, see Figure 4.31f) and provides accurate angular and rotational information.
- *Linear inductosyns* (Figure 4.31g): Has a movable *slider* that has two secondary rectangular-waveform-winding traces (one trace is shifted one-quarter of a cycle relative to the other) and a fixed primary winding trace, exactly the same pattern as the slider, printed on a PCB (called *scale*). A linear inductosyn provides a translation measurement. The slider and the scale remain separated by a small air gap of about 0.2 mm.
- *Rotary inductosyns*: Is similar to a linear inductosyn but the scale is printed on a circular stator, and the slider's trace pattern is printed on a circular rotor.

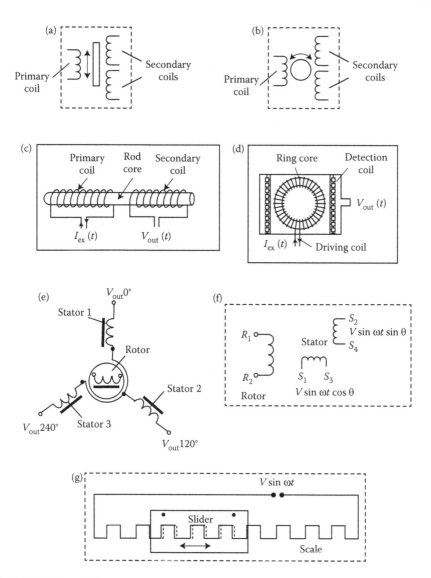

FIGURE 4.31   (a) LVDT; (b) RVDT; (c) rod-core fluxgate; (d) ring-core fluxgate; (e) synchro; (f) resolver; (g) linear inductosyns.

Among these transformer-type sensors, LVDTs, RVDTs, synchros, and resolvers are the most commonly used sensors and will be discussed in detail in this section.

EXAMPLE 4.10

Two wire windings are wrapped around a common iron core, so whatever magnetic flux may be produced by one winding is fully shared by the other winding

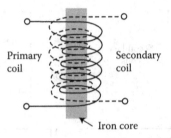

FIGURE 4.32 Two wire windings wrapped around a common core.

(Figure 4.32). Derive an expression that relates the induced voltage at the primary winding (V_p) and the secondary winding (V_s) and the numbers of turns of wires (N_p and N_s).

SOLUTION

Using Equation 4.24, $V_p = N_p \dfrac{d\Phi}{dt}$ and $V_s = N_s \dfrac{d\Phi}{dt}$. Based on the fact that the magnetic flux is equal for each winding, thus, $\dfrac{V_p}{N_p} = \dfrac{V_s}{N_s} \Rightarrow \dfrac{V_p}{V_s} = \dfrac{N_p}{N_s}$

EXAMPLE 4.11

Figure 4.33 shows a transformer circuit. Calculate V_p, V_s, I_p, and I_s (voltages and currents; p—primary, s—secondary).

SOLUTIONS

For this type of transformer circuit, the following relationship exists:

$$\frac{V_s}{V_p} = \frac{N_s}{N_p} \quad \text{and} \quad \frac{I_p}{I_s} = \frac{N_s}{N_p}$$

Thus,

$$V_s = \frac{N_s}{N_p} V_p = \frac{4000}{13000} \, 48 \text{ V} = 14.77 \text{ V}$$

$$I_s = \frac{V_s}{R_s} = \frac{14.77 \text{ V}}{150 \, \Omega} = 0.0985 \text{ A}$$

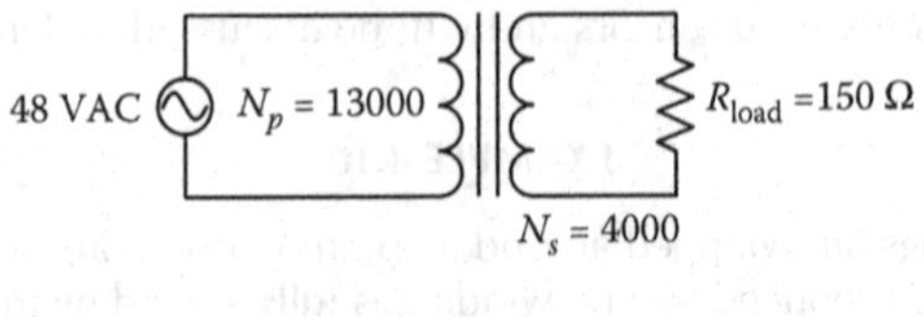

48 VAC $N_p = 13000$ $R_{load} = 150 \, \Omega$

$N_s = 4000$

FIGURE 4.33 A current transformer (CT) circuit.

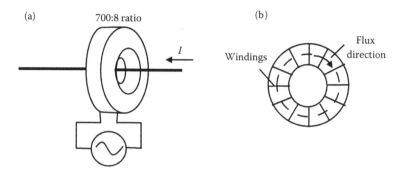

FIGURE 4.34 A current transformer.

$$I_p = \frac{N_s}{N_p} I_s = \frac{4000}{13000} 0.0985 \text{ A} = 0.0303 \text{ A}$$

EXAMPLE 4.12

A current transformer (CT) as shown in Figure 4.34a is commonly used to measure high AC currents in power systems. The CT has a "donut" shape, and a current-carrying conductor passes it. The purpose of the CT is to create a secondary current that is a precise fraction of the primary current, for easier measurement of the current in the power conductor. Given this function, would CTs be considered a "step-up" or "step-down" transformer? Also, draw how the secondary windings of a CT are arranged around its toroidal core. What is the CT's output current if the load current is 360 A and the CT ratio is 700:8?

SOLUTIONS

The terms "step-up" and "step-down" are usually referenced to voltage, thus, a CT is a step-up transformer. Its secondary windings should be wound perpendicular to the magnetic flux path (typical in all transformers) as shown in Figure 4.34b.

$$I_{output} = \frac{N_1}{N_2} I_{load} = \frac{8}{700} 360 \text{ A} = 4.11 \text{ A}$$

4.8.2 Sensing Principles of LVDTs, Fluxgate Sensors, RVDTs, Synchros, and Resolvers

4.8.2.1 LVDTs/RVDTs

All transformer-type inductive sensors operate based on changing the mutual inductance between primary and secondary coils. Figure 4.35 shows the internal design of an LVDT, its circuit diagram, and its typical output. The primary coil driven by an AC excitation current (typically several kHz) induces an AC current in each of the secondary coils (they are connected in the opposed phase). A ferromagnetic core,

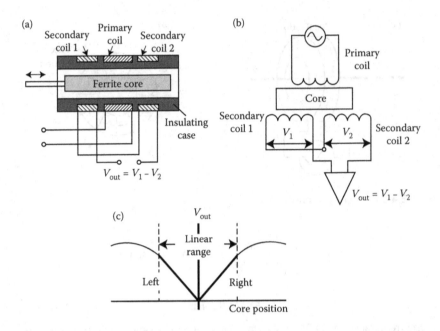

FIGURE 4.35   (a) Internal structure of an LVDT sensor, (b) its schematic diagram, and (c) its output.

usually threaded to a moving object, is inserted coaxially into the tube opening without physically touching the coils. When the core is in the center of the transformer (null position), the outputs from the two secondary coils are canceled out, and the net output voltage is zero. When the core moves away from the central position, a nonzero net output is produced due to unbalances in magnetic field in the secondary coils. There is also a phase change as the core moves. Thus, by measuring both the voltage amplitude and its phase angle, the direction and the displacement of the object motion can be determined.

An LVDT requires a circuit to generate a proper output. Figure 4.36 shows a simplified industry-standard AD598 LVDT signal conditioning circuit. Its on-chip oscillator can be set to generate a 20 Hz to 20 kHz excitation frequency through a

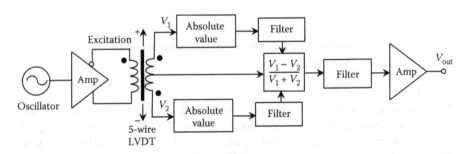

FIGURE 4.36   A simplified AD598 LVDT signal conditioner.

single external capacitor. The two secondaries' outputs—V_1 and V_2, followed by two filters—are used to generate the ratiometric function $(V_1 - V_2)/(V_1 + V_2)$. This function is independent of the amplitude of the primary winding's excitation voltage, assuming the sum of the LVDT output voltage amplitudes remains constant over the operating range. This is usually the case for most LVDTs except that the manufacturer specifies otherwise on the LVDT data sheet. This method often requires the use of a 5-wire LVDT. A single external resistor sets the AD598 excitation voltage from approximately 1 V (RMS) to 24 V (RMS) with a current drive capability of 30 mA (RMS). The AD598 can drive an LVDT at the end of 300 ft of cable, since the circuit is not affected by phase shifts or absolute signal magnitudes. The position output range of V_{out} is ±11 V for a 6 mA load and it can drive up to 1000 ft of cable. The V_1 and V_2 can be as low as 100 mV RMS.

A RVDT works in the same way that an LVDT does, except that its movable magnetic core rotates instead of translates. Thus, it measures the angular position or speed.

4.8.2.2 Fluxgate Sensors

Fluxgates are precise vector sensors of magnetic fields (can measure both magnitude and direction). They were invented by Aschenbrenner and Goubau in 1936 and improved during World War II by Victor Vacquier. They have been used for navigation, compass work, metal detection, and prospecting. The earliest investigation of Earth's magnetic field by fluxgate sensors was found to be important for terrestrial life because Earth's magnetic field forms a protective bubble with distinct regions, called *magnetosphere*, and deflects charged particle flux from the Sun and other extraterrestrial sources [25]. In 1948, the fluxgate was adapted to space magnetometry application. A three-axis fluxgate was flown in an Aerobee sounding rocket to a peak altitude of 112 km. Satellites or space probes carrying fluxgate include USSR mars probe, NASA Explorer 12, 14, and 18, Mariner 2 (Venus) the USSR earth–orbit Electron 2, and Apollo 12, 14, 15, and 16. Latest developments for fluxgate sensors are based on CMOS technology for the coils and CMOS compatible post-process technology (i.e., sputtering) for the core deposition—all to realize micro-fluxgate with very low-power consumption (in a few milliwatts).

Different from other transformer-type sensors, a fluxgate sensor uses a soft magnetic material (with a low coercive force/field and high permeability) as its core material and operates under magnetic saturation state. In saturation, the permeability of the core drops, causing the flux associated with the magnetic field decreases. The term "fluxgate" comes from this "gating" or limiting of the flux at saturation point. The core is magnetically saturated alternatively in opposing directions, normally by means of an excitation coil driven by a sine or square waveform. Thus, an alternating cycle of magnetic saturation in both direction is formed: magnetized (to the saturation point) → unmagnetized → inversely magnetized (to the saturation point in the opposite direction) → unmagnetized → magnetized (to the saturation point). This constantly changing field induces a current in the second (sensing) coil. Note that the peak of the excitation current should be big enough to ensure the core to be saturated. In the absence of an external magnetic field, the input and output currents will match. However, when the core is exposed to an external magnetic field, it will be more easily saturated in alignment with that field direction and less easily saturated in the opposite direction. Hence, the alternating magnetic field, and the induced output current, will be out of phase with the input current. Some flux-gate sensors operate with current sensing in the pickup (secondary) coil, and others with voltage sensing in the pickup (secondary) coil. No matter which mechanism is used, the magnitude of the output signal should be proportional to the strength of the external magnetic field to be measured.

4.8.2.3 Synchros and Resolvers

4.8.2.3.1 Synchros

A synchro is essentially a variable coupling transformer that uses the principle of electromagnetic induction. The magnitude of the magnetic coupling between the primary and secondary windings varies according to the position of the rotating element. Traditionally, the simplest synchro system contains two parts: *synchro transmitter* and *synchro receiver*. The synchro transmitter consists of a single-phase, salient-pole (dumbbell-shaped) rotor and three-phase Y-connected stator. As shown in Figure 4.37, the primary coil (usually driven at 400 Hz) is the rotor and has two terminals (R_1 and R_2). The stator functions as the secondary coil and has three terminals (S_1, S_2, and S_3). There are three stator coils in a 120° orientation and they are electrically Y-connected. When a synchro emitter is driven by an AC current, the stator has three output voltages V_{1-2}, V_{2-3}, and V_{1-3}. The transmitter equations show that nowhere over the entire 360° rotation of the rotor has the same set of voltages produced. Therefore, each set of voltage output (V_{1-2}, V_{2-3}, and V_{1-3}) corresponds to a unique rotor position. A receiver can take the three outputs V_{1-2}, V_{2-3}, and V_{1-3} from the emitter and transfer these three voltages into an angular position. Sometimes a receiver has its own rotor that rotates when receiving the three outputs from the emitter (V_{1-2}, V_{2-3}, V_{1-3}). The term "synchro" is an abbreviation of the word "synchronous," which came from the fact that the receiver's rotor rotates synchronously with the emitter's rotor. Today, most synchros only contain a rotor (or emitter), and they rely on other means to determine the rotor's position based on three voltage outputs.

EXAMPLE 4.13

Examine the working mechanism of a synchro transmitter (see Figure 4.38) under a 115 V, 60 Hz AC excitation voltage applied to its rotor.

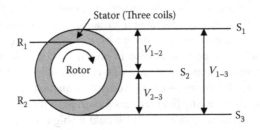

FIGURE 4.37 A synchros sensor configuration.

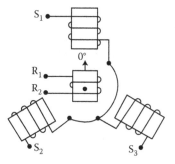

FIGURE 4.38 A synchro transmitter.

SOLUTION

When a 115 V, 60 Hz AC excitation voltage is applied to the rotor, the current in the rotor coil produces an AC magnetic field in the rotor winding and the core. The magnetic flux and force cut through the turns of the three stator windings and, by transformer action, induce voltage into the stator coils. The voltage induced in any stator coil depends upon the angular position of that coil's axis with respect to the rotor axis. When the maximum effective coil voltage is known, the effective voltage induced into a stator coil at any angular displacement can be determined.

Figure 4.39 shows a cross section of a synchro transmitter and the effective voltage induced in one stator coil as the rotor rotates to different positions. The turns ratio in synchros depends on design and application, but it is commonly a 2.2:1 step down between the rotor and a single stator coil. Thus, for 115 V applied voltage to the rotor, the highest value of effective voltage induced in any one stator coil is 52.27 V, which occurs whenever there is maximum magnetic coupling between the rotor and the stator coil (views a, c, and e). The effective voltage

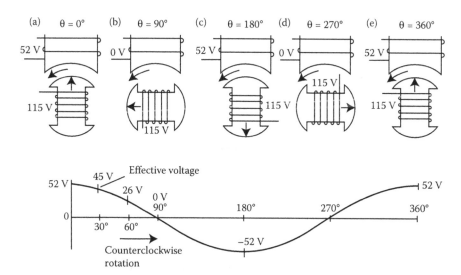

FIGURE 4.39 Stator output voltage versus rotor position.

induced in the secondary winding is approximately equal to the product of the effective voltage on the primary, the secondary-to-primary turns ratio, and the magnetic coupling between primary and secondary. Since the primary voltage and the turns ratio are constant, the secondary voltage varies with the angle between the rotor and the stator. When stator voltages are measured, reference is always made to terminal-to-terminal voltages (voltage induced between two stator terminals) instead of to a single coil's voltage. This is because the voltage induced in one stator winding cannot be measured because the common connection between the stator coils is not physically accessible. In summary, the synchro transmitter converts the angular position of its rotor into electrical stator signals (voltages).

4.8.2.3.2   Resolvers

A resolver is basically a rotating transformer with one primary winding and two secondary windings that are phased 90° (see Figure 4.40a). As the rotor turns, the amplitude of the secondary voltage changes, modulating the input carrier. This establishes two separate outputs having a sine/cosine relationship. As shown in Figure 4.40b, it accepts an AC excitation through terminals R_1 and R_2 at the rotor and produces a pair of two-wire outputs: $\sin \theta$ (between terminals S_1 and S_3) and $\cos \theta$ (between terminals S_2 and S_4), where θ is the angular position of the rotor [26]. The position of the rotor can then be calculated by

$$\theta = \arctan \frac{\sin \theta}{\cos \theta} \qquad (4.41)$$

Synchros are more difficult than resolvers to manufacture and are therefore more costly. Today, synchros find decreasing use, except in certain military and avionic retrofit applications.

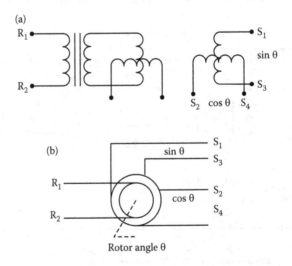

FIGURE 4.40   (a) A resolver configuration: one primary winding and two secondary windings (with 90° phase difference); (b) the resolver's input ($R_1 - R_2$) and two outputs ($S_1 - S_3$ and $S_2 - S_4$).

4.8.3 FEATURES OF TRANSFORMER-TYPE SENSORS

4.8.3.1 Features of LVDTs and RVDTs

Many LVDTs and RVDTs have virtually no friction, possess excellent null stability, and operate in broad temperature spectrum. They are often used for linear or angular displacement measurement. LVDTs have a wide measurement range typically from ±100 μm to ±25 cm. Typical excitation voltages range from 1 to 24 V (RMS), with frequencies from 50 Hz to 20 kHz. A well-designed LVDT can provide a linear output within ±0.25% over the range of core motion and with a very fine resolution (limited primarily by the ability to measure voltage changes) such as 1 part in 100,000 within the linear range. The time response depends on the moving object to which the core is connected. The slope of the transfer function is typically given in millivolts per millimeter ($mV \cdot mm^{-1}$). Some LVDT displacement sensors come with integral electronics that internally generate the alternating current and convert the measured signal into a calibrated DC output. Table 4.10 listed the specifications of an LVDT displacement sensor, made by *WUNTRONIC GmbH*, Germany [27].

Most RVDTs are composed of a wound, laminated stator and a salient two-pole rotor. The stator contains both the primary and the two secondary windings. Some secondary windings may also be connected together. RVDTs utilize brushless, noncontacting technology to ensure long-life, reliable, repeatable position sensing with infinite resolution under the most extreme operating conditions.

EXAMPLE 4.14

An LVDT's core motion range is −1.5 ~ +1.5 cm. Its linearity over this range is ±0.3%, and its sensitivity is 23.8 $mV \cdot mm^{-1}$. If used to track a workpiece motion from −1.2 to +1.4 cm, find the sensor's expected output voltage and the error in position determination due to nonlinearity. Assume a linear transfer function $V_{out} = a\, V_{in}$.

TABLE 4.10
Specifications of *WUNTRONIC* AC LVDT Displacement Transducers Series S

Range	±2.5 to 250 mm
Linearity	0.25%
Resolution	Virtually infinite
Excitation, nominal	5 V RMS @ 2.5 kHz
Excitation, nominal	0–10 V RMS @ 0.4–20 kHz
Output impedance	Less than 10 Ω
Operating temperature	−30°C to +90°C
Storage temperature	−55°C to +110°C
Frequency response	1–100 Hz, ±3 dB
Shock survival	500 g, 11 ms
Vibration	20 g, up to 2 kHz
Housing material	Stainless steel, 400 series
Lead wires	#28 AWG plated stranded copper

SOLUTIONS

Using a linear transfer function, the output voltages can be easily found ($a =$ 23.8 mV · mm^{-1})

$$V(-1.2 \text{ cm}) = (23.8 \text{ mV} \cdot \text{mm}^{-1}) \, (-12 \text{ mm}) = -285.6 \text{ mV}$$

and

$$V(1.4 \text{ cm}) = (23.8 \text{ mV} \cdot \text{mm}^{-1}) \, (14 \text{ mm}) = 333.2 \text{ mV}$$

The linearity deviation shows up in deviations of the transfer function. Thus, the transfer function has an uncertainty of

$$(\pm 0.003) \, (23.8 \text{ mV} \cdot \text{mm}^{-1}) = \pm 0.0714 \text{ mV} \cdot \text{mm}^{-1}$$

This means that a measured voltage, V_m (in mV), could be interpreted as a displacement that ranges from $V_m/23.73$ mm (23.73 comes from 23.8–0.0714 mV · mm^{-1}) to $V_m/23.87$ mm (23.87 comes from 23.8 + 0.0714 mV · mm^{-1}), which is approximately ±0.3%, as expected.

Common specifications of commercially available RVDTs are

- *Power input*: 3–15 V (RMS) sine wave with a frequency between 60 and 20,000 Hz.
- *Angle*: Most RVDTs have effective angle limits of ±60°, although they are capable of continuous rotational measurement.
- *Nonlinearity*: Higher accuracy in the smaller angle range: 0.25% @ ±30°, 0.50% @ ±40°, 1.50% @ ±60°.

The major advantages of RVDTs include (1) relatively low cost and robust like most inductive sensors; (2) no frictional resistance due to noncontact between the iron core and coils, and thus long service life; (3) high signal-to-noise ratio and low output impedance; (4) negligible hysteresis; (5) theoretically infinitesimal resolution (in reality, angle resolution is limited by the resolution of the amplifiers and the circuit used to process the output signal); and (6) no permanent damage of RVDTs if measurements exceed the designed range. The primary limitation of RVDTs is that the core must be in contact directly or indirectly with the measured surface, which is not always possible or desirable.

There are several factors affecting a RVDT sensor's output: the number of turns in the sensing winding, magnetic permeability of the core, sensor geometry, and the excitation frequency. Phase synchronous detection is used to convert a senor's output (harmonic signals) to a DC voltage proportional to the measurant.

4.8.3.2 Features of Fluxgate Sensors

For low magnetic field sensing, the fluxgate sensors provide the best trade-off between cost and performance. It can measure both magnitude and direction of a DC (static) or low-frequency AC magnetic field in the range of 0.1 nT–0.1 mT with the achievable resolution of 0.0001 µT. Many fluxgate sensors have a bandwidth of a few hertz to kilohertz frequencies [28].

Fluxgates can operate over a wide temperature range. Typical temperature stabilities drift less than 0.1 nT $\cdot$ °C^{-1} with a temperature coefficient around 30 ppm $\cdot$ °C^{-1}. Some fluxgates can be compensated to 1 ppm $\cdot$ °C^{-1}. In terms of dynamic range and resolution, fluxgate sensors perform better than the Hall effect sensors and are preferable to superconducting quantum interference devices (SQUIDs) because of their lower cost and size. If the fluxgate operates in feedback mode, the linearity error may be as low as 10^{-5}.

Fluxgate sensors, using soft magnetic materials as their core materials (with low coercive field H_c), have the following characteristics [29]:

- Low losses at the excitation frequency (usually in the tens of kilohertz range)
- A low saturation induction value (implies a low power consumption)
- A minimal magnetostriction effect (see Chapter 5 for details)
- Low magnetic noise due to easy reversibility of the magnetization

These important features have made fluxgate sensors ideal for a variety of applications.

4.8.3.3 Features of Synchros and Resolvers

Synchros resemble motors—consisting of a rotor, stator, and a shaft. Usually, slip rings and brushes connect the rotor to external power. A synchro can have single- or three-phase configuration. Single-phase units have five wires: two for an exciter winding (positive and negative ends) and three for the output. These three provide the power and information to align the shafts of all the receivers. Synchro transmitters and receivers must be powered by the same branch circuit. Three-phase synchro can handle more power and operate more smoothly. The excitation is often a 240 V 3-phase main power. Synchros designed for terrestrial use tend to be driven at 50 or 60 Hz, while those for marine or aeronautical use tend to operate at 400 Hz.

Multispeed synchros have stators with many poles, so that their output voltages go through several cycles for one physical revolution. *Differential synchros* have three-lead rotors and stators, and can be transmitters or receivers. A differential transmitter is connected between a synchro transmitter and a receiver, and its shaft's position adds to (or subtracts from) the angle defined by the transmitter. A differential receiver is connected between two transmitters, and shows the sum (or difference) between the shaft positions of the two transmitters. Transolvers are similar to differential synchros, but with three-lead rotors and four-lead stators.

Since synchros have three stator coils in a 120° orientation, they are more difficult than resolvers to manufacture and are therefore more costly. Today, synchros find decreasing use, except in certain military and avionic retrofit applications.

The main parameters of resolvers are as follows [30]:

- *Input voltage*: 1–26 V. Larger voltage can cause the saturation of a resolver's magnetic structure, resulting in increased error and null voltage.

- *Frequency*: 400–5000 Hz frequency. Lower frequency can result in the saturation of a resolver's magnetic structure, increase errors, and change some other parameters. Higher frequencies may result in increased magnetic flux leakage as well as changes in capacitance coupling.
- *Voltage sensitivity* or *voltage gradient*: defined by the output voltage per one degree (1°) rotor rotating angle.
- *Transformation ratio (TR)*: defined as the ratio of output voltage to input voltage when the output is at maximum coupling, that is

$$TR = \frac{\max\{V_{out}\}}{V_{in}} \tag{4.42}$$

TR is approximately proportional to the ratio of effective turns, secondary N_2 to primary N_1:

$$TR = \frac{KN_2}{N_1} \tag{4.43}$$

where K is a constant.

A higher TR is easy to achieve in one-speed resolvers. However, it is more difficult to achieve in multispeed resolvers because of increased flux leakage and increased N_2 using very fine magnetic wire that complicates the manufacturing process.

- *Phase shift*: the difference between the time phase of the primary and secondary voltage when the output is at maximum coupling.
- *Null voltage*: the residual voltage at the point of minimum magnetic coupling between the primary and secondary windings. It is measured when the "in-phase" secondary voltage is zero.
- *Number of speeds*: the number of amplitude-modulated sinusoidal cycles in one revolution of the resolver. Multiple-speed resolvers are achieved by increasing the number of magnetic poles in the rotor and stator equally. Increasing the number of speeds can increase the accuracy, but it is limited to the size of the resolver. A single-speed resolver is essentially a single-turn absolute device. By increasing the speeds of a resolver, the absolute information is lost. If space permits, mounting a single-speed resolver on top of a multiple-speed resolver will provide higher accuracy and absolute information.

4.8.4 Design and Applications of LVDTs, RVDTs, Fluxgate Sensors, Synchros, and Resolvers

4.8.4.1 Design and Applications of LVDTs/RVDTs/PVDTs

By attaching the core to a moving object or a diaphragm, an LVDT can measure position, acceleration, and pressure.

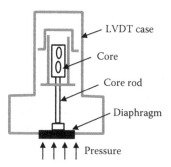

FIGURE 4.41   An LVDT pressure sensor.

4.8.4.1.1   LVDT Pressure Sensor

To use an LVDT sensor to measure pressure, the pressure must be converted to a linear displacement. Any pressure-sensitive element that changes size or shape as a pressure is applied, such as bellows, Bourdon tubes, or aneroid capsules, can be used. As a pressure is applied, the core is displaced from its null position, causing an output voltage. Figure 4.41 shows an LVDT pressure sensor in which a pressure diaphragm drives the core up and down and varies the inductive coupling between the primary and secondary windings. The LVDT is a differential device and can be made to measure absolute, relative, or differential pressures.

4.8.4.1.2   Plant Growth Monitor

An LVDT can also be used for measuring the growth of a plant (see Figure 4.42). One end of the moving core is attached to the plant, the other end of the core is connected to a spring to keep the plant straight without bending. A linear screw system utilizes the feedback data from a computer to adjust the LVDT's position as the plant

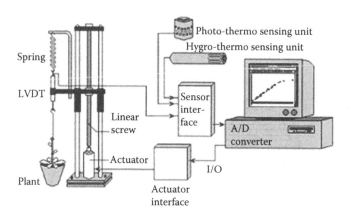

FIGURE 4.42   An LVDT sensor measuring plant growth. (After Floral stem growth measurement, Physiomatics Laboratory, University of Geneva, Switzerland. www. physiomatique.unige.ch/lvdt-e.html)

grows. The photo-thermo and hygro-thermo sensing units are used to measure light, temperature, and humidity.

4.8.4.2 Design and Applications of Fluxgate Sensors

Fluxgate sensor designs fall broadly into two styles, those employing rod cores and those using ring cores. The rod cores can be further divided into a *single-rod* (Figure 4.43a) or a *twin-core* design. Twin cores are further split into two categories: the *Forster* (Figure 4.43b) and *Vacquier* (Figure 4.43c) based. A ring core can have either a circular (Figure 4.43d) or race-track (Figure 4.43e) shape.

In a single rod-shaped core, the sense coil will not only pick up the signal voltage, but also the excitation drive, which due to its high level can be troublesome to remove electronically. A common solution for this is to use two parallel cores with the excitation winding reversed from one to the other. Thus, the sensing coil picks up only the signal voltage, while the induced the excitation voltage is cancelled by the phase reversal.

The only difference between a Forster and a Vacquier is their sensing coil configuration as shown in Figure 4.43b and c. In the Forster design, the sensing coils indicated by the darker ones in Figure 4.43b are wound such that the excitation affection is cancelled out, whilst the useful voltage (output) signals are combined. It is also possible to build a fluxgate based on the Forster using a common set of coils for excitation and sensing. The core material for the rod in the Forster design is traditionally high mu-metal wire (e.g., 77% nickel, 16% iron, 5% copper, and 2% chromium or molybdenum). The rod is typically 20 mm in length but can vary from 15 to 75 mm. The long-term stability of this type of sensor is better than 3 nT drift per year [31]. The sensor's sensitivity direction is along the core's longitudinal direction. This type is often used in geophysical observatory. The Vacquier configuration is often used in satellite instruments. Unfortunately, most of the amorphous wires exhibit large noise when a core has a long dimension. For example, Vacquier-type sensors made of stress-annealed Vitrovac 6025 amorphous tape exhibited a minimum noise of 11 pT

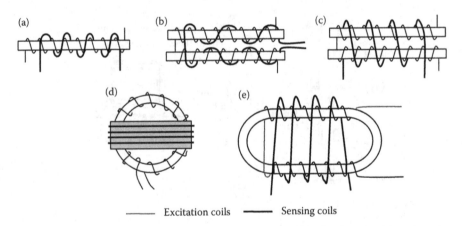

FIGURE 4.43 Core types of fluxgate sensors: (a) single-rod, (b) forster, (c) vacquier, (d) ring-core, and (e) race-track.

and 46 pT RMS (60 MHz–10 Hz) for 65 and 35 mm long sensors, respectively. It has been shown that the cores and sensing coils should be 10% shorter than the excitation coils to reduce noise caused by unsaturated ends [32].

A ring core can be considered as the connection of the tops and the bottoms of the twin rods to form a continuous magnetic loop. It normally uses a toroidal excitation winding spread evenly around the core (Figure 4.43d), with a sensing coil wound around the outside and orthogonal to the core plane, similar to the Vacquier sensing winding. The axis of the sensing or feedback winding defines the sensitive direction of the sensor, since the core itself is isotropic. It is common for a single axis sensor to have its sensing winding completely cover the ring core in order to achieve better performance. For two-dimensional (XY) sensing, two sets of sense coils are put over the toroid at 90° angle to one another to create two sensitive axes. These coils, however, are usually short to avoid cross coupling. In a tri-axis system, the sensor can be built in a similar way with two rings.

The ring can be a simple ferrite core or tape wound core using permalloy or amorphous metal tapes of soft magnetic materials. Ring core fluxgates are broadly used in modern satellite missions for mapping the magnetosphere and deep space explorations. The ring-core geometry has advantages of creating lower noise and using less power consumption than rod core sensors. Ring-core sensors also demonstrate less stability offset than other shape cores for the same length. However, ring-core sensors have low sensitivity due to large demagnetization. Race-track sensors, however, have higher sensitivity than other types due to the lower demagnetization factor. Race-track shape cores have a problem of producing large spurious signals that cannot be easily balanced as the ring cores do (by rotation to minimize the spurious components). One solution is to change the track width along the core length to adjust for unbalance caused by core or excitation winding nonhomogeneity [33].

A superior design is to add a feedback coil (or double up the sensing coil for this purpose) to feedback a magnetic field in opposition to the sensed field such that the two fields cancel one another. In this mode of operation, the fluxgate is used as a *null detector*, and the current in the feedback coil is proportional to the sensed field. This technique improves linearity of measurement, allows a much greater dynamic range to be measured, and is used by the majority of modern devices.

In recent years, PCB (printed circuit board) planar fluxgates or micro-fluxgates have increasing applications due to their low cost and easy fabrication. Figure 4.44a shows a structure for a differential single-axis planar fluxgate magnetic sensor. The ferromagnetic core is placed over the diagonal of the excitation coil. An AC current in the excitation coil saturates each half of the core periodically in opposite directions. If no external magnetic field is present, the output voltage of two sensing coils is zero. When an external magnetic field is present and parallel to the core, the magnetization in one half of the core is in the same direction as the external magnetic field, while the magnetization of the other half of the core is in the opposite direction (see Figure 4.44b). Thus, the voltage induced in the two sensing coils is not equal and the differential output voltage increases its value. Figure 4.45 shows the magnetization process and the sensor's outputs with or without an external magnetic field.

Many planar fluxgate sensors do not require electroplating or sputtering processes for ferromagnetic material deposition because very thin amorphous metals

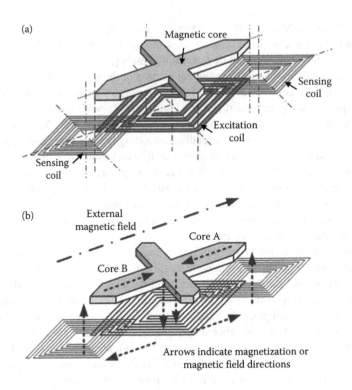

FIGURE 4.44 A single-axis planar fluxgate sensor configuration: (a) without an external magnetic field, (b) with an external magnetic field. (Adapted from Dottorato di, T.D. and Marchesi, M., *Fluxgate Magnetic Sensor System for Electronic Compass*, University of Degli Studi Di Pavia, Italy, pp. 25–26, 2013.)

are commercially available (e.g., 25 μm for Vitrovac 6025X; 20 μm for Vitrovac 6025Z; and 16 μm for Metglas 2714A) with very good magnetic characteristics.

Fluxgate sensors are commonly used for high-resolution measurements at low frequencies in applications where the sensor must operate at room temperature.

Simple fluxgate sensors can detect metal objects and read magnetic marks/labels. Fluxgate compasses are used in aircraft, land vehicle, and submersible navigation systems. The fluxgate principle is also used in electrical current sensors and current comparators. The most common use of fluxgate sensors is in static magnetic field measurement. Fluxgate sensors can also measure alternating fields up to several kilohertz, depending on excitation frequency, the core, sensing and pickup coil designs, and circuits used. Followings are some application examples.

4.8.4.2.1 DC Magnetometer

In a DC magnetometer, there are separate pickup and feedback coils [34]. The core, in most cases a ring core, is periodically saturated in both directions by excitation current in the primary coil. When an external field (to be measured) appears, it

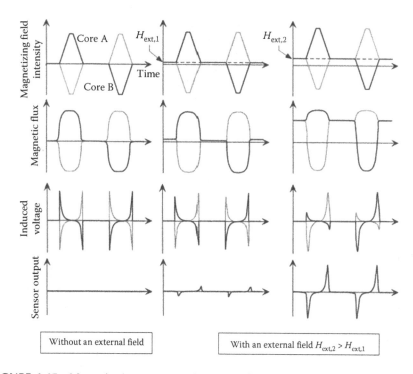

FIGURE 4.45 Magnetization process and the sensor's outputs with or without absence of an external magnetic field. (Adapted from Dottorato di, T.D. and Marchesi, M., *Fluxgate Magnetic Sensor System for Electronic Compass*, University of Degli Studi Di Pavia, Italy, pp. 25–26, 2013.)

alters the harmony and voltage (or current if it is a short-circuited mode) induced in the pickup (secondary) coil, which are measured by a phase detector and fed back as a current into a compensation coil. The integrator in the feedback loop ensures enough DC gain and makes the loop to respond to only low-frequency (DC) magnetic field [28].

4.8.4.2.2   PCB Planar Fluxgate for Electronic Compass

Figure 4.46 shows a single-axis planar fluxgate magnetic sensor using an excitation (primary) coil (30 turns, 30 µm thick, 25.12 mm × 31.3 mm) and two sensing coils (30 turns, 17 µm thick, 31.3 mm × 31.3 mm) configured in differential format with metal lines' pitch of 400 µm [29]. The excitation and sensing coils are placed on two different metal layers in a multilayer PCB structure. The ferromagnetic sheet core is a special amorphous alloy (Vitrovac 6025, $\mu_r \cong 10^5$, magnetic saturation at 0.55 T).

4.8.4.3   Design and Applications of Synchros and Resolvers

Synchros and resolvers offer extremely high accuracy and fast measurement, and are used in industrial metrology, radar antennae, and telescopes. Synchros are found in just about every weapon system, communication system, underwater detection system, and navigation systems. They are reliable, adaptable, and compact. Resolvers

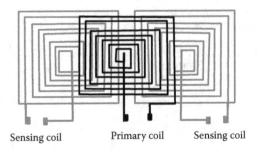

Sensing coil   Primary coil   Sensing coil

FIGURE 4.46   A single-axis PCB fluxgate sensor.

lend themselves to maximum applications because of their simple and standard components similarity to electric motors (windings, laminations, and bearings). The most popular use of resolvers is in permanent magnet brushless AC servo motors, military, and aerospace applications.

Figure 4.47 is a DC-biased eddy-current speed sensor developed by *Hood Technology Corporation*, Hood River, Oregon, USA. It can be mounted on the outside of the engine case (no holes and no interruption in the gas path) to monitor the turbine engine's blade-tip speed [35], even in the presence of contaminants and at temperatures up to 1000°F. With a permanent magnet the sensor generates a static magnetic field that penetrates nonferromagnetic engine case walls. As a conductive rotor blade passes through the sensor's magnetic field, the eddy current is induced in the blade tip, creating the secondary magnetic field that interacts with the sensor's magnetic field. The resulting perturbation in the field is detected by the sensor's pickup coil as an induced voltage.

Based on Faraday's law of electromagnetic induction (Equations 4.23 and 4.24), if N is fixed, the induced voltage is a function of only the time-rate change of the magnetic flux. Since some component, such as an engine case, is fixed to the sensor face, the penetrated flux in the material will not change as time changes. Therefore, it

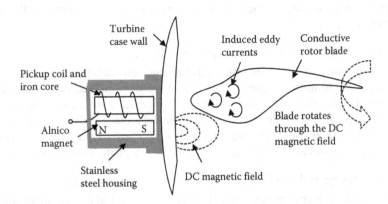

FIGURE 4.47   An inductive turbine engine speed sensor. (From von Flotow, A. and Drumm, M.J., Blade-tip monitoring with through-the-case eddy current sensors, *Sensors*, 21(6) 28–35, 2004.)

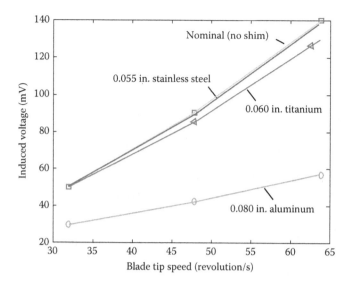

FIGURE 4.48 The sensor's response to the blade's speed. (From von Flotow, A. and Drumm, M.J., Blade-tip monitoring with through-the-case eddy current sensors, *Sensors*, 21(6) 28–35, 2004.)

will not contribute to the induced voltage. Only the moving blades on the other side of the wall change the magnetic flux, resulting in an induced voltage.

Figure 4.48 shows the sensor's output, the induced voltage, versus the speed of a two-blade titanium rotor with three different test shims between the sensor and the rotor. Here, the nominal gap between the rotor and the sensor is 0.160 in. The thickness and material of each nonferromagnetic shim are

0.080 in. aluminum
0.060 in. titanium
0.055 in. stainless steel

As can be seen in Figure 4.48, the presence of stainless steel was the same as the nominal case (no shim); titanium reduced the output voltage signal by ~10%, and aluminum reduced it by ~50%. This result proves that increased electrical conductivity in the case of material resistance changes in magnetic flux and thus attenuates the sensor signal. However, the signal amplitude increases with increased electrical conductivity and magnetic permeability of the blades.

4.9 OSCILLATOR AND SIGNAL PROCESSING CIRCUITS OF INDUCTIVE SENSORS

Many inductive sensors come with circuits inside them. One of the circuits is an oscillator. An oscillator is an inductive capacitive tuned circuit (or an *RLC circuit*, as shown in Figure 4.49) that creates a radio frequency, typically 500 kHz to 1 MHz. The electromagnetic field produced by the oscillator is emitted from the coil away

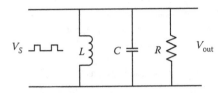

FIGURE 4.49 A typical oscillator for inductive sensors.

from the face of the sensor. The circuit usually has a feedback from the field to keep the oscillator operating properly. When a metal target enters the field, currents are induced and circulate within the target. This causes a load on the sensor, decreasing the amplitude of the electromagnetic field. The trigger circuit measures the amplitude change to determine how far the target is from the sensor (if it is for a distance measurement) or compare the amplitude change with a predetermined value (threshold) to switch the sensor's output state, e.g., ON or OFF for a proximity sensor.

When two or more sensors are mounted in close proximity to each other such that their electromagnetic fields intermix, some interference may occur. This interference can be in the form of cross-talk (see Chapter 6), resulting in beat notes whose frequency will be the difference between the frequencies of the oscillators in each unit. Synchronization is to use one oscillator unit to drive multiple electronics. The oscillator driver, or Master, is connected to other units whose oscillators have been disabled. These units are referred to as the Slave units.

There are three main types of circuits used for inductive sensors: *Colpitts oscillator*, *balanced bridge*, and *phase circuit* [17].

4.9.1 COLPITTS OSCILLATOR

A Colpitts oscillator (see Figure 4.50), named after its inventor, Edwin H. Colpitts, consists of a simple and robust LC components (C_1, C_2, and L_1), the sensor coil L_2, and a transistor (for control and amplification of the sensor output signal). When the sensor coil interacts with a conductive target, the oscillator frequency and amplitude vary. This variation is proportional to the distance between the sensor and the target, and then the sensor outputs an analog signal proportional to the distance. The Colpitts circuit provides a larger measurement range for the same size sensor than other types of inductive circuits. It responds to any conductive material, but works really well to magnetic steel and highly resistive targets except that it does not have compensations for offset, temperature, or nonlinearity of the output signal.

A single adjustment for gain control is used to raise or lower the output voltage level to a desired value. Sensor cable length is limited (or circuit will not oscillate). Typical applications are fuel injector testing, small-diameter targets that are not detected well by inductive bridge circuits, and machining and grinding.

4.9.2 BALANCED WHEATSTONE BRIDGE CIRCUIT

In Figure 4.51, an oscillator excites a Wheatstone bridge—a balanced bridge tuned to be near resonance. Slight changes in impedance of the sensor coil (due to the

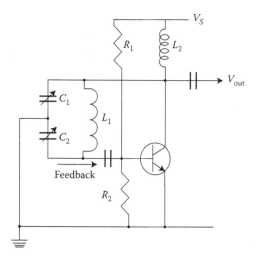

FIGURE 4.50 A Colpitts oscillator. (From *Inductive Technology Handbook*, Kaman Precision Products, a division of Kaman Aerospace Corp., Springs, Colorado, USA.)

target's movement and the interaction between the sensor and the target) will result in big shifts in the output of the bridge. This change of impedance is measured by a demodulator circuit, linearized by a logarithmic amplifier, and amplified by a final amplifier. The system output voltage is directly proportional to the target position relative to the sensor.

The bridge circuit can be used with either single- or dual-coil sensors. In the single-coil configuration, only one coil is exposed to the measuring environment. In the dual-coil design, the sensing and reference coils are on opposing sides of the bridge, providing a cancelation effect and improving the sensor's performance. This circuit has a wide variety of applications for both ferrous and nonferrous targets.

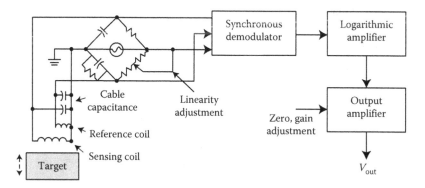

FIGURE 4.51 A balanced bridge circuit. (From *Inductive Technology Handbook*, Kaman Precision Products, a division of Kaman Aerospace Corp., Springs, Colorado, USA.)

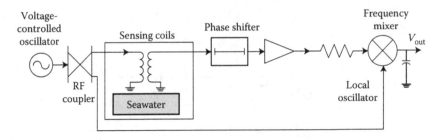

FIGURE 4.52   A phase circuit.

4.9.3 Phase Circuit

The induced currents in inductive sensors have both amplitude and phase components. A phase circuit measures a phase change (shift) rather than an amplitude
change. Since noise is amplitude sensitive, the phase-shift measurement will not be
affected by the noise. Moreover, the phase circuit uses the pulse width modulation
(PWM) signals instead of high-gain and noisy op-amps analog signals; thus, it can
produce extremely low noise outputs. The PWM method also allows for temperature
stability or linearization.

Figure 4.52 shows an inductive sensor that uses RF (radio frequency) phase detection to measure the conductivity of sea water samples containing different salinities
[36]. The sensor detects a phase change in an alternating signal that couples the two
closely-spaced coils inside the fluid. The phase shifter modifies the output voltage.

EXERCISES

Part I: Concept Problems

1. Which of the following factor(s) will *not* affect an inductive sensor's
   performance?
   A. Target material
   B. Media material between sensor and target
   C. Core material
   D. Coil material
2. Indicate the Lorentz force direction in Figure 4.53a and b.

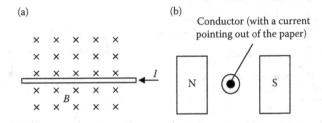

FIGURE 4.53   Electromagnetic fields.

3. Which of the following phenomena is related to Faraday's law?
   A. A changing magnetic field produces an electric field
   B. A moving charge creates a magnetic field
   C. A moving charge has a curvilinear trajectory in a magnetic field
   D. A moving charge experiences a force
4. The magnetic field strength H will increase if (choose one from below):
   A. N turns are wound in two layers around a core of total length $l/2$
   B. N turns are wound in one layer around a core of total length l
5. The MMF (magnetomotive force) will increase if (choose one from below)
   A. Using stacked layers of turns
   B. Using one layer of turns
6. Which core in Figure 4.54 will store more energy (assume same coil or core materials and dimensions)?
7. Why do some inductive sensors and actuators (e.g., motors) use laminated core?
8. State advantages and disadvantages of coil sensors with ferromagnetic cores compared to air coil sensors.
9. An inductive sensor's performance is *least* affected by which of the following factor(s)?
   A. Winding
   B. Driving current frequency
   C. Magnetic permeability of the core
   D. Core geometry
10. What is the difference between *ferrimagnetic* materials and *ferromagnetic* materials? What are their roles in the magnetic sensors?
11. Which of the following statement(s) is/are true for a fluxgate sensor?
   A. A fluxgate sensor has a permeable core wrapped by a primary coil and secondary coil.
   B. "Gate" in the term "fluxgate" comes from transistor "gate," meaning ON and OFF.
   C. LVDT is one of the fluxgate sensors.
   D. An alternating driving current is passed through the secondary coil.
12. Skin effect states that_____.
13. Which of the following materials that inductive sensors cannot detect
   A. Metallic
   B. Nonmetallic
   C. Magnetic
   D. Nonmagnetic

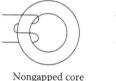
Nongapped core

Gapped core

FIGURE 4.54 Ring cores with or without a cut.

    E. Ferroelectric
    F. Nonferroelectric
    G. Conductive
    H. Nonconductive
    I. Ferromagnetic
    J. Nonferromagnetic

14. The inductance L of a moving core type of sensors is usually a function of the permeability of the core μ, the cross-sectional area A, the number of turns N, and the length of coil l. When using this type of sensors to measure a linear position or motion, which of the above parameters causes the inductance L to change during measurement?

15. Why are eddy-current sensors a type of inductive sensors?

16. A student needed a 215 mH inductor in a circuit, but he only had 550 and 350 mH inductors on hand. He decided to connect these two inductors in parallel with one another (see Figure 4.55) on a breadboard to achieve approximately 215 mH of inductance. However, upon testing, he found the total inductance was significantly *less* than the 215 mH predicted. He asked the lab assistant to help, who inspected the board and immediately suggested that the two inductors be relocated with their axes perpendicular to each other. The student did not understand why the physical orientation of the inductors should matter, because it never mattered how he located resistors and capacitors with respect to one another, as long as their connecting wires went to the right places. Explain why inductors might be sensitive to physical orientation?

17. State the difference between various transformer-type sensors, such as LVDT, fluxgate, and resolver sensors.

Part II: Calculation Problems

18. Find the inductance of a circular wire/conductive loop ($N = 1$) in Figure 4.56, given that the loop radius $R = 10$ mm, and the wire radius $r = 0.5$ mm.

19. A particle carrying a 1 μC charge is moving with the velocity $3i - 3k$ (m · s⁻¹) in a uniform field $-5k$ (T). (1) Determine the force experienced by the particle. (2) If the mass of the particle is 5.686×10^{-18} kg, find the acceleration of the particle due to this magnetic force.

20. Find (1) the inductance of a toroid (or *ring core*) and (2) the magnetic field strength at its center. Given $N = 5$ turns, $R = 4.75$ mm, $r = 1.6$ mm, $\mu_r = 2490$, and $I = 5$ A.

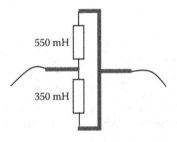

FIGURE 4.55 Two inductors in parallel with the same orientation.

FIGURE 4.56 An inductive loop.

21. An airplane with a wingspan of 39.9 m is flying northward near the North Pole at 850 km/h. The magnetic field of the Earth is $B = 5.0 \times 10^{-5}$ T. Calculate the induced emf.

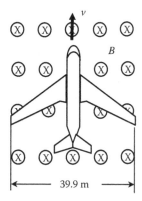

Hints: (1) use Equation 4.1 EMF = Blv; (2) the geographic North Pole is a magnetic south pole.

22. Consider a loop of wire, carrying a 0.225 A current, shown in Figure 4.57, which is partially immersed in the magnetic field. The force that the magnetic field exerts on the loop can be measured with a balance method, and this permits the calculation of the strength of the magnetic field. Suppose that the short side of the loop is $l = 10.0$ cm and the magnetic force is $F_m = 5.35 \times 10^{-2}$ N. What is the strength of the magnetic field? [Hint: The three segments of the current loop shown in the figure are immersed in the magnetic field. The magnetic forces acting on segments 1 and 3 are

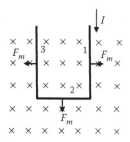

FIGURE 4.57 A wire loop in a magnetic field.

cancelled out. The magnitude of the magnetic force acting on segment 2 can be calculated using $F = BIl$].

23. An eddy-current sensor is used to detect the defects on a steel target ($\rho = 1.18 \times 10^{-7}\ \Omega \cdot m$). Knowing that μ_r for steel is 100 and the sensor driving frequency is 50 Hz, what is the minimum thickness of the steel target required to be detected by this eddy-current sensor?

24. What is the minimum thickness of an iron target to be detected by an eddy-current sensor? Given $\rho = 9.7 \times 10^{-8}\ \Omega \cdot m$, μ_r (for iron) $= 5000$, and the sensor driving frequency $f = 50$ Hz.

25. In Example 4.4, if $r_1 = 12.7$ mm, $r_2 = 76.2$ mm, $B = 0.675$ T, and $\omega = 10{,}000$ rpm, calculate the voltage generated between the brushes on a Faraday's disk.

26. An inductive sensor ($D = 5$ mm) is chosen to detect the cracks on a spindle surface. The available space in the area is 20 mm × 25 mm × 16 mm (length × depth × height). If $S_n = 3$ mm, refer to Figure 4.16, what type of sensor mounting should be used to perform the detection job?

27. A tin target is detected by an eddy-current probe. Knowing the probe-driven frequency is 500 kHz, what is its minimum thickness for an optimal measurement?

28. A moving object made of 304 SST is detected by an eddy-current sensor. If the target has a minimum thickness of 1.8 mm, what should the sensor's maximum excitation frequency be?

29. An inductive proximity sensor has $S_n = 14$ mm. If the target material is aluminum instead of iron, what is the detection range?

30. Determine a sensor's size, so that it can measure a 10 mm × 12 mm × 2 mm rectangular copper object.

31. A damping coefficient ζ is related to a quality factor Q by the equation $Q = 1/(2\zeta)$. A high Q circuit's response has a sharp peak at resonance, while a low Q circuit has a smooth broad response curve. Calculate the Q for $\zeta = 2.2$ ($\zeta > 1$ overdamped), $\zeta = 1$ (critical damped), and $\zeta < 0.7$ ($\zeta < 1$ underdamped) and indicate in which situation a circuit has a sharp peak at its response.

32. Figure 4.58 is a transformer circuit. Calculate V_p, V_s, I_p, and I_s.

33. Two wire windings are wrapped around a common iron core in Figure 4.32. Thus, if one winding produces a magnetic flux in the core, it will be fully shared by the other winding. Express the induced voltage at each winding (V_p and V_s, respectively) in terms of the instantaneous current through that winding (I_p and I_s, respectively) and the inductance of each winding (L_p and L_s, respectively). Note that the induced voltages in the two windings are related to each other by the equation $V_p/N_p = V_s/N_s$ if there is perfect

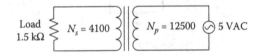

Load
1.5 kΩ $N_s = 4100$ $N_p = 12500$ 5 VAC

FIGURE 4.58   A transformer circuit.

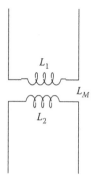

FIGURE 4.59 An inductive coupling.

"coupling" of magnetic flux between the two windings. Based on this, write two more equations describing induced voltage in each winding in terms of the instantaneous current in the other winding, that is, $V_p = ?I_s$ and $V_s = ?I_p$.

34. Determine the target size for an extended range M18 sensor with a rated sensing distance (S_n) of 8 mm.

35. *Mutual inductance* is the term given to the phenomenon where a change in current through one inductor causes a voltage to be induced in another. When two inductors (L_1 and L_2, as shown in Figure 4.59) are magnetically "coupled," the mutual inductance (L_M) relates their voltages and currents as such:

$$v_1 = L_M \frac{di}{dt} \text{ Voltage induced in coil 1 by change of current in coil 2}$$

$$v_2 = L_M \frac{di_1}{dt} \text{ Voltage induced in coil 2 by change of current in coil 1}$$

When the magnetic coupling between the two inductors is perfect ($k = 1$), how does L_M relate to L_1 and L_2 (i.e., write an equation defining L_M in terms of L_1 and L_2, given perfect coupling)? Hint:

$$v_1 = L_2 \frac{N_1}{N_2} \frac{di_2}{dt}, v_2 = L_1 \frac{N_2}{N_1} \frac{di_1}{dt}, \text{ and } \frac{L_1}{L_2} = \left(\frac{N_1}{N_2}\right)^2$$

36. The best way to analyze the inputs and/or outputs of a complex or complete system is to step by step analyze the information passing through. Figure 4.60 shows the block diagram of an LVDT output that is amplified and connected to a chart recorder. Calculate the output movement of the chart recorder pen. The equations for the subsystems are $X_{in} = 12$ mm, LVDT output $V = 0.2$ V · mm^{-1}, amplifier gain $= 4$, and chart recorder sensitivity $a = 5$ mm · V^{-1}.

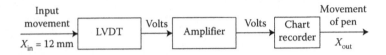

FIGURE 4.60   Block diagram of an LVDT output.

37. A resolver has voltage outputs of sin $\theta = 0.616$ and cos $\theta = 0.788$; find the rotor's angular position.

38. An *RLC* oscillator in Figure 4.49 has an input of $V_{in} = 5 \sin 3t$. If $L = 0.1$ H, $C = 0.2\ \mu F$, and $R = 750\ \Omega$, find its output V_{out}.

REFERENCES

1. Chattock, A.P., On a magnetic potentiometer, *Phil. Mag.*, 24, 94–96, 1887.
2. Rogowski, W. and Steinhaus, W., The measurements of magnetic potential, *Arch. Elektrotechnik*, 1, 141–150, 1912.
3. Werner, E., The device for testing of electrical steel magnetized by AC field, Austrian Patent No. 19, 10–15, 1957.
4. Linberger, M., *FE Review Manual*, 3rd ed., Professional Publications, Inc. (PPI), Belmont, California, USA, 2011.
5. Giancoli, D.C., *Physics: Principles with Applications*, 7th edn, Addison-Wesley Educational Publishers, Inc. Boston, Massachusetts, USA, 2013.
6. Morgan, K., *Ampere's Law*, HandiComputing, Charlotte, Michigan, USA, 2000.
7. History of eddy current testing, eddy current array tutorial, Olympus Corporation, Center Valley, Pennsylvania, USA. Available at: www.olympus-ims.com/en/ndt-tutorials/eca-tutorial/intro/history/.
8. AC resistance, skin & proximity effect, General Cable New Zealand Ltd., 2010.
9. Storr, W., Basic electronics tutorials, 2011. Available at: www.electronics-tutorials.ws/
10. Wilson, T.G. Sr., *Fundamentals of Magnetic Materials*, APEC Tutorial Seminar 1, 1987.
11. Kretschmar, M. and Welsby, S., Chapter 8 Capacitive and inductive displacement sensors. In: *Sensor Technology Handbook*. J.S. Wilson (editor), Elsevier Inc., Burlington, Massachusetts, USA, pp. 197–198, 2005.
12. Dehmel, G., Chapter 6 Magnetic field sensors: induction coil (search coil) sensors. In: *Sensors—A Comprehensive Survey*. Vol. 5. Hesse, J (Ed.), etc. VCH Publishers, Birmingham, UK, pp. 205–254, 1989.
13. Stuart, W.F., Earth's field magnetometry, *Rep. Prog. Phys.* 35, 803–881, 1972.
14. Campbell, W.H., Induction loop antennas for geomagnetic field variation measurements. ESSA Technical Report ERL123-ESL6, 1969.
15. Tumanski, S., Induction coil sensors—A review, *Meas. Sci. Technol.* 18, 31–46, 2007.
16. Harlow, J.H., *Electric Power Transformer Engineering*. CRC Press. Boca Raton, Florida, USA, pp. 2–216, 2004. ISBN 9780849317040.
17. *Inductive Technology Handbook*, P/N 860214-001. Kaman Precision Products, A division of Kaman Aerospace Corp., Springs, Colorado, USA. Available at (7/23/2013 retrieved): www.kamansensors.com.
18. Inductive (eddy current) displacement, SensorCentral.Com. Available at: www.sensorcentral.com/displacement/laser03.php.

19. *Eddy Current Distance and Displacement Transducer, Product Handbook*, WayCon, Germany. p. 3. Available at: www.waycon.de/fileadmin/pdf/Eddy_Current_Probe_TX.pdf.

20. McGraw-Hill Science & Technology Dictionary: Magnetic core, www.answers.com/topic/magnetic-core#Straight_cylindrical_rod, retrieved on 7/23/13.

21. Vetterlein, J. et al., Eddy current testing at high temperatures for controlling heat treatment processes, *International Symposium on Non-Destructive Testing in Civil Engineering*. 2003 (NDT-CE 2003), September 16–19, 2003, Berlin, Germany, p. 22.

22. Probes—mode of operation, NSF-NDT Resource Center. Available at: www.ndt-ed.org/EducationResources/CommunityCollege/EddyCurrents/ProbesCoilDesign/ProbesModeOp.htm.

23. Inductive eddy current sensors for force and stress measurements, *Chen Yang Technologies GMBH & Co. KG*, Finsing, Germany, 2014. Available at: www.chenyang-ism.com/EddyCurrent.htm.

24. Randall, H.F. and James, A., Statistical approach to the verification of thread holes. *Sensors*, 20(3), 28–32, 2003.

25. Maugh, T., Victor Vacquier Sr. dies at 101; geophysicist was a master of magnetics. *Los Angeles Times*, January 24, 2009.

26. *Synchro/Resolver Conversion Handbook*, 4th ed., Data Device Corporation, New York, USA, pp. 5–6.

27. AC LVDT Displacement Transducers Series S and M, WUNTRONIC GmbH, Germany. Available at: www.wuntronic.du/displacement/acdispac.htm.

28. Ripka, P., Primdahl, F., Nielsen, O.V., Petersen, J.R., and Ranta, A., A.C. magnetic-field measurement using the fluxgate, *Sens. Actuators A* 46–47, 307–311, 1995.

29. Dottorato di, T.D. and Marchesi, M., *Fluxgate Magnetic Sensor System for Electronic Compass*, University of Degli Studi Di Pavia, Italy, pp. 25–26, 2013.

30. *Pancake Resolvers Handbook*, Axsys Technologies, Inc., San Diego, California, USA. pp. 12–14, 2005.

31. Evans, K., Fluxgate magnetometer explained, INVASENS, Gloucester, UK, pp. 1–5, 2006.

32. Moldovanu, C., Brauer, P., Nielsen, O.V., and Petersen, J.R., The noise of the Vacquier type sensors referred to changes of the sensor geometrical dimensions, *Sens. Actuators A* 81, 197–199, 2000.

33. Ripka, P., New directions in fluxgate sensors, *J. Magn. Magn. Mater.*, Vol. 215–216, 735–739, 2000.

34. Ripka, P., Review of fluxgate sensors, *Sens. Actuators A*, Vol. 33, 129–141, 1992.

35. von Flotow, A. and Drumm, M.J., Blade-tip monitoring with through-the-case eddy current sensors. *Sensors*, 21(6), 28–35, 2004.

36. Natarajan, S.P., Huffman, J., Weller, T.M., and Fries, D.P., Contact-less toroidal fluid conductivity sensor based on RF detection. *Proceedings of IEEE Sensors*, Oct. 24–27, 2004, Vol. 1, pp. 304–307.

5 Magnetic Sensors

5.1 INTRODUCTION

On the basis of sensing principles, magnetic or electromagnetic sensors can be classified as:

- *Hall sensors*: operate based on the *Hall effect*—when current flows through a thin flat conductor in a magnetic field, a voltage will build up on the sides of the conductor. Hall sensor examples include Hall current sensors, Hall magnetic sensors, and Hall position sensors.
- *Magnetoresistive sensors*: operate based on the *magnetoresistive (MR) effect*—a current carrying magnetic material changes its electric resistance in the presence of an external magnetic field. For example, anisotropic magnetoresistive (AMR) and giant magnetoresistive (GMR) sensors are MR sensors.
- *Magnetostrictive/Magnetoelastic sensors*: operate based on the *magnetostrictive effect*—ferromagnetic (FM) materials (e.g., iron, nickel, and cobalt) exhibit a change in size and shape resulting from magnetization change, e.g., magnetoelastic torque sensors, magnetostrictive force sensors, and Terfenol-D stress sensors.
- *Nuclear magnetic resonance (NMR)/Magnetic resonance imaging (MRI) sensors*: operate based on the *nuclear magnetic resonance*—micromagnets (generated due to nuclei's spins) line up themselves in one direction and create a detectable magnetic field when a strong and uniform external magnetic field is around (e.g., resonant magnetic field sensors and MRI sensors).
- *Barkhausen sensors*: operate based on *Barkhausen effect* that relates magnetism to acoustics (e.g., impact toughness testers and Barkhausen stress sensors).
- *Wiegand sensors*: operate based on *Wiegand effect*—when the strength of a magnetic field reaches a magnetic threshold, a Wiegand wire (made of an alloy of cobalt, iron, and vanadium) nearby will switch its polarity (e.g., Wiegand speed sensors).
- *Magneto-optical sensors*: operate based on *Magneto-optical Kerr effect* (MOKE)—the light that is reflected from a magnetized surface can change in both polarization and reflected intensity—the interaction of light with a magnetic system (e.g., magneto-optic disk readers and Faraday current sensors).
- *Superconducting quantum interference devices (SQUIDs)*: operate based on *Meissner* or *Meissner–Ochsenfeld effect*—magnetic fields are expelled from superconducting materials when superconducting materials are cooled below their *critical* or *transition temperature* T_c (e.g., SQUID battery monitors and SQUID biomagnetic sensors).

TABLE 5.1

Magnetic Field Detection Ranges of Inductive and Magnetic Sensors

Inductive/Magnetic Sensors	Detectable Magnetic Field (Tesla)
	10^{-14} 10^{-10} 10^{-6} 10^{-2} 10^{2}
Coil sensors	├———————————————┤
Fluxgate sensors	├————————————┤
Hall effect sensors	├————————┤
Magnetoresistive sensors	├——————————┤
Magnetoimpedence sensors	├——————————┤
Nuclear magnetic resonance sensors	├————————————┤
Magneto-optical sensors	├————————┤
SQUIDs	├————┤

Table 5.1 summarizes the measurement ranges of inductive and magnetic sensors in detecting magnetic fields (in Tesla).

5.2 HALL SENSORS

5.2.1 HALL EFFECT

When a current flows through a thin flat conductor placed in a magnetic field, the magnetic field exerts a transverse force (i.e., Lorentz force) on the moving charge carriers and pushes them to one side of the conductor. The charge then builds up and forms a measurable voltage between the two sides of the conductor (see Figure 5.1). This voltage is called *Hall voltage*, V_H, named after Edwin Hall who discovered this phenomenon in 1879 [1]. V_H can be described by

$$V_H = \frac{IB}{qN_c t_h} \tag{5.1}$$

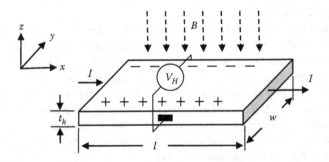

FIGURE 5.1 Hall effect.

where I is the current (in A), B is the magnetic field density (in T), q is the charge of an electron (1.602×10^{-19} C), N_C is the number of charge carriers per cubic meter (in m^{-3}, called *Charge Carrier Density*), and t_h is the thickness of the conductor (in m). The Hall voltage V_H is directly proportional to the current I and the magnetic field B, and inversely proportional to the thickness of the conductor t_h. If the applied magnetic field is along the z-axis, perpendicular to the current in the x-axis, and then the Hall voltage is measured in the y-axis.

The inverse of qN is called *Hall coefficient*, denoted as C_H (in $m^3 \cdot C^{-1}$ or $m^3 \cdot A^{-1} \cdot s^{-1}$):

$$C_H = \frac{1}{qN_C} \tag{5.2}$$

EXAMPLE 5.1

In a typical Hall sensor application, $I = 1$ mA, $B = 0.1$ T, $t_h = 10$ μm, and $C_H = 0.006$ $m^3 \cdot C^{-1}$ (or $m^3 \cdot A^{-1} \cdot s^{-1}$). Find the Hall voltage V_H.

SOLUTION

Apply Equation 5.1:

$$V_H = \frac{C_H}{t_h} IB = \frac{0.006 \, m^3 \cdot C^{-1}}{10 \times 10^{-6} \, m} (1 \times 10^{-3} \, A)(0.1 \, T) = 0.06 \, V \, or \, 60 \, mV$$

Equation 5.1 shows that the Hall voltage does not depend on the strip width, w, and the length, l, but its thickness, t_h. A thinner or smaller t_h provides a larger Hall voltage. However, a thin plate limits the maximum allowed current I, while I depends on the plate cross-sectional area $w \cdot t_h$ and the current density, ΔI, that is,

$$I = \Delta I w t_h \tag{5.3}$$

Therefore, in Hall sensor design, there is a tradeoff between choosing a small thickness t_h while still maintaining enough current I to flow through the sensor.

Besides a simple and popular rectangular shape, a Hall sensor may have other shapes. Theoretically, the characteristics of a Hall sensor should not depend on its shape. In reality, the geometry of a Hall sensor does influence the Hall voltage output. A geometry factor, G_H, is introduced to represent this influence:

$$V_H = G_H \frac{C_H}{t_h} IB \tag{5.4}$$

Ideally, $G_H = 1$. In real Hall devices, G_H has values between 0.7 and 0.9.

EXAMPLE 5.2

A Hall sensor is made of a copper foil. Its Hall coefficient is 7.42×10^{-11} $m^3 \cdot C^{-1}$, thickness is 25 μm, and geometry factor is 0.8. If a 1 A current flows through the foil and the applied magnetic field is 1 T, find the Hall voltage.

SOLUTION

Apply Equation 5.4:

$$V_H = G_H \frac{C_H}{t_h} IB = (0.8)\frac{7.42 \times 10^{-11}\, m^3 \cdot C^{-1}}{25 \times 10^{-6}\, m}(1\, A)(1\, T) = 2.37 \times 10^{-6}\, V$$

For a rectangular Hall element with length l, width w, and thickness t_h, its electrical resistance can be calculated using Equation 2.1:

$$R = \rho \frac{l}{A} = \frac{\rho l}{w t_h}$$

where l and A are the length and the cross-sectional area of the conductor, respectively; and ρ is *resistivity* or *specific resistance* of the material. For metals, ρ is the characteristic of the metallic material itself. For semiconductors, ρ is a function of doping and carrier mobility. Carrier mobility measures how fast the charge carriers move under an electric field, and it varies with respect to the type of semiconductor, the dopant concentration level, the carrier type, and temperature.

5.2.1.1 Hall Effect in Metals

In metals, the number of charge carriers per cubic meter, N_C, can be calculated by

$$N_C = \frac{C_A}{M_m} D_g \tag{5.5}$$

where C_A is the *Avogadro constant* $(6.02 \times 10^{23}\, mol^{-1})$, M_m is the *molar mass* of the metal $(g \cdot mol^{-1})$, and D_g is *specific gravity* of the metal $(g \cdot cm^{-3})$.

EXAMPLE 5.3

Find the number of charge carriers per cubic centimeter N_C and Hall coefficient C_H for a 25 μm-thick copper foil. If the applied current is 1 A and the magnetic field is 1 T, find the resultant Hall voltage, V_H, on the foil. (The molar mass of copper is 63.55 $g \cdot mol^{-1}$ and the specific gravity of copper is 8.89 $g \cdot cm^{-3}$.)

SOLUTIONS

Apply Equation 5.5:

$$N_C = \frac{C_A}{M_m} D_g = \frac{6.02 \times 10^{23}\, mol^{-1}}{63.55\, g \cdot mol^{-1}} \times 8.89\, g \cdot cm^{-3} = 8.42 \times 10^{22}\, cm^{-3}$$
$$= 8.42 \times 10^{28}\, m^{-3}$$

Apply Equation 5.2:

$$C_H = \frac{1}{qN_C} = \frac{1}{(1.602 \times 10^{-19}\, C)(8.42 \times 10^{28}\, m^{-3})} = 7.41 \times 10^{-11}\, m^3 \cdot C^{-1}$$

Apply Equation 5.1:

$$V_H = \frac{IB}{qN_c t_h} = \frac{(1\,A)(1\,T)}{(1.602 \times 10^{-19}\,C)(8.42 \times 10^{28}\,m^{-3})(25 \times 10^{-6}\,m)} = 2.97 \times 10^{-6}\,V$$

This result shows that the Hall voltage produced by a copper foil is extremely small. For this reason, it is not practical to make Hall-effect sensors with most metals.

5.2.1.2 Hall Effect in Semiconductors

In semiconductors, the charge carrier density, N_c, is usually referred to as *carrier concentration*. A material with a lower carrier density or concentration will exhibit the Hall effect more strongly for a given current and thickness as indicated in Equation 5.1. Semiconductor materials such as silicon, germanium, and gallium-arsenide, often have much lower carrier densities (see Table 5.2) than metals (e.g., copper's free electron density is $8.42 \times 10^{28}\,m^{-3}$), thus they are often used to make Hall sensors.

Semiconductor materials are rarely used in their pure forms, but are doped with other semiconductor materials to form n-type or p-type semiconductors. This allows a choice of the predominant charge carriers (electrons or holes), unlike in metals where electrons are the only charge carriers. More sensitive Hall sensors can be made using n-type semiconductors because electrons tend to move faster than holes under a given set of conditions. In addition, for pure semiconductors, the carrier concentration highly depends on temperature, while for doped semiconductors, the carrier concentration is mostly a function of the dopant concentration, which is less temperature dependent. Thus by adding adequate dopants, one can obtain more temperature-stable Hall sensors.

EXAMPLE 5.4

A Hall sensor made from an n-type silicon has been doped to a level of $4 \times 10^{15}\,cm^{-3}$. If its thickness is 25 μm, the current is 1 mA, and the magnetic field is 1 T, find the sensor's output voltage V_H.

SOLUTION

$$V_H = \frac{IB}{qN_c t_h} = \frac{(1 \times 10^{-3}\,A)(1\,T)}{(1.602 \times 10^{-19}\,C)(4 \times 10^{21}\,m^{-3})(25 \times 10^{-6}\,m)} = 0.0624\,V = 62.4\,mV$$

TABLE 5.2

Commonly Accepted Values of Intrinsic Carrier Concentrations at 300 K

Semiconductor Material	Carrier Concentration (m^{-3})
Silicon (Si)	1.4×10^{16}–1.5×10^{16}
Germanium (Ge)	2.1×10^{19}–2.4×10^{19}
Gallium-arsenide (GaAs)	1.8×10^{12}–11.0×10^{12}

The above result shows that a doped semiconductor Hall sensor can provide 21,000 times (=0.0624/(2.97 × 10⁻⁶)) of Hall voltage as a copper sensor does. Also the bias current (1 mA) applied is only 1/1000 of the bias current (1 A) used in the copper sensor. Therefore, the doped semiconductor Hall sensors provide more practical functions than metal Hall sensors.

5.2.2   OPERATING PRINCIPLE OF HALL SENSORS

Hall sensors operate based on the Hall effect. A thin sheet of metal or semiconductor material (*Hall plate*) with a current passing through it is placed in a magnetic field, and then a voltage is generated perpendicular to the field and the direction of current flow. Since the Hall voltage V_H is small, Hall sensors require amplification and signal conditioning. A distinguished feature of a Hall sensor is that it can be entirely integrated on a single silicon chip containing the Hall sensor, amplifier and signal conditioning circuits inside. This allows the low-cost and high-volume production of Hall sensors. Thus, Hall sensors are usually surface-mount types and can be mounted on a PCB (printed circuit board).

Figure 5.2a shows the cross section of a Hall sensor and Figure 5.2b shows its IC (integrated circuit) package. It is manufactured by doping different materials into a silicon substrate to form n-type or p-type carrier regions. These n- and p-type regions are formed into geometries to create various active and passive components of the IC including the Hall element. The geometries are at 100-μm, 10-μm, or even smaller scale. The circuit density can be extremely high, allowing complex circuits to be built on a very small area of silicon.

5.2.3   CHARACTERISTICS OF HALL SENSORS

The key characteristics of Hall sensors can be described as follows [1].

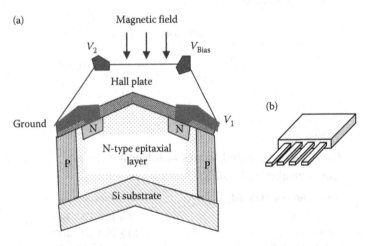

FIGURE 5.2   (a) A cross section of a Hall sensor; (b) a Hall IC sensor package.

5.2.3.1 Transfer Function

A Hall sensor's transfer function describes its input and output relationship, characterized by sensitivity, null offset, and span. Sensitivity is defined as the change in output resulting from a given change in input. Null offset is the output from a sensor with no magnetic field excitation. Span defines the output range of a sensor. Span is the difference in output voltages when the input is varied from a negative extreme to a positive extreme. Figure 5.3 illustrates the characteristics of a typical analog Hall sensor under three different supplied voltages—5, 8, and 10 V, respectively. Its transfer function can be expressed by the relationship between a magnetic field input (in Gauss, G) and a voltage output (in V) as:

$$V_{out} = (6.25 \times 10^{-4} \, V_s)B + 0.5 \, V_s \qquad (5.6)$$

Note this equation is valid only if the flux density B in the core has not reached saturation level, that is, -640 (Gauss) $< B < +640$ (Gauss). The factor of B, 6.25×10^{-4} V_s, in the equation expresses the sensitivity for the sensor, and it is the slope of each characteristic curve (a straight line in this case) in Figure 5.3. The second term in Equation 5.6, $0.5 \, V_s$, is the null offset, which is the output voltage at 0 G under a given supply voltage.

5.2.3.2 Sensitivity (or Gain)

A Hall sensor's sensitivity can be characterized in two ways

1. Volts per unit magnetic field, per unit of bias current: $V/(B \times I)$, in $V \cdot T^{-1} \cdot A^{-1}$
2. Volts per unit magnetic field, per unit of bias voltage: $V/(B \times V)$, in $V \cdot T^{-1} \cdot V^{-1}$

For example, a CMOS-compatible vertical Hall angle sensor, developed by *Microsystems Institute* in Switzerland, has a current sensitivity up to $450 \, V \cdot T^{-1} \cdot A^{-1}$ and a voltage sensitivity of $0.035 \, V \cdot T^{-1} \cdot V^{-1}$.

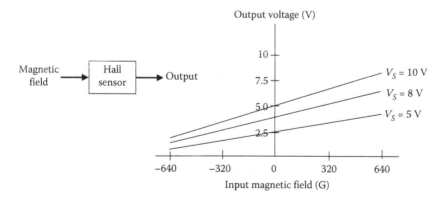

FIGURE 5.3 Characteristic curves of an analog Hall sensor under three different voltage supplies. (From *Hall Effect Sensing and Application*, Honeywell Inc., Freeport, Illinois, USA, p. 6, 2012.)

Sensitivity of a Hall sensor varies slightly with temperature. Many sensor manufacturers provide sensitivity versus temperature curves in their datasheets to indicate their products' stability with temperature variation. Signal conditioning electronics may be incorporated into Hall sensors to compensate temperature effects.

5.2.3.3 Ohmic Offset

An ohmic offset is a small voltage present in a sensor's output, even in the absence of a magnetic field. This offset appears in almost all Hall sensors. It limits the ability of a sensor in discriminating a small steady-state magnetic field. Many factors can cause ohmic offsets, including alignment error in the sensor's contacts, inhomogeneities or stresses in the sensing material. An ohmic offset can be expressed by the output voltage for given bias conditions or in terms of magnetic field units. For instance, sensor A has a 500 μV offset, while sensor B has a 200 μV offset.

5.2.3.4 Nonlinearity

Similar to many other sensors, Hall sensors are not perfectly linear over their operating ranges. They typically exhibit a nonlinearity between 0.5% and 1.5% over their operating ranges.

5.2.3.5 Input and Output Resistance and Their Temperature Coefficient

Input resistance of Hall sensors affects the design of the bias circuitry, while the output resistance affects the design of the amplifier circuitry used to detect the Hall voltage (see Figure 5.4).

The temperature coefficients of the input and output resistances of Hall sensors should be equal or very close to each other since their difference will affect the measurement accuracy.

5.2.3.6 Noise

Hall sensors also present electrical noise at their outputs, mainly *Johnson noise* and sometimes *flicker noise* (1/*f* noise). Johnson noise occurs in all conductive materials due to the random and thermally induced motion of electrons through a conductor (also called *thermal noise*). 1/*f* or flicker noise is more significant when detecting DC or near DC low-frequency signals. It depends on the sensing material used and the fabrication processes.

Johnson noise limits how small a sensor signal can be recovered from its output. It can be minimized by choosing a low impedance sensor.

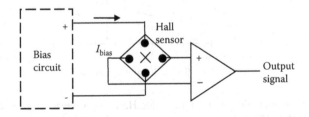

FIGURE 5.4 A simple Hall sensor system.

TABLE 5.3
Key Features of the SIEMENS KSY-46 Hall Sensor

Features	Value
Nominal supply current	7 mA
Open-circuit sensitivity	$150 - 265$ V $\cdot$ A^{-1} $\cdot$ T^{-1}
Open-circuit Hall voltage at $B = 0$ T	$105 - 185$ mV
Ohmic offset voltage at $B = 0$ T	$\leq \pm 15$ mV
Nonlinearity of Hall voltage:	
at $B = 0$–0.5 T	$\leq \pm 0.2\%$
at $B = 0$–1.0 T	$\leq \pm 0.7\%$
Input resistance at $B = 0$ T	600–900 Ω
Output resistance at $B = 0$ T	1000–1500 Ω
Temperature coefficient of the internal resistance at $B = 0$ T	$\sim 0.3\%$ K^{-1}
Temperature coefficient of Ohmic offset voltage at $B = 0$ T	$\sim 0.3\%$ K^{-1}
Noise figure	~ 10 dB
Operating temperature	$-40°C \sim +150°C$

Table 5.3 shows the key features of a SIEMENS KSY-46 Hall sensor. This sensor is made of monocrystalline GaAs material and is built on a surface-mount package. It operates with a constant current and can be used for magnetic field detection, current and power measurement, rotation and position sensing, and DC motor control.

Keep the following things in mind when using Hall sensors:

1. Using a voltage drive rather than a current drive can reduce the temperature effect on the Hall output voltage V_H, for example, from $-2.0\%°C^{-1}$ to $\pm 0.1 - 0.2\%°C^{-1}$ at room temperature.
2. Constant voltage bias has a larger signal-to-offset ratio than constant current bias.
3. Dual or quad Hall elements provide a higher magnetic sensitivity and smaller offsets.
4. Hall voltage V_H is very small, so use high amplification for Hall sensors. The DC offset of an amplifier often limits the usefulness of the V_H output.
5. The use of twisted or shielded cables can attenuate influence of static and dynamic electric and magnetic fields.
6. A Hall sensor's bandwidth is limited by magnetic core losses and eddy current-induced temperature rise.
7. Hall sensors should be orientated correctly.

5.2.4 Types and Design of Hall Sensors

On the basis of the Hall effect Equation 5.1, any changes in parameters, C_H, t_h, I, and B will cause the Hall voltage V_H to change. C_H and t_h are often fixed once the Hall element dimensions and material are chosen, thus a Hall sensor's output will depend

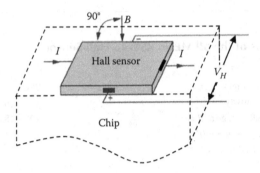

FIGURE 5.5　Traditional Hall element built in the plane of the chip surface.

on either I or B. Fixed current I, variable flux density B or vice versa allows various Hall sensors to be designed.

Hall sensors have many configurations: vertical, cylindrical, multiaxis, dual, or quad. Traditional Hall elements are built in the plane of the chip surface as shown in Figure 5.5. In this situation, the Hall element is only sensitive to the magnetic field that is perpendicular to the chip surface. The contacts for the input current and Hall output voltage are on the different faces.

5.2.4.1　Vertical Configurations

A vertical Hall sensor (VHS), as shown in Figure 5.6, responds to the magnetic field parallel to the plane of the chip and allows all electrical contacts to be placed on the top surface of the chip so that it can be easily manufactured using microelectronic fabrication processes. The Hall voltage is measured between the contacts. This configuration greatly improves both sensitivity (about 10 times greater than that of conventional Hall devices) and V_H output (about 20 times greater than that of standard Hall sensors).

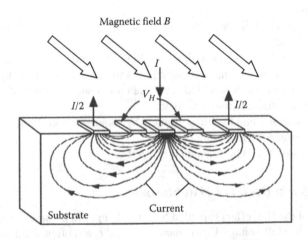

FIGURE 5.6　A vertical Hall sensor.

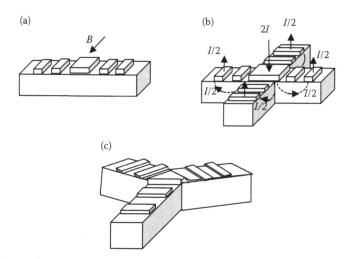

FIGURE 5.7   Three configurations of VHSs: (a) 1D VHS; (b) 2D VHS; (c) Three-branch VHS. (From Schott, C., Burger, F., Blanchard, H., and Chiesi, L., Modern integrated silicon Hall sensors, *Sensor Review*, 18(4), 252–257, 1998.)

Figure 5.7 shows three configurations of VHSs for different applications [2]. (a) is a basic one-dimensional (1D) VHS; (b) consists of two VHSs laid in a cross shape for measuring magnetic fields in both x and y (2D) directions; and (c) is a three-branch VHS for three-phase brushless micromotor control.

5.2.4.2   Cylindrical Configurations

If the magnetic field being measured has a circular geometry, as around a current carrying wire, a cylindrical configuration provides the best solution. Figure 5.8a shows a cylindrical Hall sensor (CHS) made by a conformal deformation of a vertical Hall device [3]. It can measure a circular magnetic field around a current-carrying wire or around an air gap of two FM flux concentrators as shown in Figure 5.8b. The

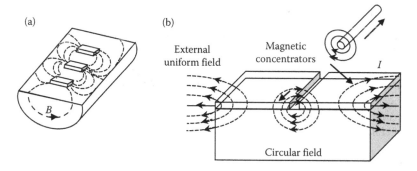

FIGURE 5.8   (a) A cylindrical Hall sensor; (b) applications of the cylindrical sensor in measuring a circular magnetic field. (From Blanchard, H., Chiesi, L., Racz, R., and Popovic, R.S., Cylindrical Hall device, EPFL-*Swiss Federal Institute of Technology*, Lausanne, Switzerland. IEDM 96 pp. 541–544.)

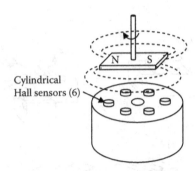

FIGURE 5.9 Operating principle of a magnetic angular position sensor.

sensitivity of this cylindrical Hall device can be as high as 2000 V · A⁻¹ · T⁻¹ due to the integrated magnetic flux concentrators. Its resolution is 70 nT.

Figure 5.9 shows a magnetic angular position sensor and its operating principle. It contains six high-sensitivity CHSs mounted on a ceramic substrate.

A cylindrical Hall device equipped with FM field concentrators is very sensitive. However, its sensitivity decreases when the field value reaches the saturation value of the concentrator material (e.g., 20 mT). The drawback of a CHS is its nonlinearity.

5.2.4.3 Multiaxis Configurations

A multiaxis (2D or 3D) Hall sensor is used to measure more than one component of a magnetic field simultaneously. Such a sensor can be achieved by merging two or three mutually orthogonal vertical Hall devices.

Hall sensors that contain two or four Hall elements in one package are called *dual* or *quad* Hall sensors, respectively. This arrangement makes Hall sensors more stable and predictable than using several individual Hall sensors. It also minimizes the effects of mechanical or thermal stress on the output. In addition, its rail-to-rail operation over a full voltage range provides a more usable signal with higher accuracy.

5.2.5 Applications of Hall Sensors

Hall sensors are widely used in automobile, security, brushless DC motors, damper control, various instrumentation, or any applications that involve electric current or magnetic field measurements. Table 5.4 lists the main applications of Hall sensors. High-quality Hall sensors can be constructed inexpensively with the standard IC processes used in the microelectronics industry, and integrate ancillary signal-processing circuitry on the same silicon die. Following are some application examples of Hall-effect sensors.

5.2.5.1 Hall Position Sensor

A Hall position sensor designed by the *Sensing and Control* group, *Honeywell*, Freeport, Illinois, USA, is shown in Figure 5.10a. It consists of a magnet and a Hall sensor—both are mounted rigidly in a fixture made of a nonmagnetic material. A ferrous vane can move in and out through the gap, which alters the magnetic flux lines

TABLE 5.4
Main Applications of Hall Sensors

Magnetic flux density/ magnetic field	Flow rate
	Temperature
Current	Torque
Pressure diaphragms	Encoded switches
Position	Voltage regulation
Occupancy	Ferrous metal
Proximity	Vibration
Sequence	Magnetic toner density
Speed/Velocity	Tachometers
	Computer keyboards

in the gap. The Hall sensor then detects the presence, absence, or position of the vane. Figure 5.10b shows a Hall angular speed sensor designed by Allegro Microsystems, LLC, Worcester, Massachusetts, USA.

5.2.5.2 Hall Current Sensor

Hall sensors measure current via the intensity of the magnetic field generated by current flow. A larger current produces a stronger magnetic field. A Hall sensor's output voltage is therefore directly proportional to the current. Hall sensors can measure both AC and DC currents, and pulsed waveforms (e.g., PWM—pulse width modulation signals). Figure 5.11 illustrates a Hall current sensor. A C-core of soft magnetic material is placed around a conductor to concentrate the field. The Hall sensor, placed in the small air gap, delivers a voltage that is proportional to the current in the conductor. Hall current sensors are usually surface-mount types that can be mounted on a PCB to measure the current in the traces. Hall current sensors have advantages

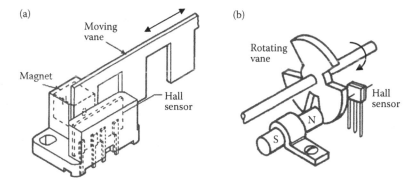

FIGURE 5.10 (a) A Hall position sensor (From *Hall Effect Sensing and Application*, Honeywell Inc., Freeport, Illinois, USA, Page 33. 2012.); (b) a Hall angular speed sensor. From Dooge, M. and Thomas, M., Integrating Hall-effect magnetic sensing technology into modern household appliances, Allegro Microsystems, LLC, Worcester, Massachusetts, USA, p. 2, 2013.)

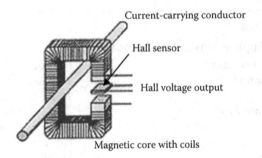

FIGURE 5.11   A Hall current sensor.

of maintaining galvanic isolation between the sensor and the measuring circuits, and measuring the current without interrupting the circuit.

5.2.5.3 Door Security System

Figure 5.12 shows a door interlock security system. It consists of a Hall sensor, a magnetic card, an actuator, and a circuit. When the keycard slides by, the Hall sensor detects the magnetic embedded in the card, and then sends an output signal to the microprocessor where the analog signal is converted into digital pulse to pull in the relay to open the door. If additional security is required, a series of magnets

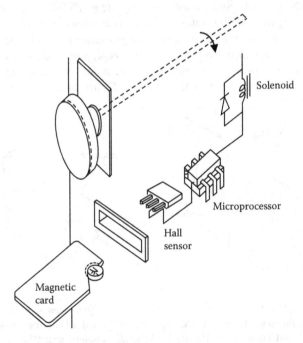

FIGURE 5.12   A door interlock security system. (From *Hall Effect Sensing and Application*, Honeywell Inc., Freeport, Illinois, USA, p. 74, 2012.)

(form a magnetic pattern) can be molded into the card, and an array of Hall sensors or quad Hall sensors can be used to generate a series of pulses to trigger the solenoid.

5.2.5.4 Flow Rate Meter

Hall sensors can also be used to measure flow rate (see Figure 5.13). Two Hall sensors measure the rotational velocity of the metal rotor that relates to the flow rate of a liquid transmitted through a pipe. The rotor is modified to have four recessed areas that have magnetic properties on their surfaces. The two Hall sensors pickup the magnetic field changes due to the rotor's rotation, and the signals are converted into the flow rate.

5.2.5.5 Motor Control

Hall sensors are broadly used in brushless DC motor control due to their following features:

- Fast response time (<5 μs)
- Capability to detect high speed (theoretically can detect maximum of 16×10^5 rpm)
- Over-current sensing and stepper motor stall detection
- Support for closed-loop regulation

Hall sensors can provide two-phase, three-phase, and quad-phase motor control, implemented by using either two or more discrete Hall sensors or a single multiaxis Hall sensor. For two-phase brushless motor control, one can use two individual Hall sensors that are oriented 90° apart (or a single two-axis Hall sensor) on the end of a brushless DC motor and a magnet attached on the end of the motor shaft. The advantage of using a 2-axis Hall sensor, instead of two discrete Hall sensors, is to eliminate the physical mounting tolerances of the discrete Hall sensors.

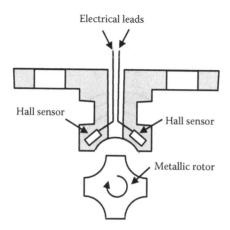

FIGURE 5.13 Flow-rate meter using two Hall sensors. (Based on US Patent: 6,397,686 B1.)

EXAMPLE 5.5

An electromagnetic system (shown in Figure 5.14) consists of a soft magnetic toroidal core with an air gap and a coil wrapped around the core. Design a linear
Hall current sensor that has a transfer function is shown in Equation 5.6 for this
magnetic system, so that the voltage output of the Hall sensor, positioned in the
air gap, is proportional to the flux density B at that point.

SOLUTIONS

If the air gap is narrow (compared to the cross-sectional area of the core), the flux
can be considered to be concentrated in the air gap, and the flux density in the
core should be equal to that in air. Applying Ampere's law yields

$$B_a = \frac{0.4\pi\mu_c NI}{l_c + \mu_c l_a} \tag{5.7}$$

or

$$B_a = C_M I \left(\text{where } C_M = \frac{0.4\pi\mu_c N}{l_c + \mu_c l_a} \right)$$

where B_a is the flux density (in Gauss, G) in the air gap; μ_c is the relative permeability of the core; l_a is the length of the air gap in centimeters; l_c is the mean length
of the core in centimeters; I is the current flowing through the coil with N turns.

Note that this equation is valid only if the flux density B_a, has not reached the
core's saturation flux density, B_{sat} (= ±640 Gauss), that is,

$$-640 \text{ (Gauss)} < B_a < +640 \text{ (Gauss)}.$$

For reliability, we choose: -400 (Gauss) $< B_a < +400$ (Gauss).
Plug B_a from Equation 5.7 into the linear output Hall sensor's transfer function:

$$V_{out} = (6.25 \times 10^{-4} V_S)B + 0.5V_S = (6.25 \times 10^{-4} V_S)\frac{0.4\pi\mu_c NI}{l_c + \mu_c l_a} + 0.5V_S$$

Thus, the design of the Hall current sensor becomes to

1. Choose a core such that B_{sat} is much greater than 400 Gauss;
2. Choose l_a and N so that $C_M I_{max} < 400$ Gauss and $C_M I_{min} > -400$ Gauss;

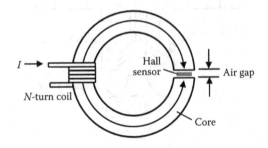

FIGURE 5.14 A magnetic system.

3. Choose V_S so that C_M (6.25×10^{-4} V_s) yields the desired overall sensitivity if no additional output signal conditioning circuit is to be used.

5.3 MAGNETORESISTIVE SENSORS

5.3.1 MAGNETORESISTANCE EFFECTS

In magnetoresistance (MR) effect, a conductor changes its electric resistance in the presence of an external magnetic field. The MR effect was discovered by William Thomson (*Lord Kelvin*) in 1856 [4]. He found that iron and nickel exhibited a small increase in electrical resistance along the direction of an applied magnetic field and a similar decrease in resistance in the transverse direction. The magnitude of the effect, M_R (unitless), depends on the material of the conductor as

$$M_R = \frac{\rho_B - \rho_0}{\rho_0} \times 100\% \tag{5.8}$$

where ρ_B (in $\Omega \cdot m$) is the resistivity of the conductor under an applied magnetic field B, and ρ_0 (in $\Omega \cdot m$) is the resistivity of the conductor without an applied magnetic field B. The value of the effect M_R is in the order of few percentage points.

EXAMPLE 5.6

Under an applied magnetic field $B = 0.75$ T, an iron (Fe) conductor changes its resistivity from $\rho_0 = 10$ ($\mu\Omega \cdot cm$) to $\rho_{0.75} = 10.53$ ($\mu\Omega \cdot cm$); while a nickel (Ni) conductor (with the same size and shape as the iron conductor) changes its resistivity from $\rho_0 = 6.99$ ($\mu\Omega \cdot cm$) to $\rho_{0.75} = 7.07$ ($\mu\Omega \cdot cm$). Which material displays a stronger magnetoresistance effect?

SOLUTION

For Fe:

$$M_R = \frac{10.53 - 10}{10} = 0.053 \quad \text{or } 5.3\%$$

For Ni:

$$M_R = \frac{7.07 - 6.99}{6.99} = 0.011 \quad \text{or } 1.1\%$$

Thus, iron demonstrates a stronger MR effect.

It was not until recent decades that advances in microelectronic thin film techniques (can achieve as thin as 10–50 nm) and the utilization of *anisotropic magnetoresistive* (AMR) materials have allowed practical applications of MR effect. The discovery of *giant magnetoresistance* (GMR) by Albert Fert and Peter Grünberg in 1988 raised MR sensors' maximum M_R value from 2–5% to 10% or more [4,5], a feat honored by the 2007 Nobel Prize in Physics.

On the basis of the materials and structures, MR effects can be classified as [6]:

1. *Ordinary magnetoresistance* (OMR) effect in nonmagnetic metals;
2. *Anisotropic magnetoresistance* (AMR) effect in ferromagnetic alloys;
3. *Giant magnetoresistance* (GMR) effect in multiple alternating ferromagnetic-alloy and metallic layer structures;
4. *Tunneling magnetoresistance* (TMR) effect in multiple alternating ferromagnetic-alloy and thin-insulating layer structures;
5. *Ballistic magnetoresistance* (BMR) effect in multiple alternating ferromagnetic-alloy-layer and nonferromagnetic-point structures; and
6. *Colossal magnetoresistance* (CMR) effect in $La_{1-x}M_xMnO_{3+\delta}$ (M = Ca or Sr) perovskite structures.

The following sections discuss each of these effects and their characteristics with a primary emphasis on AMR and GMR effects due to their practical and wide applications.

5.3.1.1 Ordinary Magnetoresistance (OMR) Effect

OMR effect is present in normal (nonmagnetic) metals. It arises from the effect of Lorentz force acting on an electron in a magnetic field, causing a circular or helical motion of the electron. The resistivity ρ (in $\Omega \cdot m$) to determine the M_R value for the OMR effect is [7]:

$$\rho = \frac{B}{nqc\omega_c\tau_r} \tag{5.9}$$

where B is the applied magnetic field strength (in T); n is the *electron density* of the metal (in m^{-3}); q is the electron charge (1.602×10^{-19} C); c is the speed of light (3×10^8 m $\cdot$ s^{-1}); ω_c is the *cyclotron frequency* (in rad $\cdot$ s^{-1}); and τ_r is the *electron relaxation time* (the mean time between collisions, in s). ω_c can be found by

$$\omega_c = \frac{qB}{m^*c} \tag{5.10}$$

where m^* is the *effective mass* of an electron (in kg). Thus, the resistivity becomes

$$\rho = \frac{m^*}{nq^2\tau_r} \tag{5.11}$$

In Equation 5.9, $\omega_c\tau_r$ is the dominant factor for the OMR effect. In metals, such as Cu, Ag, and Au, the dependence of $\omega_c\tau_r$ value on magnetic field can be approximately described by $\omega_c\tau_r \approx 0.005B$. Thus, their OMR value is smaller than 1% under 1 T. To have a substantial M_R, $\omega_c\tau_r$ should be at least of order 1. Research has found that for some metals (e.g., Bi) the $\omega_c\tau_r$ value can be increased up to 100 times [7].

EXAMPLE 5.7

A nonmagnetic metallic material has the electron density $n = 6 \times 10^{28}$ m^{-3}, the electron relaxation time $\tau_r = 3 \times 10^{-14}$ s, and the effective mass of $m^* = 0.13\ m_0$ ($m_0 = 9.109 \times 10^{-31}$ kg is the free-electron mass). (1) Calculate the material's resistivity. (2) Find its $\omega_c \tau_r$ value under an applied magnetic field $B\ (= 2$ T).

Solutions

1. The material's resistivity is

$$\rho = \frac{m^*}{nq^2\tau_r} = \frac{(0.13)(9.109 \times 10^{-31}\ \text{kg})}{(6 \times 10^{28}\ \text{m}^{-3})(1.602 \times 10^{-19}\ \text{C})^2 (3 \times 10^{-14}\text{s})}$$
$$= 2.56 \times 10^{-9} (\Omega \cdot \text{m})$$

2. From Equation 5.9, the $\omega_c \tau$ value is

$$\omega_c \tau_r = \frac{B}{nqc\rho} = \frac{2\text{T}}{(6 \times 10^{28}\text{m}^{-3})(1.602 \times 10^{-19}\text{C})(3 \times 10^8 \text{m})(2.56 \times 10^{-9}\Omega \cdot \text{m})}$$
$$= 2.71 \times 10^{-10}$$

For nonmagnetic metals, the MR effect under low magnetic fields is very small, although the effect can become larger for higher magnetic fields. The change in resistivity, $\Delta \rho$, is positive for the magnetic fields both parallel ($\rho_{//}$) and transverse ($\rho_\perp$) to the current direction with $\rho_\perp > \rho_{//}$.

5.3.1.2 Anisotropic Magnetoresistance (AMR) Effect

AMR, discovered in 1857 by William Thomson, is a typical effect in ferromagnetic (FM) materials. The term *anisotropic* is from the dependence of the resistivity on the orientation of the magnetic field relative to the current direction Θ. Mathematically it is expressed by:

$$\rho(\Theta) = \rho_0 + \Delta \rho \cos^2 \Theta \tag{5.12}$$

where ρ_0 is the resistivity of the material without an applied magnetic field, and $\Delta \rho\ (= \rho(0°) - \rho(90°))$ is the resistivity difference between the parallel ($\Theta = 0°$) and perpendicular ($\Theta = 90°$) relationship between the current direction and the applied magnetic field direction. Thus, the resistance is at maximum when the current and the magnetic field directions are parallel and is at minimum when their directions are perpendicular. The AMR effect reflects the change of electron scattering in the atomic orbitals due to a magnetic field.

To understand AMR sensors better, consider a strip of ferromagnetic material, *permalloy* (Ni$_{81}$Fe$_{19}$), as shown in Figure 5.15. During the deposition process of the strip manufacturing, a strong magnetic field is applied to magnetize and define the *preferred (internal) magnetization direction* (parallel to the length of the trip—the x-axis) for the strip. After the manufacturing process is finished,

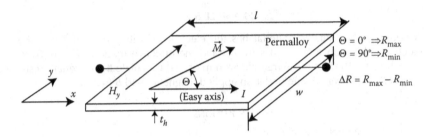

FIGURE 5.15   The MR effect in permalloy.

this magnetic field is removed, and the strip (sensor) will maintain its magnetization in the x direction (called *easy axis*) due to its ferromagnetic properties. If no
external magnetic field is present, the sensor's magnetization vector $\vec{M}$ is in its
preferred direction and the angle between $\vec{M}$ and the sensor's current I (also in
the x-direction) is zero (i.e., $\Theta = 0°$). In this case, the sensor's resistance R is at
its maximum value, R_{max}. If a magnetic field to be measured, H_y, is applied in the
y-direction, the sensor's magnetization vector $\vec{M}$ will be affected by H_y and rotate
an angle Θ. If H_y is very strong, $\vec{M}$ tends to align itself parallel to H_y and $\Theta \approx 90°$.
In this case, R reaches its minimum value, R_{min}. Thus, the angle Θ will be directly
related to the magnetic field strength of H_y.

The dependence of the sensor resistance R on the angle Θ is described by [8]:

$$R(\Theta) = \underbrace{\rho_\perp \frac{l}{wt_h}}_{R_0} + \underbrace{(\rho_{//} - \rho_\perp)\frac{l}{wt_h}\cos^2\Theta}_{\Delta R} = R_0 + \underbrace{(R_{max} - R_{min})}_{\Delta R}\cos^2\Theta \quad (5.13)$$

where $\rho_\perp$ and $\rho_{//}$ are the resistivities when $\vec{M}$ is perpendicular and parallel to the
easy axis, respectively; l is the length of the strip, w is its width, and t_h is its thickness. In general, $l > w \gg t_h$. R_0 (=R_{min}) is the resistance when $\vec{M}$ is perpendicular
to the easy axis. Both R_0 and ΔR are material and geometry dependent. For the
permalloy, ΔR is in the range of 2–3% of R_0. The AMR effect can be evaluated by
the quotient $(\rho_{//} \rho_\perp)/\rho_\perp = \Delta\rho/\rho_\perp$ as described in Equation 5.8. The typical M_R value
of the AMR effect is 1% at room temperature. Equation 5.13 can also be written as

$$R(\Theta) = R_{min} \sin^2\Theta + R_{max}\cos^2\Theta \quad (5.14)$$

The angle Θ directly relates to the strength of the external magnetic field H_y:

$$\cos^2\Theta = 1 - \left(\frac{H_y}{H_{max}}\right)^2 \quad \text{or} \quad \sin^2\Theta = \left(\frac{H_y}{H_{max}}\right)^2 \quad (5.15)$$

Thus,

$$R(H_y) = R_0 + (R_{max} - R_{min})\left[1 - \left(\frac{H_y}{H_{max}}\right)^2\right] \quad (H_y \le H_{max}) \qquad (5.16)$$

where H_{max} is the maximum field strength that an AMR sensor can sense before saturation, and it is a parameter of material and geometry of the sensor. For $H_y > H_{max}$, R equals R_0.

EXAMPLE 5.8

Find the resistance change of an AMR sensor when an external magnetic field strength $H_y = 3$ kA · m^{-1} is applied perpendicular to the AMR's easy axis, knowing $H_{max} = 10$ kA · m^{-1}, $R_{max} - R_{min} = 400$ Ω.

SOLUTION

When no external magnetic field is applied ($H_y = 0$):

$$R(0) = R_0 + (R_{max} - R_{min})\left[1 - \left(\frac{0}{H_{max}}\right)^2\right] = R_{max}$$

When the external magnetic field is applied ($H_y \ne 0$):

$$R(H_y) = R_0 + (R_{max} - R_{min})\left[1 - \left(\frac{H_y}{H_{max}}\right)^2\right] = R_{min} + (400\ \Omega)\left[1 - \left(\frac{3 \times 10^3\ A \cdot m^{-1}}{10 \times 10^3\ A \cdot m^{-1}}\right)^2\right]$$

$$= R_{min} + 364\ \Omega$$

The resistance change is:

$$R(0) - R(H_y) = R_{max} - (R_{min} + 364\ \Omega) = 400\ \Omega - 364\ \Omega = 36\ \Omega$$

5.3.1.3 Giant Magnetoresistance (GMR) Effect

Baibich et al. [9] and Binasch et al. [10] are the first who reported "Giant" magnetoresistance measured on Fe/Cr/Fe thin multilayers. They demonstrated that the electric current was strongly influenced by the relative orientation of the magnetizations of the magnetic layers. The cause of this giant change in resistance is attributed to the scattering of the electrons at the layers' interfaces. Thus, any structure with metal–FM interfaces is a candidate to display the GMR effect. Since then, a huge effort has been carried out on improving structures to maximize the effect. Today above 200% of M_R value can be achieved at room temperature, which is much higher than the M_R value of either OMR or AMR effect.

Figure 5.16, taking a three-layer structure (FM-Metal-FM) as an example, shows the characteristics of a GMR sensor under various magnetic field strengths. A big

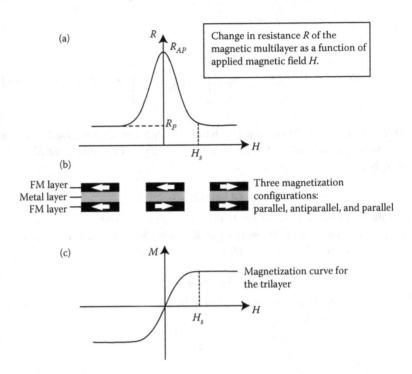

(a)

Change in resistance R of the magnetic multilayer as a function of applied magnetic field H.

(b)

FM layer
Metal layer
FM layer

Three magnetization configurations: parallel, antiparallel, and parallel

(c)

Magnetization curve for the trilayer

FIGURE 5.16 Characteristics of a three-layer GMR sensor. (From Tsymbal, E.Y. and Pettifor, D.G., Perspectives of giant magnetoresistance, in *Solid State Physics*, 56, Academic Press, London, 113–237, 2001.)

change in resistance R of the trilayer occurs when the magnetization alignment changes from parallel to antiparallel, and from antiparallel to parallel (indicated by the arrows). H_S is the saturation field. R_P and R_{AP} are the resistance in the parallel and antiparallel configuration, respectively. Figure 5.16c is the magnetization curve for the trilayer [4]. The GMR effect in % can be found by

$$M_R \text{ (for GMR)} = \frac{R_{AP} - R_P}{R_P} \times 100\% \qquad (5.17)$$

The orientation of the magnetization of the FM layers (indicated by the arrows) can be changed from antiparallel to parallel, or from parallel to antiparallel, by applying an external magnetic field. The change in electrical resistance can be made to over 200% of M_R, depending on the number and thickness of layers, materials, and manufacturing processes.

5.3.1.4 Tunneling Magnetoresistance (TMR) Effect

If the non-FM layer in a GMR sensor is replaced by a thin insulating layer (usually aluminum oxide), the TMR effect can be observed. This insulating layer is so thin that electrons can "tunnel through" the barrier if a bias voltage is applied between the two metal electrodes. The TMR effect was first investigated by Julliere [11] and Maekawa [12]. Tunneling conductance or the TMR effect depends strongly on the relative *densities of states* (DOS) of spin species (e.g., electrons) at the Fermi energy level. Julliere was the first to describe the cause of the TMR effect: the spin splitting of the Fermi level in the magnetic metals causes an unequal distribution of up- and down-spin electron states. In the classical model of tunneling, the overall tunneling conductance is proportional to the product of the densities of states of the two FM layers [13]. Assume that spin is conserved in the tunneling process, then:

$$M_R(\text{for TMR}) = \frac{\Delta R}{R} = \frac{R_{AP} - R_P}{R_{AP} + R_P} = \frac{2P_1P_2}{1 + P_1P_2} \qquad (5.18)$$

where R_P and R_{AP} are the electrical resistances of the tunnel junction (insulating layer) when the magnetic moments of the adjacent electrode layers are aligned parallelly and antiparallelly to each other, respectively. P_i ($i = 1, 2$) is the spin polarization of the FM electrode layer i defined as

$$P_i = \frac{N_i^\uparrow - N_i^\downarrow}{N_i^\uparrow + N_i^\downarrow} \quad (i = 1,2) \qquad (5.19)$$

N_i is the density of states of the electrode layer i defined at the Fermi level. Arrow $\uparrow$ represents the spin-up electrons, whereas $\downarrow$ represents the spin-down electrons. Figure 5.17 illustrates the mechanism of a TMR sensor.

One of the most promising TMR structures is Fe/MgO/FeCo, where very large M_R values have been found at room temperature [14]. The typical magnitude of the effect is near 100%. The importance of the TMR effect lies in its applications such as mesoscopic scale magnetic sensors and magnetic random-access memory (MRAM). TMR sensors are also used to determine the spin polarization P at Fermi level as indicated in Equations 5.19.

5.3.1.5 Ballistic Magnetoresistance (BMR) Effect

In a GMR device, if the non-FM layer shrinks to a point (atomic scale), the "ballistic" magnetoresistance (BMR) effect occurs (see Figure 5.18), resulting in a very

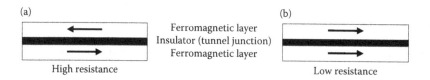

(a)   Ferromagnetic layer   (b)
Insulator (tunnel junction)
Ferromagnetic layer

High resistance   Low resistance

FIGURE 5.17 Schematic of a TMR sensor: (a) high-resistance TMR; (b) low resistance TMR.

FIGURE 5.18   Comparison of GMR and BMR structures.

large MR effect (over 3000% of M_R value can be achieved at low magnetic fields such as a few hundred Oersteds, 1 Oersted is equal to 79.58 amperes per meter) [15]. BMR is still in the experimental phase before its practical applications. The barrier to bring BMR to its practical applications includes the challenge to create nano- or atomic-scale point contacts between FM electrodes.

5.3.1.6   Colossal Magnetoresistance (CMR) Effect

The *Colossal Magnetoresistance* (CMR) effect, discovered in 1993, is an MR phenomenon that has *Perovskite structures*—$A_{1-x}B_xMnO_3$, where A = (La, Pr, Nd, or Sm); B = (Ca, Sr, or Ba) [16]. The term "colossal" came from the *huge* MR effects observed (~100,000%) when the resistivity of the material undergoes a *phase transition* at a low temperature from an insulating (high resistivity) phase to a metallic (low resistivity) phase. Research has shown that the CMR effect can be raised up to room temperature. *Hewlett Packard* company has produced high-quality CMR films in room temperature with a resistance change of about 95%. The major challenge associated with CMR sensors is to achieve useful, reproducible behavior at room temperature.

Among these effects, BMR and CMR are still under study phase. OMR has not found much practical applications yet due to its very low MR effect (<1%). TMR, however, has been applied in magnetic random access memory (MRAM). AMR and GMR effects by far have found most applications in sensors. A summary of the above MR effects is shown in Table 5.5.

5.3.2   AMR Sensors and the Barber-Pole Structure

A standard AMR sensor is composed of a nickel–iron (permalloy) resistive thin film deposited as a strip on a silicon wafer. Its *R–H* (resistance vs. magnetic field strength) curve is shown with a dashed line in Figure 5.19, which is drawn based on Equation 5.16. When H_y is small, H_y/H_{max} is nearer to zero and the sensitivity is very low and nonlinear (indicated by the zero slope of the curve at $H_y/H_{max} = 0$ with sharp bending curvature). This *R–H* curve, however, cannot distinguish the sign of H_y because both $+ H_y$ and $- H_y$ result in the same resistance values as shown in Equation 5.16.

An improved structure is to deposit aluminum stripes (called *barber poles*) on the top of the permalloy strip at an angle of 45° to the strip's easy axis. Since the conductivity of aluminum is much higher than permalloy, the electrons tend to flow as much as they can along the aluminum strips and as less as they can along the permallery strips (see Figure 5.20), resulting an effect of the current's rotating 45°. Thus, the angle between the magnetization and the electrical current is changed from Θ to

TABLE 5.5

Comparison of MR Effects

MR Effect	$\Delta R/R$ (%)	B (T)	Mechanism	Comments
OMR	<1	1 T (metal)	Lorentz force	There is no saturation at large magnetic field.
AMR	1–5	0.5 mT–1 T (depend on bulk or wire permalloy)	Electron spin–orbit interaction (leads to scattering of conducting electrons).	R is directly related to the orientation of magnetization M relative to current I.
GMR	10–200	1–10 T	Spin-dependent electron transport. In FM-metal-FM alternating layer structure.	Interface quality is crucial. They are broadly applied in sensors and read-heads of magnetic hard disks.
TMR	≈ 100	≈ 4 T	Spin-polarized tunneling. In FM-insulator-FM alternating layer structure.	Interface quality is crucial. They are temperature independent and applied in magnetic random access memory (MRAM).
BMR	>3000	<0.1 T	Spin scattering across very narrow magnetic domain walls trapped at nano-sized constrictions.	BMR has nano- or atomic-scale point contacts between FM electrodes. They are still in experimental phase.
CMR	≈ 100000	≈ 3 T	Insulation to conduction (metal) phase transition at Curie temperature.	CMR is extremely temperature and doping dependent. It is challenging to get useful, reproducible behavior at room temperature. They are still in study.

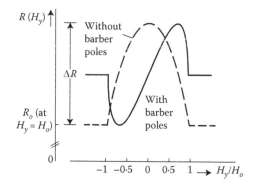

FIGURE 5.19 The R–H characteristics of a standard sensor and a Barber-pole sensor. (From Ripka, P. and Závěta, K. Chapter Three, Magnetic Sensors: Principles and Applications, *Handbook of Magnetic Materials*, Vol. 18, pp. 347–420, 2009.)

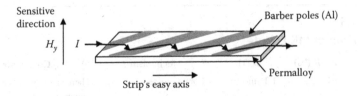

FIGURE 5.20 A Barber-pole sensor structure.

$\Theta - 45°$, the sensitivity of the sensor, indicated by the solid curve in Figure 5.19, is much higher, a linear behavior for a small H_y is achieved, and the sign of H_y can be distinguished.

The characteristic of a barber-pole AMR sensor can be modeled as [17]

$$R(H_y) = R_0 + \frac{\Delta R}{2} \pm \Delta R \frac{H_y}{H_{max}} \sqrt{1 - \left(\frac{H_y}{H_{max}} \right)^2} \tag{5.20}$$

The sensor displays a linear characteristic at about $H_y^2/H_{max}^2 = 0$. The $\pm$ sign in the above equation is determined by the inclination of the barber poles ($\pm45°$) to the strip's easy axis.

EXAMPLE 5.9

Figure 5.21 shows a contactless potentiometer that consists of two AMR sensors serially connected at B and current flowing between A and C. The sensors are identical but their preferred magnetization vectors are set perpendicular to each other. An external magnetic field is applied to this 3-terminal device in a direction perpendicular to the paper as shown. Find the output voltage V_{out}.

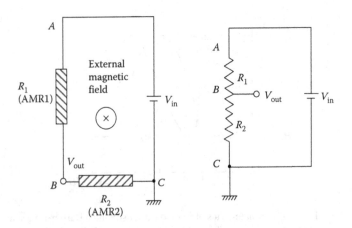

FIGURE 5.21 An MR-type potentiometer circuit.

SOLUTION

Since the two AMR sensors' preferred magnetization directions are perpendicular to each other, we have:

$$R_1(\Theta) = R_{min} \sin^2 \Theta + R_{max} \cos^2 \Theta$$

$$R_2(\Theta + 90°) = R_{min} \sin^2(\Theta + 90°) + R_{max} \cos^2(\Theta + 90°) = R_{min} \cos^2 \Theta + R_{max} \sin^2 \Theta$$

Applying the voltage divider formula, yields:

$$V_{out} = \frac{R_2(\Theta + 90°)}{R_1(\Theta) + R_2(\Theta + 90°)} V_{in} = \frac{R_{min} \cos^2 \Theta + R_{max} \sin^2 \Theta}{(R_{min} \sin^2 \Theta + R_{max} \cos^2 \Theta) + (R_{min} \cos^2 \Theta + R_{max} \sin^2 \Theta)} V_{in}$$

$$= \frac{R_{min} \cos^2 \Theta + R_{max} \sin^2 \Theta}{(R_{min} + R_{max})} V_{in}$$

The above equation can be rewritten as

$$V_{out} = \frac{V_{in}}{2} - \frac{(R_{min} - R_{max}) \cos 2\Theta}{2(R_{min} + R_{max})} V_{in} \tag{5.21}$$

where the first term, $V_{in}/2$, is the output voltage at a demagnetized state (i.e., at $\Theta = 45°$); the second term indicates the direction of the magnetic field:

$$\cos 2\Theta = \left(1 - \frac{2V_{out}}{V_{in}}\right) \frac{R_{min} + R_{max}}{R_{min} - R_{max}} \tag{5.22}$$

Note that the total resistance in the circuit, $R_1(\Theta) + R_2(\Theta + 90°) = R_{min} + R_{max}$ remain constant, which does not depend on Θ.

EXAMPLE 5.10

In Equation 5.22, if $V_{out} = 2.3$ V, $R_{min} = 800$ Ω, $R_{max} = 2$ kΩ, $V_{in} = 5$ V, find the angle Θ.

SOLUTION

$$\cos 2\Theta = \left(1 - \frac{2V_{out}}{V_{in}}\right) \left(\frac{R_{min} + R_{max}}{R_{min} - R_{max}}\right) = \left(1 - \frac{2(2.3\,V)}{5\,V}\right) \left(\frac{800\,\Omega + 2000\,\Omega}{800\,\Omega - 2000\,\Omega}\right)$$

$$= -0.1867$$

$$2\Theta = \cos^{-1}(-0.1867) = 100.76°, \quad so\ \Theta = 50.38°$$

5.3.3 AMR SENSOR MATERIALS AND CIRCUIT CONFIGURATIONS

The common materials used for AMR sensors are binary and ternary alloys of Ni, Fe, and Co. Table 5.6 compares the common AMR materials and their features (note

TABLE 5.6

Comparison of Common AMR Sensor Materials at 20°C

Materials	Resistivity ρ $(10^{-8}\ \Omega \cdot m)$	Resistivity Change $\Delta\rho/\rho$ (%)	Magnetic Field Strength H (A $\cdot$ m^{-1})
NiFe 81:19	22	2.2	250
NiFe 86:14	15	3	200
NiCo 50:50	24	2.2	2500
NiCo 70:30	26	3.7	2500
CoFeB 72:8:20	86	0.07	2000

that these are just approximate values as the exact values depend on a number of variables such as thickness, deposition, and post-processing).

As shown in the table, permalloys (with magnetocrystalline anisotropic structure, e.g., NiFe 81:19, NiFe 86:14) have a low H. Adding Co considerably increases H. The amorphous alloy (with noncrystalline structure, e.g., CoFeB 72:8:20) has a low $\Delta\rho/\rho$ but high H and exhibits an excellent magnetic behavior.

Like most resistive sensors, the AMR sensor circuit has either a series or a Wheatstone bridge structure depending on its application (see Figure 5.22). A series circuit with two elements is the simplest one. A Wheatstone circuit can compensate for temperature variation and provide more accurate measurements with higher sensitivity; and the resistance variation ΔR due to an applied magnetic field is linearly converted into the differential output ΔV.

5.3.4 GMR Sensors and Their Multilayer Structures

A GMR sensor has a "multilayer" structure, typically consisting of alternate stacks of ferromagnetic (FM) (such as Fe, Co, Ni, and their alloys) and non-FM (e.g., Cr, Cu, Ru) metallic (conductive) layers. Each layer is like an AMR strip, but only a few atoms thick (5–10 Å). If the FM layers are aligned parallelly (i.e., their predefined magnetization direction is in the same direction as the applied magnetic field), the up-spin electrons can pass through the structure easily with less or no scattering, thus

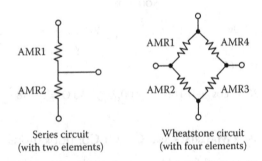

Series circuit
(with two elements)

Wheatstone circuit
(with four elements)

FIGURE 5.22   AMR circuit configurations.

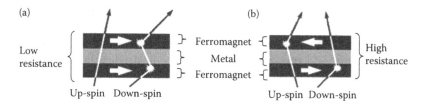

FIGURE 5.23 Microscopic illustration of GMR: spin-dependent scattering.

the resistance is low (see Figure 5.23a) [5]. "Up-spin" means that the spin orientation of the electrons is parallel to the external magnetic field; while "down-spin means that the spin orientation of the electrons antiparallel align with the external magnetic field. If the magnetic layers are aligned antiparallelly, scattering occurs for both up-spin and down-spin electrons and neither of them can go through the structure easily, thus the electrical resistance is high (see Figure 5.23b).

In the earlier GMR devices, the current flows in the plane, i.e., current-in-plane (CIP). In the latest GMR devices, GMR effects are substantial in current-perpendicular-to-plane (CPP) since the electrons move across the layers, thus the interfaces between layers are critical.

There are several structures that have been designed to achieve giant magneto-resistance as shown in Figure 5.24.

In the magnetic multilayer structure shown in Figure 5.24a, the FM or the anti-ferromagnetic (the dashed arrow) layers are separated by nonmagnetic (NM) spacer layers. If no external magnetic field is applied, these FM and AF layers are aligned antiparallel as indicated by the solid and dashed arrows, and the resistance reaches its maximum value R_{max}. If an external magnetic field is applied and at its saturation state, both FM and AF layers are aligned parallelly (as indicated by all the solid

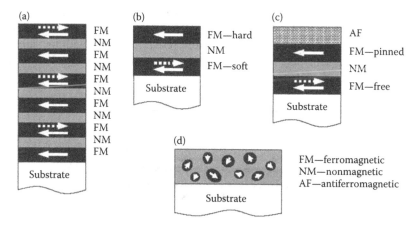

FIGURE 5.24 Various GMR structures: (a) magnetic multilayer; (b) pseudo spin valve; (c) spin valve; (d) granular thin film. (From Tsymbal, E.Y. and Pettifor, D.G., Perspectives of giant magnetoresistance, in *Solid State Physics*, 56, Academic Press, London, 113–237, 2001.)

arrows) and the resistance drops to its minimum value R_{min}. In the pseudo-spin valve structure (Figure 5.24b), different coercivities of the FM layers are introduced (i.e., FM hard and FM soft), making the hard and soft magnetic layers switch at different magnetic field strength, thus providing a field range in which they are antiparallel. In the spin valve structure shown in Figure 5.24c, the magnetization of the top FM layer is pinned by an AF layer, whereas the magnetization of the bottom FM layer is free to switch with the applied magnetic field. The advantage of this structure is that only small magnetic fields need to be applied to change the resistance. In the granular thin film structure (Figure 5.24d), FM precipitates are embedded in a nonmagnetic host metal film. The randomly oriented magnetic moments of the precipitates can be aligned by an applied magnetic field, resulting in a resistance change.

Practical GMR sensors are composed of more complex materials, such as a series of Fe/Ag/Co$_x$Fe$_{1-x}$ sandwiches, with the possible addition of niobium, silicon, tantalum, and other elements. Figure 5.25 shows an actual GMR structure that contains nine layers. The middle five layers form the GMR active region with one FM (NiFe) layer, three non-FM (Co/Cu/Co) layers, and the other FM (NiFe) layer. The magnetization of the upper FM is locked using an AF "pinning layer" (FeMn). The lower FM is made to be magnetically soft, so that its magnetization is easily realigned in small magnetic fields.

5.3.5 MR SENSOR DESIGN

Figure 5.26a shows the structure of an MR sensor that contains a NiFe sensing layer, a spacer, and a NiFeX soft adjacent layer (SAL); Figure 5.26b shows a spin valve GMR sensor that consists of an AM layer, a Co pinned film, a Cu layer, and a NiFe sensing layer. The thickness of each layer is on the order of a few nanometers, whereas the lateral dimensions can vary from micrometers to centimeters. The output of an MR sensor often requires signal processing.

MR sensors are usually produced using thin-film technology with a variety of structures or configurations. An MRX1518H series MR sensor, made by *AnaSem, Inc.* (Japan), uses an FeNi FM thin film whose resistance varies with the strength of an external magnetic field (the maximum resistance variation is 3%). Figure 5.27 shows an AA005-02 GMR current sensor, made by the *NVE Corporation* (Eden

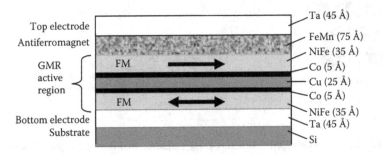

FIGURE 5.25 An actual GMR device structure.

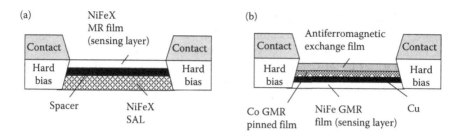

FIGURE 5.26 Structures of (a) a soft adjacent layer (SAL) MR sensor; (b) a spin valve GMR sensor.

Prairie, Minnesota) with its IC package, pin labels, and mounting diagram. This sensor detects and measures an electrical current perpendicular to the axis of sensitivity. Its typical operating point is 40 Oe. The saturation field is 100 Oe and sensitivity is 0.9–1.3 mV/V/Oe.

5.3.6 MR SENSOR APPLICATIONS

MR sensors are widely used in disk drives, automotive engine and transmission systems, vehicle traction and stability controls, electronic compasses, precision positioning, and angle/rotational speed/current/magnetic field measurement. Examples of MR sensors and their applications are described as follows.

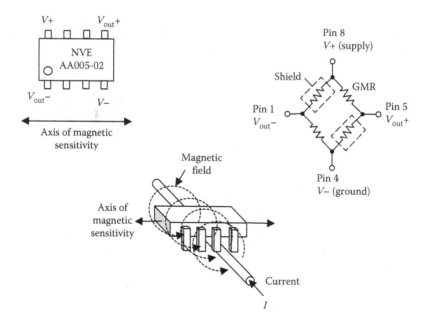

FIGURE 5.27 An NVE AA005-02 current sensor.

5.3.6.1   Magnetic Recording System

Figure 5.28 shows a magnetic recording system. The GMR read sensor is flying over the recorded media and senses the vertical component of the magnetic field of the media and varies its resistance. The inductive write element then alters the magnetization of each element on the recording track. This system can achieve GHz data recording rates with 10 year retention.

5.3.6.2   MR Biochips

The MR sensing technique can be used to detect and analyze magnetically labeled nucleic acids, proteins, whole cells, or microorganisms. An MR biochip (e.g., a DNA chip and protein chip) consists of one or multiple biosensing elements, designed and fabricated on-chip to facilitate multiprobe or multianalyte-based detection. For instance, a bead array counter (BARC III) biochip has 64 serpentine GMR elements and two reference elements. Each circular GMR element is a serpentine sensor, providing a high percentage sensing area beneath each DNA probe spot ($\sim 100 \times 100 \ \mu m^2$). The sensor's output—DNA concentration—is expressed as a voltage change that is proportional to the resistance change of the GMR elements and the percentage of sensor bead coverage.

5.3.6.3   Currency Counter

Figure 5.29 shows an *NVE*'s MR sensor array in a typical currency detection application. The sensor assembly is positioned approximately 1 mm from the currency path. The bank note is typically magnetized with a permanent magnet before it reaches the sensor array. The residual magnetization in the magnetic ink or magnetic strip in the currency is detected by the sensor array. This information is then analyzed to determine if the currency is genuine.

Figure 5.30a shows a measurement setup that uses an *NVE*'s AA GMR sensor to measure the current in a PCB trace by measuring the magnetic field created by the PCB trace at a known distance d from the trace. The axis of magnetic sensitivity

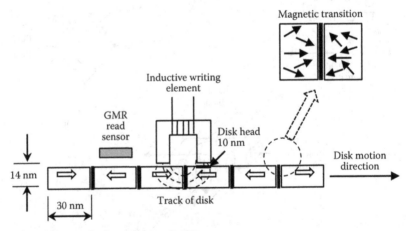

Disk width (perpendicular to the paper): 210 nm

FIGURE 5.28   A magnetic recording system using a GMR sensor.

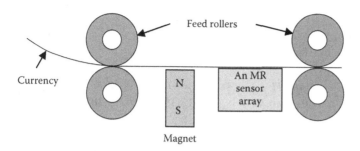

FIGURE 5.29 An *NVE*'s MR sensor array for currency detection.

must be placed at 90° to the conducting trace in order to achieve the maximum sensitivity (see Figure 5.30b). A larger view of the PCB trace layout is shown in Figure 5.30c. Note that distances L_1 and L_2 are not necessary to be equal. If $L_1 \neq L_2$ the resulting field strength at the sensor will not be uniform along the same axis. The field strength at the sensor position can be calculated by (refer to Figure 5.30d)

$$H = \frac{(\cos\theta_A + \cos\theta_B)I}{4\pi d} \tag{5.23}$$

EXAMPLE 5.11

In Figure 5.30, if $d = 2.75$ mm, measured $H = 7$ Oe, $L_1 = L_2 = 10$ mm, find the current in the PCB trace.

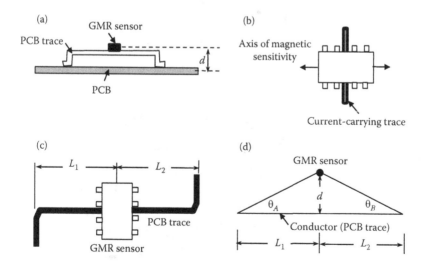

FIGURE 5.30 (a) A GMR sensor measuring the current of a PCB trace; (b) the orientation of the GMR sensor relative to the trace; (c) the sensor and the PCB trace with dimensions; (d) the geometry drawing of the trace position relative to the sensor.

<small>SOLUTION</small>

Since $L_1 = L_2$, $\theta_A = \theta_B = \tan^{-1}(2.75/10) = 15.38°$. 7 Oe $= 7 \times 79.58$ A · m^{-1}(in air) $=$ 557.04 A · m^{-1}. From Equation 5.23, the current I is:

$$I = \frac{4\pi d\, H}{\cos\theta_A + \cos\theta_B} = \frac{4\pi(2.75 \times 10^{-3}\,\text{m})(557.04\,\text{A} \cdot \text{m}^{-1})}{\cos(15.38°) + \cos(15.38°)} = 9.98\,\text{A}$$

5.4 MAGNETOSTRICTIVE/MAGNETOELASTIC SENSORS

5.4.1 MAGNETOSTRICTIVE EFFECTS

5.4.1.1 Joule and Villari Effects

The magnetostrictive effect relates a material's elastic state to its magnetic state. Nearly all FM materials such as iron, nickel, cobalt, and their alloy (alnico = **al**uminum-**ni**ckel-**co**balt alloy), exhibit a change in size and shape resulting from magnetization change. This effect was first discovered by James P. Joule in 1842, thus it is also known as *Joule effect*. Joule effect can be understood by considering an FM material consisting of many tiny, oval shaped, permanent magnets. When the material is not magnetized, the domains are randomly arranged. If the material is magnetized, the domains are oriented with their axes approximately parallel to one another. As a result, the overall dimension of the material changes (expands or contracts, depending on the direction of the magnetic field applied) as shown in Figure 5.31.

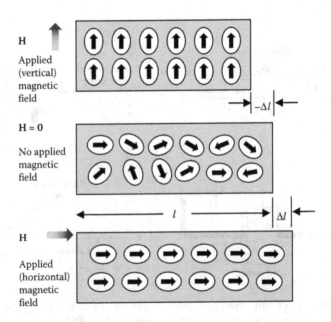

FIGURE 5.31 The principle of the magnetostrictive effect.

Inversely, applying a force to an FM material will cause a change in its magnetization. This effect is called *Villari effect*. The relationship between magnetization and mechanical stress or strain can be described by two coupled linear equations:

$$\varepsilon = \frac{\sigma}{E_y^H} + d^*H \quad \text{(Joule Effect)} \tag{5.24}$$

$$B = \mu^\sigma H + D^*\sigma \quad \text{(Villari Effect)} \tag{5.25}$$

The first equation indicates that the mechanical strain ε is due to both the mechanical stress σ and the applied magnetic field H. d^* is the strain coefficient (unit: $m \cdot A^{-1}$), $d^* = \partial\varepsilon/\partial H|_\sigma$. E_y^H is the Young's modulus under the magnetic field H. The second equation describes how the magnetic flux density B responds to the applied field H and to the mechanical stress σ. μ^σ is the permeability under the stress σ, and D^* is a coefficient (unit: $T \cdot Pa^{-1}$) which is equal to $\partial B/\partial\sigma|_H$. Both ε and B depend on σ and H. In sensor design, the first equation is often used with a fixed prestress σ, thus the strain output will be directly related to H. Both magnetostrictive position sensors and magnetic field sensors are based on this principle. Magnetostrictive force, stress, and torque sensors as well as accelerometers are often designed based on the second equation, in which H is often fixed and consequently the output of B directly relates to stress, σ, caused by external forces or torques.

EXAMPLE 5.12

Terfenol-D, an alloy of the formula $Tb_xDy_{1-x}Fe_{1.92}$ $(x = 0.3)$, is a magnetostrictive material [18]. Calculate the mechanical strain ε of a Terfenol-D sample under an applied magnetic field $H = 80\,kA \cdot m^{-1}$, given that its mechanical stress $\sigma = 20\,MPa$, Young's modulus $E_y^H = 25\,GPa$, and the strain coefficient $d^* = 75 \times 10^{-9}\,m \cdot A^{-1}$.

SOLUTION

From Equation 5.24

$$\varepsilon = \frac{\sigma}{E_y^H} + d^*H = \frac{20 \times 10^6\,Pa}{25 \times 10^9\,Pa} + (75 \times 10^{-9}\,m \cdot A^{-1})(80 \times 10^3\,A \cdot m^{-1})$$

$$= 0.0068 \text{ or } 0.68\%$$

5.4.1.2 Wiedemann and Matteuci Effects

Additional magnetostrictive effects are *Wiedemann* and *Matteuci effects*. When a current carrying FM rod is placed in a longitudinal magnetic field, the rod experiences twisting. This phenomenon is called *Wiedemann effect*, named after the German physicist, Gustav Wiedemann, who discovered the effect in 1858. Inversely, twisting a magnetostrictive element or a magnetized wire causes a change in magnetization (*Matteuci effect*).

Wiedemann effect is one of the manifestations of magnetostriction in the combined field of a longitudinal magnetic field and a circular magnetic field created by

an electric current. If the current is alternating, the rod will begin torsional oscillation. The twisting angle of rod α can be approximated by [19]

$$\alpha = j\frac{h_{15}}{2G^*} \tag{5.26}$$

where j is the current density; h_{15} is the magnetoelastic parameter, proportional to the
longitudinal magnetic field value H; and G^* is the shear modulus, a magnetoelastic
property of the rod.

Torsional twist in an FM wire allows measuring both circular and longitudinal
magnetic fields. Inversely, measuring changes in magnetization can be used to determine the applied torque. Over 50 patents have been issued in the last decade for
magnetoelastic torque sensors using the Matteuci effect.

5.4.2 OPERATING PRINCIPLES OF MAGNETOSTRICTIVE SENSORS

A magnetostrictive sensor operates based on *Joule, Villari, Wiedemann,* or *Matteuci*
effect. On the one hand, the stress created by an applied force or torque alters the
permeability of a magnetostrictive material, causing its magnetization change; on
the other, an applied magnetic field changes the magnetic state of a magnetostrictive
material, resulting in a variation in its elastic constant.

Depending on whether an excitation is used in the measurement, magnetostrictive sensors can be classified as active, passive, or hybrid sensors. Active sensors
use an internal excitation of the magnetostrictive element to facilitate the measurement. Passive sensors "passively" rely on the change in the magnetostrictive material properties to perform the measurement. Hybrid sensors use a magnetostrictive
element to actively excite or change another magnetostrictive element to make the
measurement.

5.4.3 MATERIALS AND CHARACTERISTICS OF MAGNETOSTRICTIVE SENSORS

The change in strain ε due to magnetization is usually very small. A magnetostrictive coefficient Λ is introduced to describe the fractional change in length as the
magnetization increases from zero to its saturation value. Λ may be positive or negative and is usually on the order of 10^{-5}.

The material that demonstrates a giant magnetostriction is *Terfenol-D*, an alloy
of terbium (Tb), dysprosium (Dy), and iron (Fe) – with the formula $Tb_xDy_{1-x}Fe_z$.
Terfenol-D has the highest magnetostriction of any alloy, up to 0.002 m/m at saturation. It expands and contracts in a magnetic field and has a large strain, e.g., bulk
saturation strain, over 2000×10^{-6} (at moderate magnetization levels and room temperature) and large force output, which has led its wide applications in displacement
measurement, load cells, torque and stress sensors, accelerometers, ultrasonic transducers, and many others. Its initial application was in naval sonar systems developed in the 1970s by the *Naval Ordnance Laboratory* in the United States. The
technology for manufacturing the material efficiently was developed in the 1980s

TABLE 5.7
Properties of Terfenol-D Magnetostrictive Material

Nominal composition	$Tb_{0.3}Dy_{0.7}Fe_{1.92}$
Mechanical properties	
Young's modulus	25 GPa ~ 35 GPa
Sound speed	1640 m s^{-1} ~ 1940 m s^{-1}
Tensile strength	28 MPa
Compressive strength	700 MPa
Mechanical properties	
Coefficient of thermal expansion	12 ppm $^{\circ}$C^{-1}
Specific heat	0.35 kJ kg^{-1} K^{-1}
Thermal conductivity	13.5 W m^{-1} K^{-1}
Electrical properties	
Resistivity	58×10^{-8} $\Omega \cdot$ m
Curie temperature	380°C
Magnetostrictive properties	
Strain	800 ppm ~ 1200 ppm
Energy density	14 kJ m^{-3} ~ 25 kJ m^{-3}
Magnetomechanical properties	
Relative permeability	3–10
Coupling factor	0.75

Source: Terfenol-D Data Sheet, *Etrema Products, Inc.*, Ames, Iowa, USA. www.etrema-usa.com/documents/Terfenol.pdf.

at *Ames Laboratory* under a U.S. Navy funded program. *"Terfenol-D"* is named after *ter*bium, iron (*Fe*), *N*aval *O*rdnance *L*aboratory (*NOL*), and the *D* comes from dysprosium.

Table 5.7 shows the properties of a magnetostrictive material Terfenol-D ($Tb_{0.3}Dy_{0.7}Fe_{1.92}$) [18]. Terfenol-D is currently produced in a variety of forms—solid (monolithic), powder (GMPC, giant magnetostrictive powder composite), and thin films.

5.4.4 DESIGN AND APPLICATIONS OF MAGNETOSTRICTIVE SENSORS

The earliest applications of the magnetostrictive effects include the telephonic receiver tested by Philipp Reis in the 1860s and the first magnetostrictive force sensor reported by James Alfred Ewing in 1900. The latest magnetostrictive sensor development focuses on using giant magnetostrictive material (such as Terfenol-D), magnetostrictive amorphous wire and thin films. These developments have resulted in successful sensor designs, including hearing aids, load cells, accelerometers, motion and proximity sensors, torque sensors, stress or force sensors, vibration sensors, magnetometers, flow meters, and many more. The following sections present several design and application examples of magnetostrictive sensors.

5.4.4.1   Magnetostrictive Position Sensor

A magnetostrictive position sensor usually consists of a magnetostrictive wire (or called *waveguide*, serving as the measuring element), a permanent magnet and a pickup coil (converting the mechanical pulse into an electric signal). The pickup coil can be a single coil, a dual coil, an FM recording tape, or a piezoelectric element. Figure 5.32 shows a magnetostrictive position sensor. A current pulse is applied to the FM wire where it generates a circular magnetic field around the wire. The permanent magnet (rectangular or ring shape) is connected to a moving object whose position is to be measured. This permanent magnet's magnetic field interacts with the magnetic field of the current pulse, resulting in an elastic deformation of the waveguide in a form of mechanical pulse (Joule effect). This wave travels along the waveguide in both directions with sound speed. At one end of the guide, a pickup coil detects this wave and converts the strain change into either an electrical signal or a magnetic field change (Villary effect). At the other end of the guide, the unused pulse is attenuated by a damping module to prevent further interference with returned/reflected pulse. The position of the object thus can be accurately determined by using the time elapsed, t, from emitting the current pulse to receiving the strain pulse as follows:

$$\text{Position of the Object} = \frac{v_S t}{2} \qquad (5.27)$$

where v_S is the speed of sound ($v_S = 340 \text{ m} \cdot \text{s}^{-1}$). The sensors that are based on the time elapsed are often called *magnetostrictive delay lines* (MDL). To ensure the accuracy of the output, the actual speed of sound in the waveguide is usually determined or calibrated using a laser interferometer for each sensor at the final stage of production.

To extend the measurement range or to track multiple target positions, a multi-position magnetostrictive sensor is developed (see Figure 5.33 [19]). It can monitor multiple targets in a one-time setup and can measure a stroke range greater than 5 m with a resolution down to 1 mm.

5.4.4.2   Magnetostrictive Level Transmitter

Figure 5.34 shows a *KSR's* magnetostrictive level transmitter. It consists of a magnetostrictive wire (held under tension inside a guide tube), a float (fitted with permanent

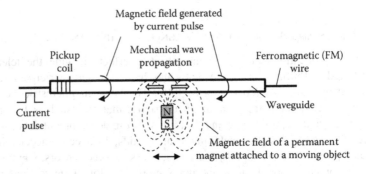

FIGURE 5.32   Working mechanism of a magnetostrictive position sensor.

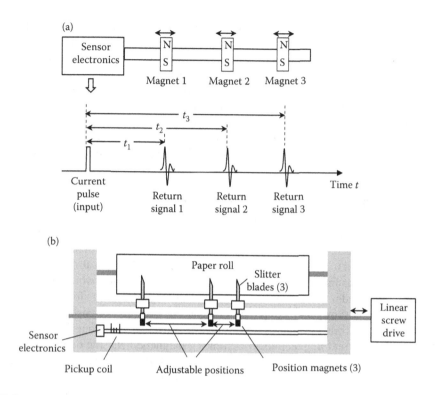

FIGURE 5.33   (a) A multiposition magnetostrictive sensor and (b) its application in monitoring the positions of the three paper slitter blades.

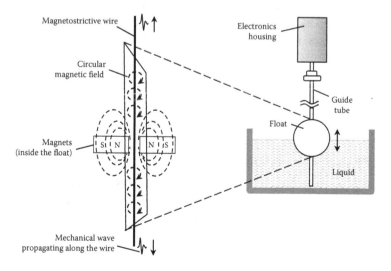

FIGURE 5.34   Operating principle of a *KSR's* magnetostrictive level transmitter.

magnets) that can move along the guide tube as the liquid level changes, and the sensor electronics. A current pulse is released to the wire and generates a circular magnetic field along the length of the magnetostrictive wire. When the pulse reaches the float, the pulse' circular magnetic field interacts with the float's magnetic field, generating a torsional stress pulse in the wire. A piezoceramic converter at the end of the wire converts this stress into an electrical signal. By measuring the elapsed transit time from the starting point of the current pulse to the receiving point of the resultant stress wave, the float position can be determined with high accuracy. This system works only if the auxiliary column and chamber walls are made of nonmagnetic material.

5.4.4.3 Magnetostrictive Force/Stress Sensors

Many magnetostrictive force and torque sensors have been developed based on Villary effect, whereby the force or torque being measured acts on the MDL and interrupts the emitted acoustic signal. The change in the acoustic wave captured by a receiver coil is thus related to the force or torque applied. The delay line can also be embedded in a fiber glass channel. The channel bends under an applied force, which changes the optical path or index that can be measured (optical-magnetostrictive force sensor).

Figure 5.35a shows an example of a magnetostrictive force sensor. The sensor consists of two coils (each surrounds a magnetostrictive core): one for excitation by passing a current to generate an oscillating magnetic field; the other for detection. Both cores are held in place by rigid end pieces. An applied force will cause a strain change in the magnetostrictive cores, resulting in a change in the core's magnetization that can be converted into a change in the output voltage by the detection coil. In Figure 5.35b, the sensing element is a magnetoelastic tape. When a force, pressure, wave, or vibration presents, the permeability of the tape changes, causing the bias magnet flux lines change. The output voltage generated in the detecting coil is proportional to the applied parameter. This method can detect strains less than 10^{-10} and it is suitable for vibration or shock measurement, as well as for very weak seismic activity monitoring. Compared with conventional force transducers, such as strain gages, magnetostrictive force sensors are simpler, more rugged, and relatively inexpensive.

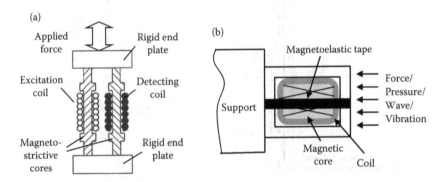

FIGURE 5.35 The structures of two magnetostrictive force sensors.

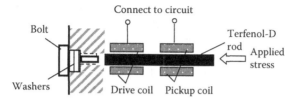

FIGURE 5.36   A magnetostrictive Terfenol-D force/stress sensor.

Figure 5.36 is a Terfenol-D stress sensor. It has a 2″ long and 1/4″ diameter laminated Terfenol-D rod ($Tb_{0.3}Dy_{0.7}Fe_{1.9}$) placed inside two coils. One coil is a multilayer
1100 turn drive coil, the other is a single-layer 110 turn pickup coil. A magnetic field
is produced by passing a current to the drive coil. Additional magnetic field is provided by a slit, cylindrical permanent magnet surrounding the coils. A mechanical
prestress is provided by inserting washers in series, and this stress can be adjusted
by turning a threaded bolt to meet the need of the measurement. The stress to be
measured is applied on the right end of the rod that will change the magnetic flux
density B. The pickup coil will detect this change and convert it into a measurable
voltage that is proportional to the time derivative of B. Notice that this Terfenol-D
sensor measures the time derivative of the load rather than the load itself.

5.4.4.4 Magnetostrictive Torque Sensors

Many torque sensors are designed based on magnetoelastic phenomenon because it
offers noncontact measurement, high sensitivity, small size, and durability. A typical magnetostrictive torque sensor includes an FM shaft supported in a housing, a
detecting coil, and a thin layer of *magnetostrictive* material coated on the peripheral
surface of the shaft. The detecting coil is usually fixed to the inner wall of the housing (stator), and a predetermined air gap exists between the stator and the shaft, as
shown in Figure 5.37.

Several designs of magnetostrictive torque sensors are shown in Figure 5.38. In
Figure 5.38a, the drive coil and the detection coil are wound around a C-shaped FM
core placed near the shaft (without contact) and oriented to 45° from the shaft axis [20].

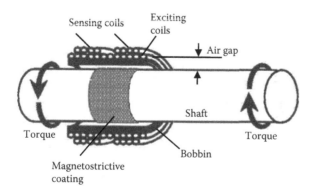

FIGURE 5.37   A typical layout and major components of a magnetostrictive torque sensor.

The current passing through the drive coil generates a magnetic field that penetrates the shaft. If a torque is applied to the shaft, stress will be developed in the shaft at 45° from the shaft axis, resulting in a change in permeability of the shaft, hence a change in the magnetic field. The changing magnetic field induces a current in the pickup coil that is proportional to the applied torque. Two factors can affect the sensor output: the air gap between the shaft and the sensor core and current in the drive coil.

To eliminate the influence of the variation in the air gap due to eccentric shaft motion, three sensors are placed equally around the circumference of the shaft Figure 5.38b [21]. In this design, the drive coil (with a laminated silicon steel core to prevent eddy currents) and the pickup coil (with a laminated permalloy core) are mounted at 90° angle to each other in a ceramic holder—one coil is parallel and the other coil is perpendicular to the axis of the shaft. The shaft is magnetized in circumference direction under the magnetic field generated by the exciting coil. If no torque is applied, the direction of the magnetization vector coincides with that of the shaft circumference. When a torque is applied, stress is generated in the shaft, which reorients the magnetization vector toward the direction of the stress. This magnetization vector change is detected by the pickup coil through electromagnetic induction.

The torque sensor shown in Figure 5.38c has a magnetostrictive material (e.g., Ni, FeNi, or FeCoNi) rigidly coated to the shaft. The coating is magnetized by passing a pulsed current through the shaft. When a torque is applied to the shaft, the magnetization in the coating reorients and becomes increasingly helical as the torque value

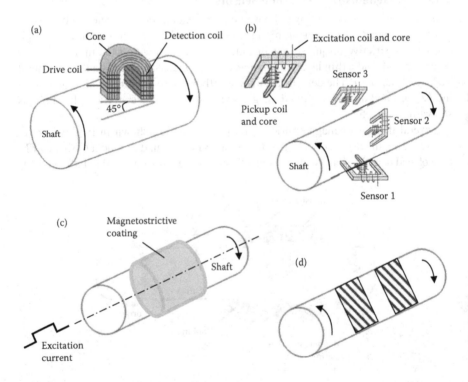

FIGURE 5.38 Several layouts of magnetostrictive torque sensors.

increases. This reorientation is detected by the sensing coil. Some shaft materials require knurl grooves in oblique direction to the shaft axis (see Figure 5.38d, US Patent 5,491,369), so that a magnetostrictive layer can be rigidly deposited (by plating or thermal spraying) in the outer periphery of the shaft and adhere to the shaft's surface to enhance the output signals.

5.4.4.5 Magnetic Field Sensor

If a magnetoelastic thin film or coating is placed on a SAW (surface acoustic wave) device, any magnetic fields nearby will cause stress to occur across the boundary layer due to the magnetostrictive effect. This stress will alter the velocity of the SAW. Hanna has reported a resonator-type SAW magnetic sensor based on SAW propagation in magnetoelastic films [22]. The relationship between resonant frequency and magnetic field of the sensor was nonlinear but a high sensitivity was achieved. Xiao et al. developed a similar sensor. They used a giant magnetostrictive material Terfenol-D and a SAW resonator. The resonant frequency of the SAW resonator varies with the stress induced by an applied magnetic field. Thus, the strength of the magnetic field can be determined through the variation of the resonant frequency. The static sensitivity of the sensor can reach $341.6 \text{ Hz} \cdot \text{Oe}^{-1}$; the resolution of the magnetic field can reach 10^{-6} T; the resolution of frequency detection is 1 Hz, and the frequency stability of the SAW resonator is 0.1×10^{-6}.

5.4.4.6 Magnetostrictive Fiber-Optic Magnetometers

Numerous magnetostrictive and fiber-optic combined magnetic field sensors have been developed. Wang et al. used an FM polymer (WCS-NG1) and the magnetostrictive film coated on an optic fiber for magnetic field detection [23]. An applied magnetic field causes the magnetostrictive film to deform and strain the optic fiber. As a result, the length of the optical path of the laser changes. An interferometer is used to measure the phase changes. A magnetostrictive fiber-optic sensor has a number of advantages:

- The optical technique provides high sensitivity to its measurand and is relatively simple to fabricate magnetostrictive coating on the optical fiber. Therefore, magnetostrictive effect is often used for the fiber optics magnetometer.
- The optical technique also prevents RF (radio frequency) interference that is common in typical electromagnetic-type sensors. Some magnetostrictive fiber-optic sensors are reported of having the resolution of 3×10^{-11} Oe at DC to low frequency (<1 Hz).
- Compared to other conventional magnetostrictive sensors, a magnetostrictive fiber-optic sensor is much less complex and relatively smaller in size.

The magnetostrictive fiber-optic method is specially useful in systems that already have optical fibers, for example, measuring the magnetic field along optical-fiber cables. Some sensitive magnetostrictive sensors also use a piezoelectric element to detect the length variation. A schematic diagram of using a magnetostrictive fiber-optic sensor to measure small-amplitude RF magnetic field distributions is shown in Figure 5.39.

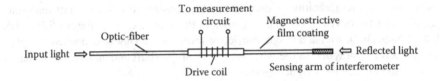

FIGURE 5.39 A magnetostrictive fiber-optic sensor for measuring a RF magnetic field.

5.5 NUCLEAR MAGNETIC RESONANCE (NMR)/MAGNETIC RESONANCE IMAGING (MRI) SENSORS

5.5.1 NUCLEAR MAGNETIC RESONANCE

As atomic nuclei spin, they create micro-magnetic fields (called *magnetic moments*) around themselves (Figure 5.40a), thus each nucleus resembles a tiny bar magnet with the north and south poles along the axis of spin. Ordinarily, nuclei are oriented randomly, so there is no net magnetic field (Figure 5.40b). However, if a strong and uniform external magnetic field is around, these nuclei will line up and create a detectable magnetic field (Figure 5.40c).

The Larmor equation expresses the relationship between *Larmor frequency*, f_L, of an individual nucleus spin and strength of an applied magnetic field, B:

$$f_L = \gamma_g (1 + \delta_C)B \tag{5.28}$$

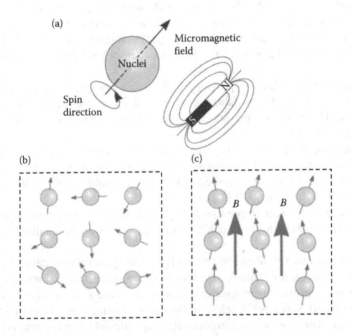

FIGURE 5.40 Principle of nuclear magnetic resonance.

where δ_C is the *chemical shift* (unitless, the difference between the resonant frequency of a nucleus and a standard). γ_g is the *gyromagnetic ratio* or *gyromagnetic constant* of the nucleus (in Hertz per tesla, $Hz \cdot T^{-1}$). For protons, $\gamma_g = 42.60 \times 10^6$ Hz T^{-1}. f_L is a specific number for each nuclear species under a specific magnetic field. For example, hydrogen nuclei under a specified magnetic field will spin at a predictable frequency. If the magnetic field changes, the Larmor frequency also changes.

EXAMPLE 5.13

A hydrogen nucleus has a gyromagnetic ratio of 4257 Hz $\cdot$ Gauss^{-1}. If the applied magnetic field is 1.5 T and the chemical shift is zero, find the Larmor frequency f_L of the hydrogen nucleus.

Solution

Apply Equation 5.28 (1.5 T = 15000 Gauss):

$$f_L = \gamma_g B = 4257 \,(Hz \cdot Gauss^{-1})\,(15000\,Gauss) = 63.855\,MHz$$

When atomic nuclei are exposed to an RF that is equal to Larmor frequency, they absorb energy from the RF wave and become excited. This phenomenon is referred to as resonance. When the RF wave stops, the atoms relax back to the equilibrium state and release the absorbed energy to the environment as RF wave emitters. The emitted RF waves can then be detected by sensors. The principle of NMR is mainly applied to obtain 2D or 3D MRI that contain the internal information of structures, materials, or human organs. Nuclear magnetic resonance was independently discovered by Bloch and Purcell in 1946. They both received the Nobel Prize in 1952.

5.5.2 Magnetic Resonance Imaging

MRI is a nondestructive or noninvasive means for obtaining high-resolution images of inside structures, materials, and human organs by mapping the distribution of hydrogen nuclei. It is based on similar principles as nuclear magnetic resonance (NMR).

The first MR spectrum of a human tumor was obtained by Griffiths et al. [24]. The first 2D (two-dimensional) and 3D MRI images were demonstrated by Lauterbur [25].

A number of schemes have been developed to combine magnetic field gradients and RF excitations for creating an MRI image:

1. 2D or 3D reconstruction from projections, similar to computed tomography (CT)
2. Formation of the image point-by-point or line-by-line
3. Use of gradients in the RF field rather than the static field

The majority of MR images today is created either by the 2D Fourier transform (2DFT) or spin-warp technique with slice selection, or by the 3D Fourier transform (3DFT) technique.

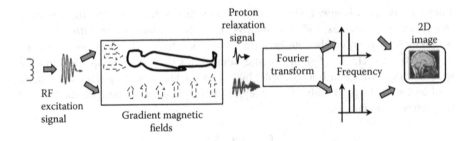

FIGURE 5.41   Procedure of forming a magnetic resonance image.

Figure 5.41 shows how a 2D MRI is formed. A patient is exposed to an external magnetic field in the z-axis (from head to feet). The magnetic field has a variable strength (gradient), that is, it is weakest at the top of the head and strongest at the bottom of the feet. Thus, the atomic nuclei located near the head have a different Larmor frequency than those near the feet (NMR principle). Then a series of RF waves that match the Larmor frequencies of atoms along the length of the body are applied (RF excitation signals must be broad-banded enough to cover all resonance values). The head region would receive RF waves of lower frequency than those at the feet. After the RF waves are stopped, sensors detect the frequencies of waves that are emitted back (proton relaxation). By selecting a specific frequency, one can view a specific section of the body (or a slice) along the z-axis. To create a visible 2D image (reconstruction process), another gradient magnetic field, perpendicular to the first one, is then applied. The second gradient may be weakest near the front of the body and strongest at the back of the body. Once again RF waves are applied and the results are recorded. A 2D image can then be obtained and displayed by a computer for any given point in the body through the 2DFT.

5.5.3 Operating Principles of Magnetic Resonance Sensors

A resonant structure presents an "amplified" response when excited by a frequency equal to the resonant frequency of the structure. This amplification is the result of an efficient transfer of the energy from the excitation source to the structure. Thus, a resonant structure based sensor can achieve larger output signals that increases its sensitivity.

5.5.4 Design and Applications of Magnetic Resonance Sensors

Magnetic resonance sensors are used to determine the composition, water content, temperature, and viscosity of cells, tissues or bones, as well as the mechanical properties of fluid and solid materials.

5.5.4.1 Resonant Magnetic Field Sensors

Resonant magnetic field sensors detect external magnetic fields through the displacement of resonant structures. The displacement change can be measured with optical, capacitive, or piezoresistive techniques.

Herrera-May et al. developed an MR magnetic field sensor using a silicon resonant microplate (400 μm × 150 μm × 15 μm) and four microbeams (130 μm × 12 μm × 15 μm) [26]. When an AC excitation current flows through the aluminum loop under an external magnetic field, the four microbeams experience a Lorentz force, which causes them to bend. To measure the deflection, the four piezoresistors (two active piezoresistors on the sensor and two passive piezoresistors on the silicon substrate) are connected in a Wheatstone bridge with an output voltage proportional to the magnetic field strength. This sensor was designed for measuring residual magnetic fields in welded steel tubes. Its main characteristics are:

- *Resonant frequency*: 136.52 kHz
- *Quality factor*: 842
- *Sensitivity*: 0.403 μV/μT
- *Resolution*: 143 nT with a frequency variation of 1 Hz
- *Power consumption*: <10 mW

A similar sensor developed by Sunier et al. [27] uses a cantilever structure embedded in an electrical oscillator. The device can be used for Earth magnetic field applications (e.g., an electronic compass). The sensor has a sensitivity of 60 kHz/T at its resonance frequency (175 kHz) and a short-term frequency stability of 0.025 Hz corresponding to a resolution below 1 T. Kádár et al. [28] developed a resonant torsional microplate sensor (2800 μm × 1400 μm × 12 μm of silicon). The external magnetic field deflects the microplate, causing a change in oscillation magnitude that is measured through the capacitance variation between the polysilicon electrode located on the microplate surface and the aluminum electrode placed under a glass package. The sensor was fabricated using a combination of bipolar processing, micromachining, glass processing, and glass-to-silicon anodic bonding. The sensor has a sensitivity of 500 μV/μT, a resonant frequency of 2.4 kHz, a quality factor of 700, a resolution of 1 nT (when it is vacuum-packaged) and a power consumption of a few milliwatts.

An MR magnetic field sensor developed by Emmerich and Schöfthaler [29] contains a movable comb-shape conducting microbeam (functions as one electrode/plate of a capacitor) and a fixed finger-shape electrode (functions as the other plate of a capacitor). This structure allows a big sensor output and a higher sensitivity. The Lorentz force is generated on the conducting beam when an AC current flows through the beam under an external magnetic field. As a result, the movable electrode changes its distance with respect to the fixed electrode and the capacitance is altered. Thus, the magnetic field is detected by means of the capacitance variation. The sensor works at resonance frequency and is micromachined in a vacuum ambient to increase its sensitivity. The main characteristics of the sensor include:

- *Sensitivity*: 820 μV/μT under an excitation current of 930 μA
- *Quality factor*: ~30 at 101 Pa
- *Resonant frequency*: 1.3 kHz
- *Resolution*: ~200 nT with a frequency increment of 10 Hz

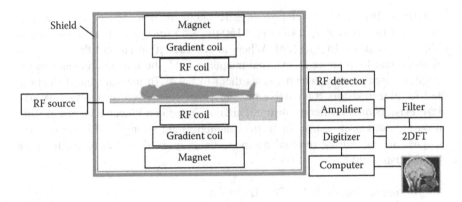

FIGURE 5.42   An MRI device and its main components.

5.5.4.2 MRI Devices

The application examples of MRI devices include obtaining clinical images and studying tissue metabolism in vivo; investigating molecular structures and molecular motion in solids and liquids, obtaining biochemical information from cells and tissues, evaluating vug (a small, unfilled cavity in rock), and controlling food production lines.

Figure 5.42 shows a typical MRI device used in hospitals and its principal components.

5.6 BARKHAUSEN SENSORS

5.6.1 BARKHAUSEN EFFECT

The *Barkhausen effect* relates magnetism to acoustics. In 1919, a German scientist Heinrich Barkhausen found that whenever he moved a magnet close to an iron-cored wire, an audible roaring sound (called *Barkhausen noise* or *Barkhausen emission*) was heard through an amplified speaker. The sound reflects a sudden (instead of a smooth or gradual) shifting or realignment of magnetic domains in the iron under an external magnetic field (see Figure 5.43). The sudden change in the magnetic

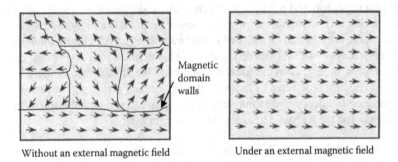

FIGURE 5.43   Microscopic magnetic domains in FM materials without and with an external magnetic field.

field around the iron induces a current pulse in the coil and creates a clicking sound in the speaker. These sudden jumps are interpreted as discrete changes in the size or orientation of FM domains that occur during magnetization or demagnetization. Crystalline imperfections temporarily hold up the movement of the domain walls causing the abrupt movements.

Two types of crystalline imperfections can interrupt the movement of domain walls: inclusions and residual stresses [30]. Inclusions, such as impurities and voids, are the regions in a material with different magnetization. Large inclusions impede the domain wall motion because they provide the areas to lower their magnetostatic energy. Small inclusions lower the overall surface energy, and thus delay the movement of walls. The causes of residual stress are crystalline imperfections, such as dislocations. When a domain wall moves through a region of varying residual stress, there is an energy increase in the wall. This is because the residual stress is trying to create a restoring force when the domain wall moves. If there is sufficient energy to overcome the restoring force, the domain wall can take irreversible jumps to equivalent energy levels.

Barkhausen effect involves a large number of very complicated interrelated mechanisms, making it difficult to express the effect analytically. In addition, magnetic behavior is inhomogeneous and does not "scale" easily as dimensions change. Williams and Shockley demonstrated the direct relation between changing magnetization and domain wall motion in single crystals of iron with 3.8% silicon, and showed that the wall position is linearly proportional to the magnetization [31].

5.6.2 Operating Principle of Barkhausen Sensors

In typical Barkhausen sensors, the domain arrangement in an FM specimen is controlled by a varying, externally applied magnetic field; the Barkhausen jumps in the specimen are detected as voltage pulses induced into a coil near to the surface of the FM material or wound around it (see Figure 5.44). Because Barkhausen jumps or noises are sensitive to the microstructural discontinuity of the FM materials such as impurities, dislocations, grain boundaries, precipitated carbides, and stresses, Barkhausen sensors can detect these defects thus. Existing methods of examining the FM material defects include fast Fourier transform (FFT), averaging over several cycles of magnetization of the sample, or using maximum entropy spectral estimation.

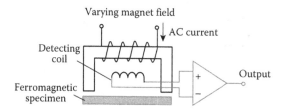

FIGURE 5.44 A typical layout of a Barkhausen effect sensor.

5.6.3 Design and Applications of Barkhausen Sensors

Barkhausen effect sensors are broadly used for nondestructive detection of structure defects. Application examples include examining stresses in a material, fatigue testing of steels, evaluating piezoelectric properties of metallic glasses, and pointing defects, dislocations, grain boundaries, and impurities of metallic materials.

5.6.3.1 Impact Toughness Tester

Korean Heavy Industries & Construction Company developed a Barkhausen noise (BN) sensor to obtain information on the impact toughness of the forged shell of SA 508 class 3 steel for a pressurized water reactor vessel. The sensor consists of a ferrite core and a 2000-turn pickup coil (see Figure 5.45a). The varying magnetic field is generated by a 1000-turn yoke magnet excited by 10 V, 0.5 Hz triangle waves from a function generator. The sensor is placed vertically on the surface of the specimen to measure the variation of a magnetic flux developed. The measured BN signal after passing an amplifier and a band-pass filter is displayed in Figure 5.45b.

5.6.3.2 Barkhausen Stress Sensor

Figure 5.46 is a Barkhausen sensor, developed by *Technical University of Szczecin* in Poland, to detect material changes introduced by stresses. It contains an 800-turn excitation coil, two 2000-turn pickup coils, and an aluminum case (shielding). The excitation coil with a 0.25 mm diameter is wound on a C-shaped ferrite core and is driven by a 30 Hz sinusoidal current source. The pickup coils, made of 0.02 mm diameter copper wires, are connected in series and placed one over the other separated by pole pieces of the ferrite core. The bottom coil is directly affected by the specimen property. The top coil picks up nonmaterial related signals. The inner distance between poles of ferrite core is 10 mm.

The sensor's output is fed into a signal conditioning circuit including a high-pass filter (500 Hz cutoff frequency, 50 dB gain), followed by a low-pass filter (30 kHz cutoff frequency and 10 dB gain). An A/D (analog-to-digital) converter with a sampling frequency of 500 kHz is used. Several measurements are taken at

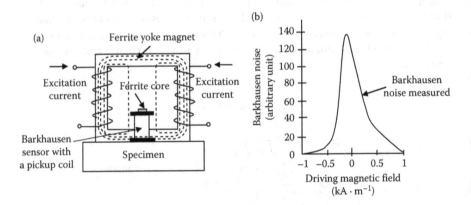

FIGURE 5.45 A Barkhausen noise sensor for impact toughness measurement of a steel shell.

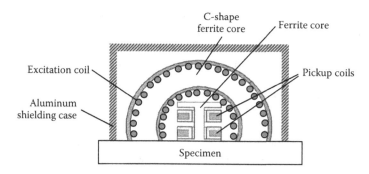

FIGURE 5.46 A Barkhausen sensor detecting material changes due to a stress.

each point of interest on the sample, and the median energy value of the signal is obtained.

5.7 WIEGAND SENSORS

5.7.1 WIEGAND EFFECT

The *Wiegand effect* relates to the nonlinear magnetization behavior of a *Wiegand wire* that switches its polarity under a strong magnetic field. A Wiegand wire is a low-carbon Vicalloy (a family of cobalt–iron–vanadium alloys) wire, typically consisting of 52% cobalt, 37.37% iron, 10% vanadium, 0.4% manganese, 0.2% silicon, and only 0.03% carbon. The features of a Wiegand wire are that it is short in length, about 0.010 in. in diameter, and magnetically "soft" in its core and magnetically "hard" on its shell. The soft-core, hard-shell Wiegand wire is obtained through a series of unique processes—a fully annealed process followed by a patented *cold-working process* (a series of twisting and untwisting operations to cold-work on the outer shell of the wire). Thus, the Wiegand wire has a very low magnetic coercivity within the wire and a very high magnetic coercivity on its outer shell. When a magnetic field is present near the wire, the high coercivity outer shell excludes the magnetic field from the inner soft core until a magnetic threshold is reached, whereupon the entire wire—both the outer shell and inner core—rapidly switches its magnetization polarity. This switchover occurs in a few microseconds, and is called the *Wiegand effect*. Once the Wiegand wire has flipped magnetization, it will retain that magnetization until a large magnetic field is applied in the opposite direction to flip it again.

5.7.2 OPERATING PRINCIPLE OF WIEGAND SENSORS

The significance of the Wiegand effect is its big pulse output generated during the magnetization switching of a Wiegand wire. A typical output of a Wiegand sensor is 2–6 V (with a typical 10 μs pulse width), which is several-orders-of-magnitude increase over a similar coil with a non-Wiegand core. This high output voltage plus the high magnetic field threshold that controls when to switch, making a Wiegand sensor one of the most sensitive and reliable sensors.

If an alternating magnetic field (generated by an AC excitation current or by one rotating magnet), instead of a static magnetic field, is around a Wiegand wire, an alternating switching with alternating pulse output can be created. There are two modes of switchings: *symmetric switching* and *asymmetric switching*. In the symmetric mode, the Wiegand wire is magnetized and triggered by an alternating positive and negative magnetic field of equal strength. In the asymmetrical mode (the optimum driving condition), the Wiegand wire is magnetized and triggered by an alternating magnetic field of opposite polarity but with unequal strength, resulting in a large pulse generated in only one direction of the field. In this situation, the magnetization direction of the magnetically soft core reverses, while the magnetization of the magnetically hard outer shell remains unchanged. A Wiegand sensor can also be made to have the magnitude of the output pulse proportional to the switching rate of magnetization.

Figure 5.47a shows a Wiegand sensor in an alternating magnetic field (created by an AC excitation current). Its two terminals are for the pulse output. The pulse is caused by a large Barkhausen jump when the core's magnetic field switches its polarity, and this voltage pulse is independent of the rate of the field change. An eddy current damper determines the pulse width. This Wiegand sensor was patented by John R. Wiegand in 1981 [32]. A typical shape of the Wiegand pulse is shown in Figure 5.47b. The magnitude of the pulse is about 3.7 V and the pulse width is about 10 μs.

5.7.3 DESIGN AND APPLICATIONS OF WIEGAND SENSORS

Wiegand effect sensors can be used in speed measurements, antilock brakers, position indicators, and anemometers. They can also be used for security by attaching them to access cards or antitheft labels.

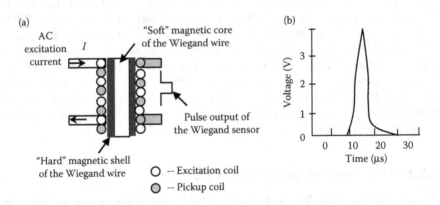

FIGURE 5.47 (a) A Wiegand sensor (From Dlugos, D.J., Wiegand effect sensors: theory and applications. *Sensors Magazine.* Questex Media Group, 1998.); (b) a Wiegand pulse (Data from Siemens).

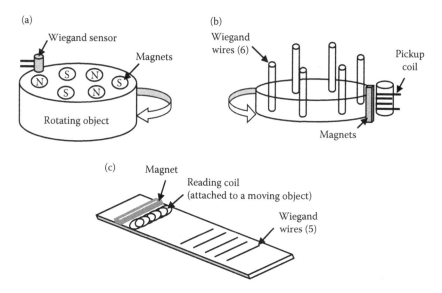

FIGURE 5.48 Wiegand speed sensors. (From Dlugos, D.J., Wiegand effect sensors: theory and applications. *Sensors Magazine.* Questex Media Group, 1998.)

Figure 5.48 shows three Wiegand sensor designs for speed measurements [33]. Figure 5.48a is a rotating object with several magnets placed on its surface (the number of the magnets depends on the measurement accuracy requirement). When the object rotates, the Wiegand sensor will produce a positive or negative voltage pulse depending on which magnetic pole (north or south) triggers the sensor. By counting how many pulses are generated, the speed of the object can be determined. The air gap between the sensor and the magnets can be as much as 1 in., depending on the strength of each magnet. In Figure 5.48b, the pulses are generated by a rotating drum (with Wiegand wires embedded in it) that passes through a fixed magnetic field. The output of this sensor is about 2 V with a pulse width of 10 μs, which has a lower pulse voltage than design (a). This is because the pickup coil is wound around the read head instead of the Wiegand wires. The advantage of design (b) is that it can indicate the direction of rotation with all positive pulses in one direction and all negative pulses in the other direction. Figure 5.48c is a linear design of Wiegand wires. The sensor can detect the linear motion of an object (attached by a reading coil) and can produce positive or negative pulses depending on the direction in which the object moves.

EXAMPLE 5.14

The Wiegand sensor shown in Figure 5.48a is used to measure the speed of a rotating object. The sensor's output is amplified and filtered and then sent to a full wave rectifier as shown in Figure 5.49. If an oscilloscope reading of the output pulse from the rectifier is 24 kHz, what is the speed of the object in rpm? Knowing that there are six magnets uniformly distributed on the surface of the object.

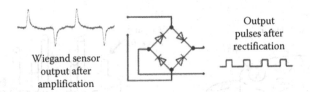

FIGURE 5.49 A full-wave rectifier rectifying the Wiegand sensor output.

Solution

The speed of the object in rpm is

$$\frac{2400\ s^{-1}}{6(\text{Pulse/Revolution})}\,60(s \cdot min^{-1}) = 24000\ \text{Revolution/Minute}$$

$$= 24000\ \text{rpm}$$

5.8 MAGNETO-OPTICAL SENSORS

Light, as an electromagnetic wave, can interact with either electric or magnetic fields, leading to numerous electro-optical and magneto-optical effects. This section focuses on magneto-optical effects and their sensors.

5.8.1 MAGNETO-OPTICAL EFFECTS

5.8.1.1 Faraday Effect

Faraday effect (also known as *Faraday rotation*), discovered by Michael Faraday in 1845, is a magneto-optical phenomenon that deals with the rotation of the polarization of light* when the light passes through certain materials in a magnetic field that is parallel to the direction of propagation (see Figure 5.50). If the light passes through ferro- or ferri-magnetic materials, the Faraday rotation, θ (the angle of the plane of polarization rotates), is proportional to the magnetization of the material:

$$\theta = K_K M l \tag{5.29}$$

where K_K, the *Kundt constant*, is material, wavelength, and temperature dependent. K_K can be as much as 350 degrees per gauss per centimeter. M is the magnetization, and l is the light path length in the material. In paramagnetic and diamagnetic materials (e.g., MgCe), the polarization rotation is proportional to the magnetic field density B. Thus, Equation 5.29 becomes

$$\theta = \int K_V B\, dl = K_V B l \tag{5.30}$$

---

* Light propagates via a sinusoidal oscillation of an electric field and a sinusoidal oscillation of a magnetic field. The direction in which the electric field oscillates as the light propagates is known as the *polarization*, and the plane that oscillates is the *plane of polarization*. If the polarization is along a single direction (e.g., the *x* axis), the electric field is linearly polarized. The magnetic field of light is perpendicular to the electric field and to the direction of propagation.

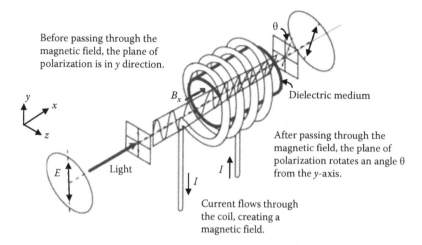

Before passing through the magnetic field, the plane of polarization is in y direction.

Dielectric medium

After passing through the magnetic field, the plane of polarization rotates an angle θ from the y-axis.

Light

Current flows through the coil, creating a magnetic field.

FIGURE 5.50 Illustration of the Faraday effect.

where K_V, the *Verdet constant*, varies from material to material, depending on the MO-material specific refraction index as well as the light wavelength. B is the magnetic field parallel to the direction of light propagation. Faraday effect, in general, is weak (small θ), and only large magnetic fields or large currents can be measured. Thus, Faraday effect sensors are often used with an optical fiber to provide a much longer light path than that of a bulk optical material. The increase in light path increases the angle of rotation and hence the sensitivity of the sensor.

EXAMPLE 5.15

Light travels through a 125 mm long SF-59 glass rod ($K_V = 19$ rad · T^{-1} · m^{-1}). How much does the polarization of light rotate when it goes through a 0.02 T magnetic field that is parallel to the light's path?

SOLUTION

Apply Equation 5.30,

$$\theta = K_V Bl = (19\,\text{rad} \cdot \text{T}^{-1} \cdot \text{m}^{-1})(0.02\,\text{T})(125 \times 10^{-3}\,\text{m}) = 0.0475\,\text{rad or } 2.72°$$

5.8.1.2 Voigt and Cotton–Mouton Effects

Voigt effect is a magneto-optical phenomenon that results in *magnetic birefringence* or *magnetic double refraction*. In this effect, the rotation of the plane of polarization occurs when an external magnetic field is applied perpendicular to the path of the light. The amount of rotation ϑ is proportional to the square of the field density B:

$$\vartheta \propto B^2 \tag{5.31}$$

The Voigt effect was discovered by Woldemar Voigt in 1902. The term "Voigt" is usually reserved for describing the polarization shift when the transparent medium is a vapor. If a liquid is the medium, the effect is much stronger (i.e., the proportion-ality to the square of the magnetic field is greater), and the effect is known as the *Cotton–Mouton effect*.

5.8.1.3 Malus' Law

Malus' law states that when a linearly polarized light is incident on an analyzer, the intensity I_L of the light transmitted by the analyzer is directly proportional to the square of the cosine of angle ϑ between the transmission axes of the analyzer and the polarizer:

$$I_L = I_0 \cos^2(\vartheta) \tag{5.32}$$

When $\vartheta = 0°$ (or $180°$), $I_L = I_0 \cos^2 0° = I_0$, meaning that the intensity of light transmitted by the analyzer is at its maximum when the transmission axes of the analyzer and the polarizer are parallel. When $\vartheta = 90°$, $I_L = I_0 \cos(90°) = 0$, mean-ing that the intensity of light transmitted by an analyzer is at its minimum when the transmission axes of the analyzer and polarizer are perpendicular to each other.

Normal light (from the Sun, a candle, or a laser) is unpolarized and can be thought of as a light containing a uniform mixture of linear polarizations at all possible angles. Since the average value of $\cos^2\vartheta$ is 1/2, the intensity of an unpolarized light transmitted by a polarizer is $0.5\,I_0$.

EXAMPLE 5.16

A vertically polarized light with intensity 0.57 W · m⁻² passes through a polar-izer oriented at $\vartheta = 63.5°$ from the vertical. Calculate the intensity using Malus' Law.

SOLUTION

Apply Equation 5.32,

$$I_l = I_0 \cos^2(\vartheta) = (0.57 \text{ W} \cdot \text{m}^{-2})\cos^2(63.5°) = 0.11\,\text{W} \cdot \text{m}^{-2}$$

EXAMPLE 5.17

An unpolarized light incidents upon two polarizers whose transmission axes are oriented at some angle with respect to each other as shown in Figure 5.51. The intensity after the first polarizer is equal to half of the intensity before. Calculate the intensity after the second polarizer, and determine the relative intensity by dividing the result by the initial intensity.

SOLUTIONS

The intensity after the first polarizer:

$$I_{L1} = \frac{1}{2}I_0$$

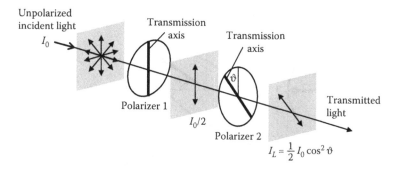

FIGURE 5.51 A two polarizer setup.

The intensity after the second polarizer:

$$I_{L2} = I_{L1} \cos^2 \vartheta = \frac{1}{2} I_0 \cos^2 \vartheta$$

Divide the final intensity by the initial intensity:

$$\frac{I_{L2}}{I_0} = \frac{1}{2} \cos^2 \vartheta$$

5.8.1.4 Magneto-Optical Kerr Effect

The Kerr effect is similar to the Faraday effect except that the Faraday effect is a measurement of the transmitted light, while the Kerr effect is a measurement of the reflected light. There are two types of Kerr effects: *electro-optical Kerr effect* and *magneto-optical Kerr effect*. To eliminate the confusion, "magneto-optical" is inserted in Kerr effect here. The magneto-optical Kerr effect (MOKE), discovered by John Kerr in 1876, describes the change of the polarization states of light when reflected at a magnetic sample subjected to a magnetic field. Thereby, linearly polarized light[*] experiences a rotation of the polarization plane (denoted by the *Kerr rotation* θ_k) and a phase difference between the electric field components perpendicular and parallel to the plane of the incident light (denoted by the *Kerr ellipticity* ε_k). As shown in Figure 5.52, the ellipticity of the light is defined as the ratio of the major axis length a to minor axis length b. This ratio can be seen as the ratio of light polarized in the y direction to light polarized in the x direction. The arrow in the diagram points in the direction of the electric field. As time increases, the electric field will change direction, but the tip of the arrow will always stay on the ellipse. θ_k and ε_k indicate the strength of the Kerr effect, governed by the nature of the magnetic material and degree of magnetization.

---

[*] Polarized light is described by a polarization ellipse. The shape of the ellipse mathematically defines the direction of the electric field vector as a function of time. For linearly polarized light, the ellipse is essentially stretched out until it becomes a straight line. For circularly polarized light, the ellipse becomes a circle. This is a visual way to describe the behavior of the electric field as a function of time.

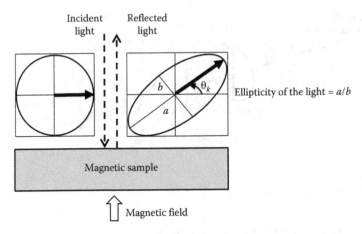

FIGURE 5.52 A diagram of Kerr rotation and the ellipticity of polarized light.

The MOKE can be observed by the three setups (see Figure 5.53): *polar, longitudinal* and *transverse*, based on the directions of the magnetic field relative to the plane of incidence and the magnetic sample surface. In the polar setup (Figure 5.53a), the magnetic field H is parallel to the plane of incidence and perpendicular to the sample surface. In the longitudinal setup (Figure 5.53b), H is parallel to both the plane of incidence and the surface of the sample. In the transverse setup (Figure 5.53c), H is perpendicular to the plane of incidence and parallel to the surface of the sample [31]. The polar Kerr effect is the strongest of the three. It may be of 20 minutes of arc. The longitudinal Kerr effect can be of 4 minutes of arc. It is the strongest when the angle of incidence is of 60°. The transverse effect is of about the same order of magnitude as the longitudinal effect.

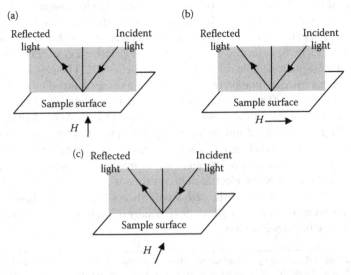

FIGURE 5.53 Three setups for observing the MOKE: (a) *polar*; (b) *longitudinal*; (c) *transverse*.

5.8.2 Operating Principles of Magneto-Optical (MO) Sensors

The operating principles of magneto-optical sensors are based on Faraday, Voigt, Cotton–Mouton, and/or Kerr effects. These effects describe the change in rotation of the polarization plane of linearly polarized light and variation in ellipticity of the light beam when passing a transparent (glass) medium or hitting a magnetic sample (or an MO film) in a magnetic field. Through measuring the Faraday rotation θ, the magnetization M, or the magnetic field strength B can be determined as indicated by Equations 5.29 and 5.30. Based on the sensing mechanism employed and the materials used, MO sensors can be classified as four types [34]:

- **All-fiber sensors:** In this type of sensor, the fiber itself acts as a transducer. The Faraday effect is used to induce a rotation of polarization of the light propagating in the fiber. The rotation angle is proportional to the magnetic field to be measured. Usually, the fiber is coiled around the electrical conductor, making it immune to external currents and magnetic fields.
- **Bulk optic sensors:** These sensors use a piece of glass or crystal with a high Verdet constant as the transducer, and place it near to (or around) an electrical conductor to measure the magnetic field. These sensors are usually less expensive, robust, and more sensitive.
- **Magnetic sample/film sensors:** These sensors utilize the MOKE effect by analyzing the reflected light from a magnetic sample (or a sensor film) to map the sample's magnetic properties (or measure the applied magnetic field or current). The change in the optical permittivity of the sample depends on the orientation of magnetization. Using various polarizing filters, different strengths of local magnetic fields can be demonstrated through different optical contrasts due to different angles of the magnetization rotation. This type of MO sensors can generate an MO image that enables a direct, real-time visualization of magnetic fields over the entire sensor surface.
- **Hybrid sensors:** These sensors employ both electromagnetic and magneto-optical technologies. Usually, the sensing part uses the conventional electromagnetic principles, while the information transportation is implemented by an optical fiber system. These hybrid sensors take advantage of the high level of electrical isolation offered by optical fibers and avoid difficulties associated with birefringence. For example, a fiber Bragg grating (FBG) sensor is combined with a magnetostrictive element to measure the mechanical strain changes in the material through detecting magnetic field change.

Figure 5.54 illustrates how an MO sensor works for mapping a magnetic sample. The incident light has different reflected rotation angles due to the specific magnetic features at local spots. Thus, an image with different intensities provides the accurate mapping of the sample's magnetic properties.

The range of detectable magnetic field strength for this type of sensors is 0.05–500 kA · m^{-1} (0.6 – 6000 Oe). Some MO–sensor systems, such as CMOS-MagView, can perform lateral resolutions up to 1 μm [35].

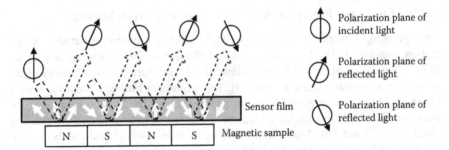

FIGURE 5.54 An MO sensor for magnetic field mapping.

EXAMPLE 5.18

An MO current sensor consists of a metallic wire and an SF2 glass rod placed perpendicular to each other (see Figure 5.55). An electrical current I flowing through the wire generates a magnetic field B. Assume that the glass rod length l is small enough, so that the magnetic field inside the glass can be considered constant. Find an expression for current I, knowing the Faraday rotation θ and the distance between the glass and the wire axis R.

Solution

The magnetic field B produced by the current in the metallic wire is

$$B = \frac{\mu_0 I}{2\pi R}$$

Plug B into Equation 5.30, yields

$$\theta = K_V B l = K_V \frac{\mu_0 I}{2\pi R} l$$

Thus,

$$I = \frac{2\pi R \theta}{\mu_0 K_V l}$$

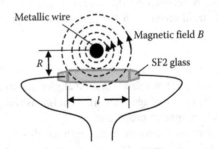

FIGURE 5.55 An MO current sensor.

Two techniques can be used to measure the Faraday rotation: *polarimetric* and *interferometric*. Polarimetric technique uses two polarizers, one at the input of the sensor and the other at the output (see Figure 5.56). The first polarizer defines the initial polarization state of the light, while the second polarizer is used to adjust the sensor's sensitivity and transforms the polarization rotation into a light intensity modulation that can be measured using a photodetector.

Malus' law can be used to determine the intensity of the output light:

$$I_{out} = I_{P1} \cos^2(\vartheta_2) = (I_{in} \cos^2(\vartheta_1))\cos^2(\vartheta_2) = I_{in} \cos^2(\vartheta_1)\cos^2(\vartheta_2),$$

where I_{P1} is the light intensity after polarizer 1. ϑ_1 and ϑ_2 are the angle of the transmission axes of polarizers 1 and 2, respectively.

To achieve the maximum sensitivity, the polarizers can be adjusted so that the sensor has a transfer function of [36]

$$I_{out} = \frac{I_{in}}{2}[1 + \sin(2\vartheta_{1-2})] \tag{5.33}$$

Where ϑ_{1-2} is the relative angle between the transmission axes of polarizers 1 and 2. Thus, the maximum sensitivity is achieved when this angle is 45°. This means that with the polarizers having a relative angle of 45°, any small change in the plane of polarization of the light caused by an external magnetic field B, will be transformed in a larger change of intensity at the output. To eliminate the dependence of the sensor's output on the input light intensity fluctuations, the final sensor output signal, S, can be delimited by its AC component only (S_{AC}) [36], that is

$$S_{AC} = \sin(2\vartheta_{1-2}) \tag{5.34}$$

If the rotation is relatively small, one can include the linear birefringence β in the sensing element, thus Equation 5.34 becomes

$$S_{AC} = \begin{cases} 2\vartheta_{1-2}\dfrac{\sin\beta}{\beta} & \beta \gg 2\vartheta_{1-2} \\ 2\vartheta_{1-2} & \beta \ll 2\vartheta_{1-2} \end{cases} \tag{5.35}$$

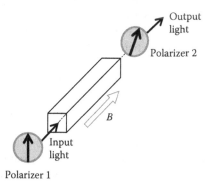

FIGURE 5.56 Polarimetric technique for detecting Faraday rotation.

Another technique to improve a sensor's output signal is to split the output light into two orthogonal polarizations (through the use of a Wollaston prism or a fiber coupler at ±45° with respect to the input polarizer) and detect the two signals (S_1 and S_2) using two independent photodetectors. The two output signals have an opposite phase. Each output signal is then sent to an analog circuit that computes the final output signal S by

$$S = \frac{S_1 - S_2}{S_1 + S_2} \qquad (5.36)$$

where S_1 and S_2 can also be determined using Equation 5.34. This dual-quadrature scheme provides better common noise rejection than the previous one. Polarimetric detection scheme, however, does not take into account the linear birefringence effect and its temperature dependence, which may affect the accuracy of the sensor. Interferometric scheme, on the other hand, overcomes this insufficiency by analyzing the rotation in terms of circular polarization, which corresponds to a phase difference between the two circular orthogonal modes (left-handed and right-handed circular polarization). In an interferometric detection, the optical phase variation (indicated by time interval between arms of the interferometer) contains the sensing information. The phase carrier can be an unbalanced Michelson or Mach–Zehnder interferometer, or other interferometer, such as Sagnac interferometer [34].

5.8.3 MATERIALS AND CHARACTERISTICS OF MO SENSORS

MO sensors offer noncontact sensing with high accuracy, high level of electrical isolation, high temporal and spatial resolution, and very fast response. MO sensor devices, such as the MO-objective adapter for commercial polarization microscopes, can perform lateral resolutions up to 1 μm. The main characteristics of MO sensors depend on the materials used, especially those that provide large Faraday rotation and Verdet constants. The magnitude of the rotational angle of the Kerr effect for a ferromagnetic material is generally between 10^{-4} and 10^{-3} degrees. The angle of rotation is more pronounced when the incident angle is increased. CeSb offers the largest MOKE, with a Kerr angle of 14°, comparing with a couple of degrees for uranium compounds and less than 0.5° for Fe, Co, and Ni. The stoichiometry formula $RE_{3-x}Bi_xFe_5O_{12}$ (abbreviation: RE = rare earth) also provides high Faraday rotations.

Optical fibers with high Verdet constants include: terbium-doped silica fiber (Verdet constant: 1.2×10^{-5} rad · A^{-1}); and doped plastic fiber (Verdet constant: 2×10^{-5} rad · A^{-1}) [37]; BGO ($Bi_{12}GeO_{20}$), BSO ($Bi_{12}SiO_{20}$), and ZnSe (Verdet constants: about 7×10^{-5} rad · A^{-1}).

$Cd_{1-x}Mn_xTe$ provides considerably higher Verdet constant, about 2×10^{-3} rad · A^{-1} [38]. Note that the Verdet constant has positive values for diamagnetic materials (such as the SF2 glass) and negative values for paramagnetic materials [39]. YIG, *Yttrium iron garnet* (a ferrimagnetic garnet crystal with the composition of $Y_3Fe_5O_{12}$), offers larger polarization rotations than that of $Cd_{1-x}Mn_xTe$ by about an order of magnitude.

YIG is transparent for light with a wavelength longer than 1.1 mm. At 1.3 and 1.5 mm wavelengths, reliable light sources and detectors are commercially available and the optical loss is very low. YIG has a substantial Faraday rotation in large parts of the visible light, infrared, and microwave spectrum [40].

A low optical absorption coefficient is also crucial in order to minimize the ellipticity of the polarized light. Other important terminologies that describe MO sensors include:

Figure of merit (FOM)—an important indicator for the maximum obtainable signal-to-noise ratio (SNR) of MO sensors, and it is defined as

$$Figure\ of\ Merit\ =\ R_r \theta_K^2 \tag{5.37}$$

where R_r is the *effective reflectivity* and θ_K is the Kerr rotation angle.

Measurement bandwidth—an indication of the dynamic measurement range of MO sensors, described by

$$Measurement\ Bandwidth = 0.44 \cdot \frac{c}{n} \cdot \frac{1}{2\pi r} \cdot \frac{1}{N} \tag{5.38}$$

where c is the speed of light in the fiber; n is the *refractive index* (unitless); N is the number of turns in the fiber coil; and r is the radius of the fiber coil. There is a tradeoff between sensitivity and bandwidth. The measurement bandwidth of bulk glass or fiber coil sensing elements is limited by the transit time of the light in the sensing element.

MO sensors have technical advantages of obtaining the magnetic field distribution directly above the surface of the magnetic material by an optical image according to the size of the MO sensor. Thus, a real-time magnetic field mapping can be implemented without the time-consuming point-to-point scans as needed when using Hall sensors for field mapping.

EXAMPLE 5.19

A Faraday sensor is made of a fiber coil. If the coil's radius is 0.1 m, the number of turns is 100, and $n = 1.5$, find the sensor's measurement bandwidth. (Note: the speed of light is 3×10^8 m · s^{-1}.)

SOLUTION

From Equation 5.38,

$$Measurement\ Bandwidth = 0.44 \cdot \frac{3 \times 10^8\ m \cdot s^{-1}}{1.5} \cdot \frac{1}{2\pi(0.1\ m)} \cdot \frac{1}{100} = 1.4\ MHz$$

5.8.4 Design and Applications of Magneto-Optical Sensors

MO sensors have attracted considerable interests in recent years because of their wide applications in MO recording devices, magnetization detection in very small objects, forensic investigations, diagnosis of electronic structure defects, nondestructive

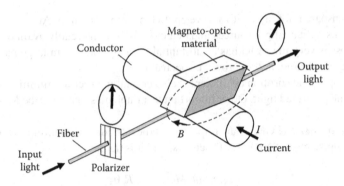

FIGURE 5.57   A Faraday current sensor.

testing, monitoring and sensing of magnetic thin films and nanostructures, observation of magnetic domains, and study of the couplings between ferromagnetic layers and nonmagnetic metallic layers. Some design and application examples of MO sensors are listed as follows.

5.8.4.1   Faraday Current Sensor

In the Faraday current sensor shown in Figure 5.57, the current flowing through a conductor induces a magnetic field that affects the propagation of light traveling through the optical fiber. By measuring how much the polarization rotates, the current in the conductor can be determined.

5.8.4.2   MO Disk Reader

In an MO disk reader shown in Figure 5.58, reading is achieved by detecting changes in the polarization direction of linearly polarized light reflected from the surface of the MO film. These changes are due to the different directions of the local magnetization. To obtain a high-quality readout signal, the magnitude of the remaining magnetization of the MO film is set equal to its saturation magnetization and the direction is set to be perpendicular to the surface of the disk (a polar setup).

5.8.4.3   MOKE Sensor for Magnetization Study

An MOKE sensor is often used to study the magnetization of an FM sample since both the Kerr rotation and ellipticity depend linearly on the sample's magnetization in first order. A laser beam can be used to strike incident on the surface of the FM

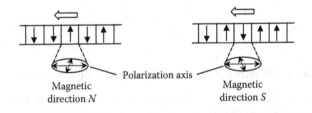

FIGURE 5.58   An MO disk reader using Kerr effect principle.

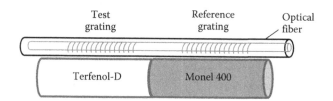

FIGURE 5.59 Schematic diagram of the fiber-Terfenol-D Hybrid sensor.

sample. Then, the reflected beam is detected to determine the Kerr rotation and ellipticity as a function of magnetization, and a hysteresis loop of the sample is formed and recorded for further analysis and characterization [31].

5.8.4.4 Fiber-Terfenol-D Hybrid Sensor

Mora et al. [41] combined an optical fiber with a Terfenol-D sensor to measure a static magnetic field (see Figure 5.59). The temperature compensation of this hybrid sensor was composed by two different alloys (Terfenol-D and Monel 400) with similar thermal expansion coefficients. The mechanical expansion of both materials due to temperature increase and magnetic field variations was detected by the two FBG (Fiber Bragg Grating) sensors attached. The spectral difference between the two Bragg wavelengths is proportional to the amplitude of the magnetostriction, and, the wavelength shift produced by the grating bonded to the nonmagnetic alloy (Monel 400) is proportional to the temperature variation.

5.9 SUPERCONDUCTING QUANTUM INTERFERENCE DEVICES (SQUIDs)

SQUIDs, or superconducting quantum interference devices, are the most sensitive detectors of magnetic flux known. They can measure extremely small variations in magnetic fields (e.g., 10^{-15} T). The SQUID has an equivalent energy sensitivity that approaches the quantum limit. Superconductivity was discovered in 1911 by H. Kamerlingh Onnes in Holland during study of the electrical resistance of frozen mercury as a function of temperature. When cooling Hg to the temperature of liquid helium, he found that the resistance vanished abruptly at approximately 4 K (= −269.15°C). Onnes was the first to liquify helium in 1908. In 1913 he won the Nobel Prize for his liquification of helium and discovery of superconductivity.

5.9.1 SUPERCONDUCTOR QUANTUM EFFECTS

5.9.1.1 Meissner Effect

The *Meissner* or *Meissner–Ochsenfeld effect*, discovered in 1933, states that magnetic fields are expelled from superconducting materials when the materials are cooled below their *critical* or *transition* temperature, T_c. This feature differs from that of normal conducting materials in which magnetic flux lines can penetrate through the material as shown in Figure 5.60.

The transition between normal-conducting and superconducting phases is comparable to different thermodynamic phases. Figure 5.61 depicts a ring made of superconducting

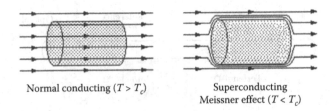

Normal conducting $(T > T_c)$

Superconducting
Meissner effect $(T < T_c)$

FIGURE 5.60 Meissner effect.

material in a magnetic field. When the temperature is above the material's transient or critical temperature, T_c (Figure 5.61a), the ring behaves as a normal (non-superconducting) diamagnetic material and the magnetic flux can penetrate through the ring. If the magnetic field is removed or turned off, an eddy current is induced and circulates around the ring, trapping or keeping the magnetic flux inside (constant). The induced current will decay exponentially with a time constant that relates to the resistance R and inductance L by

$$I(t) = I_0 e^{-(R/L)t} \tag{5.39}$$

When the temperature is below T_c (Figure 5.61b), the ring is in the superconducting state (i.e., resistance $R = 0$), and the magnetic flux is expelled from the ring (Meissner effect), but the flux, Φ, through the ring's opening remains unchanged. If the external field is switched OFF (Figure 5.61c), the induced eddy current appears and never decays. This persistent, constant, circulating current creates a constant magnetic field around the ring (i.e., the flux, Φ, unchanged).

5.9.1.2 Josephson Effect

Josephson effect, named after the British physicist—Brian D. Josephson who predicted the effect in 1962 and won the Nobel Prize in physics in 1973, is a prediction

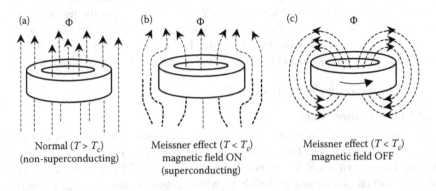

(a) Φ

Normal $(T > T_c)$
(non-superconducting)

(b) Φ

Meissner effect $(T < T_c)$
magnetic field ON
(superconducting)

(c) Φ

Meissner effect $(T < T_c)$
magnetic field OFF

FIGURE 5.61 A ring made of superconducting material (a) under a constant magnetic field for $T > T_c$; (b) under a constant magnetic field for $T < T_c$; (c) after the magnetic field is turned off for $T < T_c$.

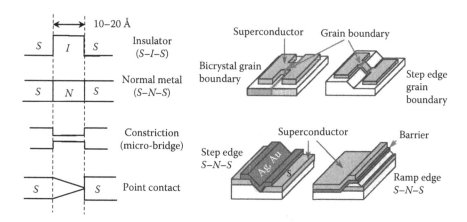

FIGURE 5.62 Examples of Josephson junctions. (From Leggett, A.J., PHYS598 Lecture Notes 13: The Josephson Effect, p. 1.)

of current flowing across two weakly linked superconductors. The weak link, called *Josephson junction*, can be a thin insulating layer (known as a superconductor–insulator–superconductor junction, or *S–I–S*), a short section of non-superconducting metal (*S–N–S*), a physical constriction (e.g., a micro-bridge), or a point of contact (*S-s-S*). The first Josephson junction was made by John Anderson at *Bell Labs* in 1963. Some examples of Josephson junctions are shown in Figure 5.62. The current that flows through the junction, $I_J(t)$, is called *Josephson current*.

5.8.1.2.1 DC Josephson Effect

The electrical properties of a Josephson junction between two superconductors change due to the phase difference, $\delta(t) = \theta_1(t) - \theta_2(t)$, of the two superconductors (recall that the two superconductors are characterized by $\psi_1(t) = \psi_0 e^{i\theta_1}$ and $\psi_2(t) = \psi_0 e^{i\theta_2}$, respectively). The *DC Josephson equation* relates the Josephson current I_J to the relative phase $\delta(t)$ and the *critical current* of the junction I_C as

$$I_J(t) = I_C \sin[\delta(t)] \qquad (5.40)$$

I_C is defined as the maximum current that a superconductor can carry without dissipating energy. It is affected by temperature and the applied magnetic field. If $I_J = 0$, it means $\delta = 2\pi n$ ($n = \pm 1, 2, 3, \ldots$). Thus, a Josephson current must flow through the junction for a physically meaningful phase difference to exist, even though no external electrical source is applied.

EXAMPLE 5.20

The phase difference of two superconductors on both sides of a Josephson junction is 8°. If the critical current and the temperature of the material (same for both superconductors) is 0.2 mA and 7.16 K, respectively, find the current flows through the junction.

SOLUTION

Apply Equation 5.40,

$$I_J(t) = I_C \sin[\delta(t)] = (0.2 \text{ mA})\sin 8° = 0.0278 \text{ mA}$$

5.8.1.2.2 AC Josephson Effect

If the Josephson current exceeds the Josephson junction's critical current I_C, the Josephson voltage appears across the junction. The *AC Josephson equation* relates the voltage across the junction, $V_J(t)$, to the temporal derivative of δ as

$$V_J(t) = \frac{K_P}{2q}\frac{\partial \delta}{\partial t} \tag{5.41}$$

where K_P is *Plank's constant* (6.626176×10^{-34} J · s); q is the charge of an electron. An AC current should flow across the junction under $V_J(t)$. Josephson also predicted that the *I–V* curve of the junction should show spikes at intervals of $q\omega/K_P$, where ω is the frequency of the AC current. The constant, $\Phi_0 = K_P/(2q) = 2.06783367 \times 10^{-15}$ Wb (or T · m²), is the *magnetic flux quantum*. Its inverse is the *Josephson constant*, $K_J = 1/\Phi_0 = 2q/K_P = 483597.89 \times 10^9$ Hz · V⁻¹.

EXAMPLE 5.21

Find the phase change rate when the voltage across the Josephson junction is 2 mV.

SOLUTION

Apply Equation 5.41:

$$\frac{\partial \delta}{\partial t} = \frac{2q}{K_P}V_J(t) = \frac{2\,(1.602 \times 10^{-19}\text{C})(2 \times 10^{-3}\text{ V})}{6.626176 \times 10^{-34}\text{ J} \cdot \text{s}} = 967 \text{ GHz}$$

5.9.2 OPERATING PRINCIPLE OF SQUIDs

The critical part of an SQUID sensor is the Josephson junction. An SQUID is in essence a superconducting loop interrupted by one or two Josephson junctions. The loop that has only one Josephson junction is called RF (radio frequency) SQUID, while the loop that has two Josephson junctions is called DC (direct current) SQUID. Both were invented by Robert Jaklevic et al. at *Ford Research Labs* in 1964–1965.

5.9.2.1 DC SQUID

A DC SQUID is based on DC Josephson effect. Figure 5.63a shows a typical DC SQUID. It has a superconducting loop interrupted by two Josephson junctions [42]. In the absence of an external magnetic field, the input current, *I*, splits into the two branches equally. If a small external magnetic field is applied to the superconducting loop, an induced current (called *screening current*, I_S) begins circulating in the

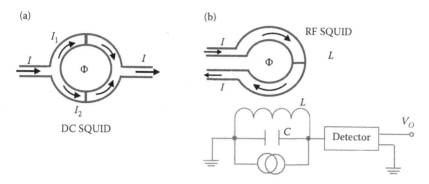

FIGURE 5.63 Two types of SQUIDs: (a) DC SQUID; (b) RF SQUID.

loop and generates a magnetic field against the applied external flux. Thus, I_S is in the same direction as I in one of the branches of the superconducting loop, and is opposite to I in the other branch. The total current, therefore, becomes $I/2 + I_S$ in one branch and $I/2 - I_S$ in the other. As soon as the current in either branch exceeds the junction's critical current I_C, the Josephson voltage appears across the junction.

If the external flux is further increased and exceeds $\Phi_0/2$ (half of the magnetic flux quantum), due to the fact that the flux enclosed by the superconducting loop must be an integer number of the flux quantum, the SQUID now energetically prefers to increase it to $n\Phi_0$ ($n = 1, 2, ...$) instead of screening the flux. The current I_S now flows in the opposite direction. Thus, the screening current changes its direction every time the flux is over $(n + 1/2)\Phi_0$, resulting in the current oscillation as a function of the applied flux. If the input current I is larger than I_C, the SQUID always operates in the resistive mode. The AC Josephson voltage in this case is therefore a function of the applied magnetic field and the period is equal to one flux quantum Φ_0. Since the current–voltage characteristics of a DC SQUID is hysteretic, a shunt resistance, R, is connected across the junction to eliminate the hysteresis (in the case of copper oxide-based high-temperature superconductors, the junction's own intrinsic resistance is usually sufficient). The screening current is the applied flux divided by the self-inductance of the ring. Thus, $\Delta\Phi$ can be estimated as the function of ΔV (flux to voltage converter) as follows (L is the self-inductance of the superconducting ring) [43,44]:

$$\left.\begin{array}{l} \Delta V = R\Delta I_S \\ I_S = \Delta\Phi/L \end{array}\right\} \Rightarrow \Delta V = \Delta\Phi R/L \qquad (5.42)$$

5.9.2.2 RF SQUID

An RF SQUID is based on the AC Josephson effect and has only one Josephson junction. The RF SQUID is inductively coupled to a resonant tank or LC (inductor–capacitor) circuit (see Figure 5.63b). When an external magnetic field is applied, the conductivity of SQUID alters, causing a change in the effective inductance and thus the resonant frequency of the LC circuit. The variation in the resonant frequency can be easily measured and converted to an output voltage that is proportional to the strength and frequency of the applied magnetic field

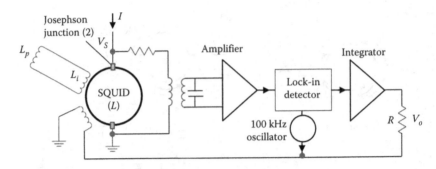

FIGURE 5.64   Schematic of a DC SQUID. (From Jenks, W.G., Thomas, I. M., and Wikswo, J. P., SQUIDS, Department of Physics, Vanderbilt University, Nashville, Tennessee, USA.)

[45]. An RF SQUID is less sensitive compared to a DC SQUID, but it is more cost efficient and easier to manufacture in smaller quantities.

A practical DC SQUID is shown in Figure 5.64. It contains a pickup coil, an input coil, the SQUID, and the associated circuit. A field change at the pickup coil, (with inductance L_p) causes a change in field at the input coil (with inductance L_i). The SQUID (with inductance L) is inductively coupled to the input coil, and it detects a change in the magnetic flux. The voltage response of the SQUID, V_s, is then fed into the flux-locked loop, which uses a modulating flux scheme to maintain the SQUID at an extremum point on the V–Φ curve. The change in the output voltage V_o is proportional to the feedback from the loop and to the change in the magnetic field at the pickup coil. To enhance the SQUID's capabilities, the SQUID does not expose to the magnetic field of interest, rather it employs a multiple-turn pickup coil (L_p); while the input coil (L_i) and the SQUID (with inductance L) are shielded from the ambient field by a superconducting niobium canister. Typical values for the inductances are $L_p = L_i = 1\ \mu H$ and $L = 0.1\ nH$. The pickup coil can be a wounded superconducting wire or a thin film fabricated on the same chip as the SQUID or an adjacent "flip chip" using integrated circuit technologies.

5.9.3   MATERIALS AND CHARACTERISTICS OF SQUID SENSORS

Traditional superconducting materials for SQUIDs are pure niobium or a lead alloy with 10% gold or indium, as pure lead is unstable when its temperature is repeatedly changed. To maintain superconductivity, the entire SQUID device needs to be operated within a few degrees of absolute zero, cooled with liquid helium.

There are two types of SQUID sensors: low-temperature SQUID (LTS) and high-temperature SQUID (HTS) sensors. These sensors need a cryogenic environment to become superconducting at very low temperature. LTS SQUIDs, with a sensitivity of 1–10 fT (f—femto, 10^{-15}), are more sensitive than the HTS SQUID, with a sensitivity of 10–40 fT [46]. Superconductivity occurs in many materials, from simple elements (e.g., aluminum, Al), to metallic alloys (e.g., MgB_2, $PuCoGa_5$), to heavily doped semiconductors. Superconductivity does not occur in noble metals like gold and silver, or in pure samples of FM metals. Table 5.8 lists several LTS and HTS materials and their critical temperature [47].

TABLE 5.8
Superconducting Materials and their Critical Temperature

LTS Materials	T_c (K)	HTS Material	T_c (K)
Al	1.2	$Bi_2Sr_2Ca_2Cu_2O_{8+X}$	85
Hg (solid)	4.2	$YBa_2Cu_3O_7$	93
$LaNi_2B_2C$	16.5	$Bi_2Sr_2Ca_2Cu_3O_{10}$	110
PuCoGa5	18.5	$TlBa_2Ca_2Cu_3O_{10}$	122
Nb_3Ge	23	$Ta_2Ba_2Ca_2Cu_3O_{8+X}$	125
MgB_2	39	$HgBa_2Ca_2Cu_3O_{10}$	133

Source: Adapted from Krasnov, V.M., *Superconductivity and Josephson Effect: Physics and Applications*, Stockholm University, p. 15, 2012.

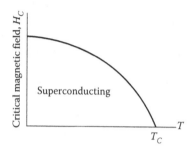

FIGURE 5.65 Temperature dependence of the critical field H_C. (From Krasnov, V.M., *Superconductivity and Josephson Effect: Physics and Applications*, Stockholm University p. 15, 2012.)

Tin (Sn) is also a popular material to make superconducting rings that can carry circulating DC currents for years without any power source and without any measurable current decay. This is because the resistivity of superconducting tin is at least 17 orders of magnitude less than that of copper at room temperature. Figure 5.65 depicts the temperature dependence of the critical field H_C for several materials [47].

SQUID are operated as either DC or AC SQUIDs depending on whether the Josephson junction(s) is biased with a direct (DC) or an alternating (AC) current. The noise mechanisms of the DC and AC SQUIDs are inherently different, with the DC SQUID offering an advantage of more than two orders of magnitude. With modern thin film fabrication techniques and improvements in control electronics design, the DC SQUID offers clear advantages over the AC SQUID for neuromagnetic applications. The demand for improved signal-to-noise requires the use of DC SQUIDs.

5.9.4 SQUID Noise

SQUID noise can be described in by many ways: power spectral density in terms of energy $(J \cdot Hz^{-1})$, magnetic flux $(\Phi_0 \cdot Hz^{-\frac{1}{2}})$, field $(fT \cdot Hz^{-\frac{1}{2}})$, or field gradient (fT $\cdot$ cm^{-1} $\cdot$ Hz$^{-\frac{1}{2}}$). These distinctions are important as one system may have a poor sensitivity, although its energy sensitivity in the SQUIDs is superior. The energy and flux noise descriptions are the figures of merit (FOM) for the bare SQUID, while the field and field-gradient noise descriptions are the FOMs for the complete SQUID system. For commercial DC SQUIDs typical orders of magnitude are: noise energy $\varepsilon = 10^{-31}$ J $\cdot$ Hz^{-1}; magnetic flux noise $(S_\Phi)^{1/2} = 10^{-6}$ J $\cdot$ Hz^{-1}; and magnetic field noise $B_N = 10$ fT $\cdot$ Hz$^{-\frac{1}{2}}$. The DC SQUID noise can be estimated by

$$S_\Phi = \frac{(4k_B T R_D^2 / R)[1 + 0.5(I_0 / I)^2]}{(R/L)^2}, \tag{5.43}$$

$$\varepsilon = \frac{S_\Phi}{2L}, \quad B_N = \frac{L_p + L_i}{\alpha\sqrt{L_i L}} \frac{S_\Phi^{1/2}}{A_e}, \tag{5.44}$$

where R_D is the dynamical resistance of the Josephson junctions $(=\partial V/\partial I)$; R is the resistance of the SQUID; A_e is the effective area of the pickup coil; k_B (1.38×10^{-23} J $\cdot$ K^{-1}) is Boltzmann constant; T is the temperature; and α is the coupling coefficient between the input coil and the SQUID. ε/α^2 is often used as the coupled energy sensitivity.

For the RF SQUIDs, the flux noise spectral density is given by

$$S_\Phi = \frac{(LL_i)^2}{\omega_{RF}} \left(\frac{2\pi k_B T}{I_0 \Phi_0} \right)^{4/3} \tag{5.45}$$

where $\omega_{RF} = 2\pi f_{RF}$ (f_{RF}—frequency of the RF). The energy and field noise descriptions have additional terms related to the room-temperature electronics.

$1/f$ noise exists in thin-film SQUIDs, but only when f is below 0.1 Hz $1/f$ noise becomes dominant in conventional niobium SQUIDs. Low-T_c DC SQUIDs are considerably less noisy than RF SQUIDs, except for RF SQUIDs that operate at microwave frequencies.

5.9.5   Design and Applications of SQUIDs

Most SQUID sensors are made of a superconducting ring/loop (see Figure 5.66), with one or two weak switching points (Josephson Junctions or fragile links). They

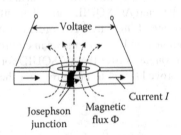

FIGURE 5.66   A typical design of a SQUID sensor.

are versatile and can measure many physical or convert them into magnetic fluxes, electric currents, voltages, or magnetic susceptibilities.

Superconductive materials are used in magnetic bearings and magnetic shields, power cables, transformers, and fault current limiters. They are also widely used in various sensors, such as Superconducting Tunnel Junction Detectors (STJs)—a viable replacement for CCDs (charge-coupled devices) in astronomy and astrophysics. SQUIDs also find many applications in medical diagnostics and have been used to monitor brain activities as a complementary method to EEG (electroencephalogram). SQUIDs have been successfully applied in the field of non-destructive evaluation (NDE).

5.9.5.1 SQUID for Biomagnetism

Since an SQUID sensor is the only magnetic sensor that can detect the most diminutive magnetic signal, it is an essential part in any type of biomagnetic device, especially in equipment used in magneto-encephalography (MEG, measuring the very weak magnetic fields generated by neuronal activity in the human brain) and in the innovative fetal magneto-cardiography (FMCG, measuring the weak magnetic fields generated by the human heart). For example, the basic components of a neuromagnetic instrument include a SQUID sensor, a field sensing coil, a cryogenic component, and electronics for control and data acquisition. The SQUID sensor can be a Josephson junction loop made of a superconducting material. The field sensing coil detects changes in the external magnetic field and transforms it into an electrical current. The cryogenic components or refrigerator maintains the SQUID and detection coil in the superconducting state. The control electronics with a negative feedback is to maintain the system operating point at a particular flux quantum. An additional signal conditioning electronics is often included to improve the signal to noise ratio. To reduce noise significantly, the sensor's operating temperature can be set to below $0.5 \, T_c$. Usually this type of systems can detect 1–10 fT (=10^{-15} T) magnetic field strength.

5.9.5.2 SQUID Battery Monitors

A lithium battery is a clean promising alternative to power electric cars. A technological challenge during the production of electric cars is to ensure that ultra-small metallic contaminants (<100 μm) do not enter lithium ion batteries during manufacturing, which could lead to malfunction of the batteries and ultimately recall of cars at great cost to the manufacturers. Hence, there is a great need to monitor lithium ion (Li-ion) batteries during manufacture to detect metallic particles, which cannot be detected by x-ray imaging or other methods. Toyohashi University of Technology (Japan) developed an ultra-sensitive system for detecting sub-100 μm magnetic contaminants in Li-ion batteries using an HTS SQUID [48]. The system consists of a permanent magnet, a conveyor, a magnetically shielded cylinder and a cryostat with a planar HTc SQUID gradiometer. It allows the SQUID magnetic field sensor to approach an object to be measured within a distance of 1 mm, and can detect iron particles as small as 50 × 50 μm located on an active material-coated sheet electrode of a lithium ion battery.

EXERCISES

Part I: Concept Problems

1. For a Hall sensor to work properly, it must be mounted in the way so that the current direction has
   A. An angle of $0°$ to the magnetic direction
   B. An angle of $45°$ to the magnetic direction
   C. An angle of $90°$ to the magnetic direction
   D. An angle of $180°$ to the magnetic direction

2. Hall sensors can NOT be made of
   A. Metals
   B. n-type semiconductors
   C. p-type semiconductors
   D. Pure semiconductors

3. A material with a lower carrier density, N, will exhibit the Hall effect more _____ (strongly or weakly) for a given current and depth.

4. Which of the following magnetoresistance (MR) effects, theoretically, produces the largest resistance change (in %)?
   A. OMR
   B. AMR
   C. GMR
   D. CMR

5. Match each MR sensor to its typical material:

OMR	Bi, Mo, Sb, or W
AMR	Fe/Cr/Fe layer
GMR	$La_{1-x}M_xMnO_{3+\delta}$ (M = Ca, Sr)
CMR	Permalloy ($Ni_{81}Fe_{19}$)

6. Write the values of the following constants, and indicate where they are used.
   Avogadro constant: _____
   Josephson constant: _____
   Magnetic flux quantum: _____
   Boltzmann's constant: _____

7. Which of the following effects relates a material's elastic state to its magnetic state?
   A. Josephson effect
   B. Hall effect
   C. MR effect
   D. Magnetostrictive effect

8. Which of the following effects results in a twisting?
   A. Wiegand effect
   B. Villari effect
   C. Wiedemann effect
   D. Barkhausen effect

9. Which of the following effects relates magnetism to acoustics?
   A. Joule effect

B. Kerr effect

C. Farady effect

D. Meissner effect

10. Which of the following materials is used to make superconducting rings that can carry circulating DC currents for a couple of years without any power source?

A. Copper

B. Silver

C. Gold

D. Tin

11. Which of the following sensors or devices can measure extremely small variations in magnetic flux:

A. SQUIDs

B. AMR sensors

C. Hall sensors

D. MDLs

12. Which of the following sensors is related to Lorentz force?

A. Resonant magnetic field sensor

B. Hall sensor

C. GMR sensor

D. MRI sensor

13. A Wiegand wire is made of which of the following materials?

A. Permalloy ($Ni_{80}Fe_{12}$)

B. Terfenol-D ($Tb_{0.3}Dy_{0.7}Fe_{1.92}$)

C. Vicalloy ($Co_{52}Fe_{38}V_{10}$)

D. Crystalline metals (FeCo)

14. The following materials are magnetostrictive materials, EXCEPT

A. Ni

B. FeNi

C. FeCoNi

D. Al

15. Which of the following effects relates magnetism to optics?

A. Wiegand effect

B. Kerr effect

C. Barkhausen effect

D. Meissner effect

16. In the SQUIDs, "superconducting" means that the material's

A. electrical resistance is zero or close to zero

B. thermal resistance is zero or close to zero

C. inductance is zero or close to zero

D. magnetic reluctance is zero or close to zero

17. In equation $V_J(t) = (K_p/2q)(\partial\delta/\partial t)$, K_p is:

A. Plank's constant (6.626176×10^{-34} m$^2 \cdot$ kg $\cdot$ s^{-1} or J $\cdot$ s)

B. Josephson constant (483597.9×10^9 Hz $\cdot$ V^{-1})

C. Gyromagnetic constant (2.675×10^8 rad $\cdot$ T$^{-1} \cdot$ s^{-1})

D. Avogadro constant (6.0221415×10^{23} mol^{-1})

18. Which of the following describe the superconductor behavior correctly?

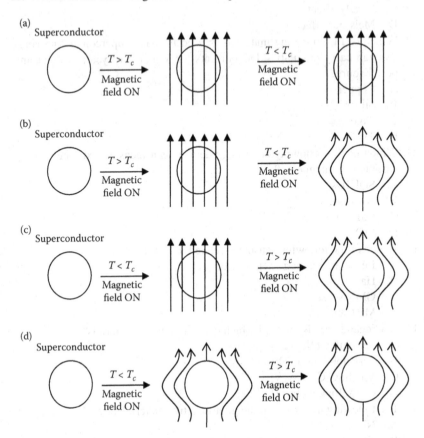

Part II: Calculation Problems

19. Calculate the Hall voltage V_H across a ribbon of copper in a magnetic field of 3 T, knowing that ribbon width: 2 mm; thickness: 0.3 mm; geometry factor G 0.94. The ribbon carries a current of 17 A. For copper: $N_c = 8.4 \times 10^{28}$ free electrons per m^3. Assume that one electron per copper atom is available for conduction.

20. An n-type silicon Hall sensor has a doped level of 5×10^{15} cm^{-3}. If the sensor is 25 μm thick, the current flowing through it is 2.5 mA, and the magnetic field is 0.9 T, what is the sensor's output voltage?

21. An n-type silicon Hall sensor has a doped level of 6×10^{15} cm^{-3}. If the sensor is 20 μm thick, the current flowing through it is 2 mA, and the magnetic field is 1 T, what is the sensor's output voltage?

22. Find the resistance of a Hall sensor made of a copper foil (1 mm long, 0.5 mm wide, and 0.025 mm thick), knowing that the resistivity of the copper is 1.72×10^{-8} Ω · m.

23. A Hall sensor is made from an n-type silicon doped to a concentration level of 4×10^{15} cm^{-3}. Its dimension is 1 mm (length) × 0.5 mm (width) × 0.025 mm

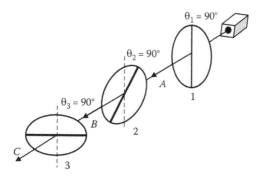

FIGURE 5.67   A three polarizer setup.

(thickness), and its resistivity is $0.0168\ \Omega \cdot$ m at room temperature. Find (1) its resistance; (2) Hall voltage if a 1 mA of current flows through the sensor.

24. A student tried to build a sensor to detect the strength of unknown magnetic fields. He used a rectangular sample of copper that is 14.2 cm wide and 0.5 cm thick. He applied a current of 2.4 A to the copper. Knowing that there was a magnetic field perpendicular to the current because he measured a Hall voltage of 0.1 μV, what was the magnitude of the magnetic field that he detected? Assume that one electron per copper atom is available for conduction. Copper has a density of 8.89 g $\cdot$ cm^{-3} and a molar mass of 63.55 g $\cdot$ mol^{-1}. 1 mol of any substance contains 6.02×10^{23} atoms (Avogadro's number).

25. Find the Larmar frequency f_L for a nucleus with the gyromagnetic constant of $\gamma_g = 2.675 \times 10^8$ rad $\cdot$ T^{-1} $\cdot$ s^{-1} under a local magnetic field of 54 μT. $\delta_c = 0.9$.

26. The measured Larmar frequency of a nucleus is 2287 Hz. Find the local magnetic field B. Given $\gamma_g = 2.675 \times 10^8$ rad $\cdot$ T^{-1} $\cdot$ s^{-1} and $\delta_c = 1.5$.

27. Figure 5.67 shows unpolarized laser light passing through three polarizers. (1) Calculate the light intensity after each polarizer (at points A, B, and C) as it passes through each of the polarizers. (Given $I_1 = 0.5\ I_0$). (2) If the second polarizer is removed, calculate the intensity at point C.

28. Find the phase difference in a DC SQUID, knowing that the critical current I_C is 0.3 mA and the Josephson current I_J is 0.041 mA.

29. If the phase difference in the superconducting materials on both sides of a Josephson junction changes in a rate of 700 GHz (i.e., $\partial \delta / \partial t = 700$ GHz), find the voltage across the Josephson junction V_J.

REFERENCES

1. Ramsden, E., *Hall-Effect Sensors: Theory and Applications*, Elsevier Inc., Burlington, Massachusetts, USA, 2006.
2. Schott, C., Burger, F., Blanchard, H., and Chiesi, L., Modern integrated silicon Hall sensors, *Sensor Review*, 18(4), 252–257, 1998.

3. Blanchard, H., Chiesi, L., Racz, R., and Popovic, R.S., Cylindrical Hall device, *Proc. of International Electron Devices Meeting*, Dec. 8–11, 1996, San Francisco, California, USA, pp. 541–544.

4. Tsymbal, E.Y. and Pettifor, D.G., Perspectives of giant magnetoresistance, in *Solid State Physics*, Vol. 56, Academic Press, London, 113–237, 2001.

5. Fert, A., Duvail, J.L, and Valet, T., Spin relaxation effects in the perpendicular magneto-resistance of magnetic multilayers, *Phys. Rev.* 52, 6513, 1995.

6. Nickel, J., Magnetoresistance overview, HPL-95-60, Hewlett Packard Company, Palo Alto, California, USA, 1995.

7. Hsu, J.H. et al., Substrate dependence of large ordinary magnetoresistance in sputtered Bi films, *J. Magnetism and Magnetic Materials*, 272, 1769–1771, 2004.

8. Ripka, P. and el Závěta, K., Chapter 3, Magnetic Sensors: Principles and Applications. *Handbook of Magnetic Materials*, Editor, Buschow, K.H.J., Elsevier. Amsterdam Holland, Vol. 18, pp. 347–420, 2009.

9. Baibich, M.N., Broto, J.M., Fert, A., Vandau, F.N., Petroff, F., Eitenne, P., Creuzet, G., Friederich, A., and Chazelas, J., Giant magnetoresistance of (001)Fe/(001)Cr magnetic superlattices, *Phys. Rev. Lett.* 61, 2472–2475, 1988.

10. Binasch, G., Grunberg, P., Saurenbach, F., and Zinn, W., Enhanced magnetoresistance in layered magnetic-structures with antiferromagnetic interlayer exchange, *Phys. Rev. B*, 39, 4828–4830, 1989.

11. Julliere, M., Tunneling between ferromagnetic films, *Phys. Lett.*, 54A, 225, 1975.

12. Maekawa, S. and Gafvert, U., Electron tunneling between ferromagnetic films, *IEEE Transaction on Magnetics*, Vol. 18(2), 707–708, 1982.

13. Simmons, J.G., Electric tunnel effect between dissimilar electrodes separated by a thin insulating film, *J. Appl. Phys.* 34, 2581, 1963.

14. Bowen, M. et al., Large magnetoresistance in Fe/MgO/FeCo(001) epitaxial tunnel junctions on GaAs(001), *Appl. Phys. Lett.* 79, 1655, 2001.

15. Chopra, H.D. and Hua, S.Z., Ballistic magnetoresistance over 3000% in Ni nanocontacts at room temperature, *Phys. Rev.*, B 66, 020403–1, 2002.

16. von Helmolt, R. et al., Giant negative magnetoresistance in perovskite like $La_{2/3}Ba_{1/3}MnO_x$ ferromagnetic films, *Phys. Rev. Lett.*, 71, 2331, 1993.

17. Hauser, H. et al., Magnetoresistive sensors, *Electro. Inform. Tech.*, 115, 382, 1998.

18. Terfenol-D Data Sheet, Etrema Products, Inc., Ames, Iowa, USA. www.etrema-usa.com/documents/Terfenol.pdf.

19. Hankinson, M., One sensor, multiple targets—A single linear sensor monitors multiple positions along its axis, *Machine Design*, Nov. 23, 2008.

20. Fleming, W.J., Magnetostrictive torque sensor performance-nonlinear analysis, *IEEE Transactions on Vehicular Technology*, 38(3), 159–167, 1989.

21. Nishibe, Y., Nonomura, Y., Tsukada, K., and Takeuchi, M., Determination of engine misfiring using magnetoelastic torque sensor, *IEEE Transactions on Instrumentation and Measurement*, 47(3), 760–765, 1998.

22. Hanna, S.M., Magnetic field sensors based on SAW propagation in magnetic films, *IEEE Trans. on Ultrasonics, Ferroelectrics, and Frequency Control*, UFFC-34(2), 191–194, 1987.

23. Wang, W.C., Hua, W.S., Lin, W.H., and Wu, W.J., Development of a polymeric magnetostrictive fiber-optic sensor system, *Proc. SPIE 6935, Health Monitoring of Structural and Biological Systems*, April 10, 2008.

24. Griffiths, J.R. et al. 31 P-NMR studies of a human tumor in situ. *Lancet I*, 1435–1436, 1983.

25. Lauterbur, P., MRI—A new way of seeing, *Nature, London*, 242, 190–191, 1973.

26. Herrera-May, A.L., Aguilera-Cortés, L.A., García-Ramírez, P.J., and Manjarrez, E., Resonant magnetic field sensors based on MEMS technology, *Sensors* 9(10), 7785–7813, 2009.

27. Sunier, R., Vancura, T., Li, Y., Kay-Uwe, K., Baltes, H., and Brand, O., Resonant magnetic field sensor with frequency output, *J. Microelectromech. Syst.* 15, 1098–1107, 2006.

28. Kádár, Z., Bossche, A., Sarro, P. M., and Mollinger, J. R., Magnetic-field measurements using an integrated resonant magnetic-field sensor, *Sensors and Actuators A*, 70(3), 225–232, 1998, ISSN 0924-4247.

29. Emmerich, H. and Schöfthaler, M., Magnetic field measurement with a novel surface micromachined magnetic-field sensor, *IEEE Trans. on Electron Devices*, 47(5), 972–977, 2000.

30. Cullity, B.D., *Introduction to Magnetic Materials*, Addison-Wesley Publishing Company, Massachusetts, 1972.

31. Time-Resolved Magneto-optical Kerr effect, (TR-MOKE), 2013 AG Aeschlimann, retrieved on June 29, 2013. www.physik.uni-kl.de/aeschlimann/?mode=methods&action=show&show=00000005

32. Wiegand, J.R., Pulse generator using read head with Wiegand wire, US Patent No.: 4263523 A. Apr. 21, 1981.

33. Dlugos, D.J., Wiegand effect sensors: theory and applications, *Sensors*. Questex Media Group, 1998.

34. Silva, R. M. et al., Optical current sensors for high power systems: A review, *Appl. Sci.*, 2, 602–628, 2012.

35. Koschny, M. and Lindner, M., Magneto-optical sensors: Visualization and analysis of magnetic field distribution of magnetic materials, *Adv. Mater. Proces.*, Feb., 13–16, 2012.

36. Ning, Y.N. and Wang, Z.P., Recent progress in optical current sensing techniques. *Rev. Sci. Instrum.*, 66, 3097–3111, 1995.

37. Muto, S., Seki, N., and Suzuki, T., Plastic fiber isolator and current sensor, *Japanese J. Appl. Phys.*, 31(3B), part 2–letters, 346–348, 1992.

38. Aikawa, E., Ueda, A., Watanabe, M., Takahashi, H., and Imataki, M., Development of new concept optical zero-sequence current/voltage transducers for distribution network, *IEEE Trans. Power Delivery*, 6(1), 414–420, 1991.

39. Berman, P.R., Optical Faraday rotation. *Am. J. Phys.* 78, 270–276, 2010.

40. Sohlström, H., Fiber optic magnetic field sensors utilizing iron garnet materials, Department of Signals, Sensors & Systems, Instrumentation Laboratory, S-100 44 Stockholm, 1993.

41. Mora, J., Diez, A., Cruz, J.L., and Andres, M.V. A magnetostrictive sensor interrogated by fiber gratings for DC-current and temperature discrimination. *IEEE Photon. Technol. Lett.*, 12, 1680–1682, 2000.

42. Fagaly, R.L., Superconducting quantum interference device instruments and applications. *Rev. Sci. Instrum.* 77, 101101, 2006.

43. de Lacheisserie, E.T., Gignoux, D., and Schlenker, M. eds. *Magnetism: Materials and Applications*, Vol. 2, Springer Science & Business Media, Inc., New York, 2005.

44. Clarke, J. and Braginski, A.I. eds. *The SQUID Handbook*, Vol. 1, Wiley-Vch Verlag GmbH & Co. KGaA, Weinheim, 2004. ISBN: 3-527-40229-2.

45. Erné, S.N., Hahlbohm, H.D., and Lübbig, H., Theory of the RF biased superconducting quantum interference device for the non-hysteretic regime, *J. Appl. Phys.*, 47(12), 5440–5442, 1976.

46. Fagaly, R.L., *Neuromagnetic Instrumentation*, Chapter 2, Advances in Neurology, Vol. 54: Magnetoencephalography, S. Sato, ed., Raven Press, New York, 1990.

47. Krasnov, V.M., *Superconductivity and Josephson Effect: Physics and Applications*, Stockholm University, P15, 2012.

48. Tanaka, S., Akai, T., Hatsukade, Y., Ohtani, T., Ikeda, Y., Suzuki, S., and Tanabe, K., High *Tc* SQUID detection system for metallic contaminant in lithium ion battery, *IEEE Transactions on Magnetics*, 45(10), 4510–4513, 2009.

6 RCIM Sensor Circuitry

6.1 INTRODUCTION

Most RCIM sensors are associated with electronic circuits. Some are delivered from manufacturers with built-in electronics (more expensive), while others need to be customized and built by users using individual electronic components. This chapter focuses on common RCIM sensor circuitry including their signal characteristics, excitation means, grounding and shielding techniques, compensation methods, measurement bridges, passive/active filters, and signal conditioning solutions. The chapter is organized as follows: Section 6.2 presents the signal levels of RCIM sensors; Section 6.3 discusses sensor noise sources and forming mechanisms; Section 6.4 reviews the grounding and shielding of RCIM sensor circuits; Sections 6.5 and 6.6 are devoted to DC and AC bridges for resistance, capacitance, and inductance measurement, respectively; Section 6.7 focuses on the output circuits of RCIM sensors; Section 6.8 provides information on analog compensation techniques, and finally, Section 6.9 covers passive and active filters and their applications in RCIM sensor signal conditioning.

6.2 RCIM SENSOR SIGNAL CHARACTERISTICS

Even with thousands of RCIM sensor designs, only a few electrical properties are handled in the sensor circuits. These electrical properties include:

- Voltage (e.g., potentiometers)
- Current (e.g., Hall effect sensors)
- Resistance (e.g., RTD, thermistors, photoresistors)
- Capacitance (e.g., capacitive displacement sensors, level sensors)
- Charge (e.g., piezoelectric sensors)
- Inductance (e.g., inductive and magnetic sensors)
- Frequency (e.g., photoresistive tachometers)

These electrical properties can be directly recorded from sensors or through some conversion mechanisms (e.g., resistance-to-voltage conversion, current-to-voltage conversion). In addition, sensor signals may also be classified as AC or DC and analog or digital signals.

To design an efficient RCIM sensor circuitry, one needs to know the RCIM sensor signal levels and ranges. Tables 6.1 through 6.4 list commonly used RCIM sensors' typical values and detecting ranges, which are compiled from multiple sources [1–8].

TABLE 6.1
Resistance Values of Common Resistive Sensors

Metal strain gauges/load cells/pressure sensors: 120–350 Ω
Semiconductor strain gauges/load cells/pressure sensors: ~3500 Ω
Relative humidity sensors: 100 Ω–10 MΩ
Photoresistors: in dark: >1 MΩ; under light: 400–9000 Ω
RTDs: 100–1000 Ω   Thermistors: 100 Ω–10 MΩ
AMR sensors: 700–1000 Ω   GMR sensors: 1000–7000 Ω

Source: From Andreev, S.J. and Koprinarova, P.D., Magnetic properties of thin film AMR sensor structures implemented by magnetization after annealing, *J. Optoelectronics Adv. Materials*, 7(1), 317–320, 2005; Kester, W., Bridge Circuit, Section 2. Seminars, Analog Devices, Inc., page 1; Introduction to NVE GMR Sensors, NVE Corporation, pp. 13–14. With permisson.

TABLE 6.2
Capacitance Values of Common Capacitive Sensors

General capacitive sensors: 10–100 pF
Capacitive position sensors: 10 pF (with plate spacing 10–20 μm)
Capacitive motion sensors: 0.75 pF (with plate spacing 250 μm)
Capacitive limit switches: 0.05–0.35 pF
Capacitive pressure sensors: 0.3–22 pF
Capacitive accelerometers: 0.004–20 pF
Capacitive proximity sensors: nominal sensing range: 20 mm. For different target materials, the detection ranges are:
Metal: 40 mm   Water: 40 mm   PVC: 20 mm   Glass: 12 mm   Cardboard: 8 mm
Wood (varies with relative humidity): 20–32 mm
Stray capacitance generated by a PC trace: ~0.5 pF

Source: From Baxter, L.K., *Capacitive Sensors: Design and Applications*, IEEE Press, New York, 51, 1997; Lepkowski, J., Designing operational amplifier oscillator circuits for sensor applications, AN866, Microchip Technology Inc., DS00866, 2003. With permission.

One major difference between magnetic sensors and inductive sensors is that inductive sensors must work under an AC source to induce an alternating magnetic field, while magnetic sensors can work under a static magnetic field generated by either a permanent magnet or a direct current (DC), or under a varying magnetic field generated by an AC source. For example, Hall sensors can measure static magnetic fields.

Most RCIM sensors have a milliampere (mA) current range and a millivolt (mV) output level before they undergo amplification and signal conditioning. A few sensors have current loads in the order of microamperes (μA) and voltage outputs in the order of microvolts (μV). Some sensors, such as potentiometers, have outputs at volt

TABLE 6.3
Inductive and Magnetic Sensor Characteristics

Excitation or driving frequency for local magnetic fields: 2–500 kHz
Inductive proximity sensor detection range 1–100 mm (for most industrial sensors,
  the maximum detection range is the same as the diameter of the detecting coils).
Measurement ranges of inductive distance sensors for various target materials:

Mild steel	$1 \times$ Nominal sensing range
Stainless steel	$(0.7–1.0) \times$ Nominal sensing range
Brass	$(0.3–0.5) \times$ Nominal sensing range
Copper	$(0.2–0.6) \times$ Nominal sensing range
Aluminum	$(0.1–0.4) \times$ Nominal sensing range

Source: From Baxter, L.K., *Capacitive Sensors: Design and Applications*, IEEE
  Press, New York, 51, 1997; *Inductive Proximity Sensors*, Allen-Bradley. With
  permission.

TABLE 6.4
Detectable Magnetic Fields (in Tesla) by Main Magnetic Sensors

SQUIDs	$1 \times 10^{-13}–10$ T
Search coils	$1 \times 10^{-12}–1 \times 10^{6}$ T
AMRs	$5 \times 10^{-11}–75 \times 10^{-3}$ T
GMRs	$1 \times 10^{-3}–1 \times 10^{4}$ T
Flux gates	$7.5 \times 10^{-11}–10 \times 10^{-3}$ T
Hall-effect sensors	$7.5 \times 10^{-3}–10$ T
Magnetic switches	$8 \times 10^{-4}–3 \times 10^{-3}$ T

Source: From Caruso, M.J. and Withanawasam, L.S., Vehicle detection and compass application
  using AMR magnetic sensors, 1999; Magnetic (AMR) sensor AS series, Murata
  Manufacturing Co., Ltd., 2008. With permission.

(V) level, where no amplification is necessary. More power is usually needed for
inductive sensors compared to capacitive sensors.

6.3 SENSOR NOISE SOURCES AND FORMING MECHANISMS

Noise from various sources often limits a sensor's performance. A good sensor system should have an effective noise attenuation means, so that the magnitude of the
useful signal is larger than that of the noise. The signal-to-noise ratio indicates the
maximum noise that a sensor can tolerate. Understanding noise sources and formation mechanisms is critical to sensor circuit design.

6.3.1 Noise Sources

Noise can arise either from internal sensor system elements or from external distur-bances. All electrical or electronic components generate noise. The main internal components that contribute to noise are:

- Electrical/electronic components (e.g., resistors, capacitors, inductors, LEDs, transistors)
- Power supplies (e.g., AC power lines, adapters)
- IC chips (e.g., amplifiers, filters, regulators)
- Lead connectors and terminals
- Circuits themselves (e.g., traces, ground loops).

External electrical noise sources that affect the output of a sensor system include:

- Signal lines nearby that interfere with a sensor system by various means—capacitive/inductive coupling, electromagnetic interference (EMI), radio frequency interference (RFI).
- 60 Hz or higher harmonics at 120 Hz, 180 Hz, and 240 Hz picked up from outlets, lights, motors, and other devices. 60 Hz noise is the most common problem and often the most difficult to attenuate by filtering.
- Radio and TV stations ($\sim 10^6$–10^8 Hz).
- Lightning, arc welders, auto ignition systems, and other "spark" generators produce wide band noise.
- Mechanical vibrations (normally at frequencies below 20 Hz) affect the output of certain sensors.

Noise is transmitted to a sensor system either by direct contact (*conductive coupling*) or interference (*capacitive coupling, inductive coupling*, and *electromagnetic coupling*). The latter usually involves three distinct parts: a source, a coupling mech-anism, and a receiver. A receiver could be the sensor itself, or a sensor's signal line, or the sensor circuit [9,10]. The following sections discuss each of these transmissions.

6.3.2 Noise Transmission Mechanisms

6.3.2.1 Conductive Coupling

Conductive coupling is the transfer of noise or electrical energy by means of physical contact via a conductive medium such as a wire, a resistor, a trace, a metallic bind-ing, or a common terminal [11]. There is a physical connection between the noise source and the receiver (e.g., sensor). Conductive coupling passes the full spectrum of frequencies including DC components. Proper grounding or filtering can elimi-nate or reduce this coupling.

6.3.2.2 Capacitive Coupling

Capacitive coupling (e.g., *electrostatic coupling*) is the transfer of noise or energy from one circuit (or component) to another by means of the mutual capacitance between the circuits (or components). There is no direct contact between the noise source and the receiver. This mutual capacitance is usually referred to as a *parasitic capacitance*. If a time-varying voltage exists between the coupling objects, a current I_P can "flow" between them [12]:

$$I_P = C_P \frac{dV}{dt} \tag{6.1}$$

where dV and dt are the voltage and time derivatives, respectively; C_P is the parasitic capacitance due to the electric field between the two objects; dV/dt is the rate of change in voltage. The *capacitive reactance*, X_C (in Ω), of a capacitor is [13]

$$X_C = \frac{1}{2\pi f C} \tag{6.2}$$

where f is frequency (in Hz) and C is capacitance (in F). The above equation indicates that as frequency increases, the capacitive reactance decreases. A smaller reactance allows more coupling action through the parasitic capacitance. Thus, a noise source that has high-frequency components is more likely to be capacitively coupled into a sensor system than a source with low-frequency components. The smaller the spacing between the circuits (or components), the greater the capacitive coupling.

6.3.2.3 Inductive Coupling

Inductive coupling requires a current loop and a change in magnetic flux. The change in magnetic flux could come from wires connected to an AC power source, a current change in the signal lines, or the earth's magnetic field. Recall that a time-varying current creates the changing magnetic flux as follows [14]:

$$\Phi = BA = \mu_0 NIA \tag{6.3}$$

where Φ is the magnetic flux (in Wb); B is the magnetic flux density (Wb m^{-2}); A is the loop area (in m^2); μ_0 is the permeability of free space ($4\pi \times 10^{-7}$ Wb $\cdot$ A^{-1} $\cdot$ m^{-1}); N is the number of turns in the loop; and I is the current. Larger loop areas mean greater self-inductance or mutual inductance. The induced voltage V is proportional to the time rate of change in current, dI/dt, and the loop area A as

$$V = d\Phi/dt = \mu_0 NA(dI/dt) \tag{6.4}$$

Reducing either dI/dt or A will reduce the induced voltage. The inductive reactance X_L (in Ω) is

$$X_L = 2\pi f L \tag{6.5}$$

where f is frequency (in Hz) and L is inductance (in H), respectively. Current flows along the path of minimum inductive reactance. Thus, at lower frequencies, inductive reactance is smaller and more current will flow, resulting in larger inductive coupling.

6.3.2.4 Electromagnetic Coupling

Electromagnetic or radiative coupling is a high-frequency phenomenon. It requires a transmitting antenna (spurious emission) in the source and a receiving antenna (spurious response) in the susceptible circuit or sensor. Electromagnetic coupling becomes a factor only when the frequency of operation exceeds 20 MHz. If the frequency is above 20 MHz and below 200 MHz, cables are the primary emitters and receivers for electromagnetic coupling; if the frequency is above 200 MHz, PCB traces begin to radiate and couple energy. Signal lines within circuits must have lengths that are an appreciable fraction of the wavelength to act as antennas. Generally the length must be longer than 5% of the wavelength λ (i.e., $>\lambda/20$) to act as an antenna [9]. A simple formula for calculating wavelength λ (in m) is as follows:

$$\lambda = \frac{v}{f} \tag{6.6}$$

where v is the velocity of the wave propagation (in m · s^{-1}); and f is the frequency of the radiation (in HZ).

Table 6.5 summarizes these coupling types and their frequency levels.

The examples of these couplings include crosstalk (a coupling from an active signal line to a passive line through either a capacitive or an inductive transmission mechanism), radio interference and crackle, and static discharge. Noise susceptible receivers are usually due to incorrect grounding and return paths or long signal lines that are not properly shielded.

Pseudo impedance (in Ω) is defined as the ratio of voltage change to current change:

$$\text{Pseudo impedance} = \frac{dV/dt}{dI/dt} \tag{6.7}$$

It is a very useful diagnostic check to determine the type of an energy coupling (thus, it is also called *Diagnostic Ratio*). Table 6.6 shows the relationship between the pseudo impedance values and their corresponding coupling types [14].

EXAMPLE 6.1

A signal line nearby a sensor circuit carries a 300 MHz clock signal. If the wavelength of the signal is 0.05 m, what is the maximum trace length in the sensor circuit allowed to avoid electromagnetic interference?

TABLE 6.5

Coupling Type and Frequency Level

Coupling Type	Frequency Level
Conductive coupling	All
Capacitive coupling	High
Inductive coupling	Low
Electromagnetic coupling	High

TABLE 6.6

Pseudo Impedance Value and Its Corresponding Coupling Type

Pseudo Impedance (Ω)	Coupling Type	Explanation
<377 Ω	Inductive	A large change in current
>377 Ω	Capacitive	A large change in voltage
~377 Ω (frequency > 20 MHz)	Electromagnetic	A midrate change in current and voltage

Source: From Fowler, K.R., *Electronic Instrument Design: Architecting for the Life Cycle*, Oxford University Press Inc., New York, Chap. 2, 1996. With permission.

SOLUTION

Since 300 MHz is above 200 MHz, the signal line and the trace in the sensor circuit begin to radiate and transmit energy. To avoid electromagnetic interference, the trace in the sensor circuit must NOT act as an antenna. Thus, the trace length in the sensor circuit must NOT be longer than 5% of the wavelength of the clock signal, that is:
The maximum trace length = 0.05 m × 5% = 0.0025 m or 2.5 mm.

EXAMPLE 6.2

At what wave propagation speed should there be concerns about energy transmission in PCB traces that carry signals with a wavelength of 175 mm?

SOLUTION

Electromagnetic transmission in PCB trances begin when the frequency of operation exceeds 200 MHz, which corresponds to a speed of

$$v = \lambda f = (0.175 \text{ m})(200 \times 10^6 \text{ Hz}) = 3.5 \times 10^7 \text{ m} \cdot \text{s}^{-1}$$

EXAMPLE 6.3

A sensor amplifier circuit is receiving noise from a nearby digital circuit when the digital circuit switches its logic levels between "0" (0.8 V) and "1" (4.5 V) within 10 ns. If its current changes from 0 to 10 mA within 100 ns, what type of noise interference is the sensor circuit experiencing?

SOLUTION

The diagnostic ratio (pseudo impedance) is

$$\frac{dV/dt}{dI/dt} = \frac{(4.5 \text{ V} - 0.8 \text{ V})/(10 \times 10^{-9}\text{s})}{(0.01 \text{ A} - 0 \text{ A})/(100 \times 10^{-9}\text{s})} = 3700 \text{ }\Omega$$

Since 3700 $\Omega \gg$ 377 Ω, the sensor circuit is experiencing the capacitive coupling.

6.3.3 TYPES OF NOISE

There are many types of noise in electrical circuits, such as common mode noise generated by ground loops, normal mode noise generated by electromagnetic fields, and electrostatic noise generated by rotating equipment, and so on. This section will only focus on five types of noise that often appear in sensors' circuits: crosstalk, popcorn, thermal, 1/*f*, and shot noise.

6.3.3.1 Crosstalk

Crosstalk is the capacitive or inductive coupling between two signal lines [15]. In the situation of capacitive coupling, crosstalk is the result of an unexpected capacitance formed between the two signal lines; while in the situation of inductive coupling, crosstalk is a transformer action between the two lines. The line that the noise signal is coupled from is referred to as the *active* or *aggressor* line, while the line that the noise signal is coupled onto is called the *passive* or *victim* line. Crosstalk becomes more significant when circuit densities increase. Some preventative design strategies can minimize crosstalk:

- Maximize allowable spacing between signal lines.
- Minimize spacing between signal and ground lines.
- Isolate critical signal lines, such as digital logic signal lines and clock signal lines from other lines to prevent malfunctions.
- Use twisted pair wiring for sensitive applications.
- Make every other line a ground line when using ribbon or flat cables.

6.3.3.2 Popcorn Noise

Popcorn or *Burst* noise was named by early users of audio amplifiers when they noticed a low-frequency noise in audio outputs that sounds like popcorn popping. Popcorn noise is found only in semiconductor components and it is associated with defects in the semiconductors' crystals [15]. It happens when there is a rapid change in a semiconductor bias current between two extremes. The level of popcorn noise can be 5–20 times of Gaussian noise level. Increasing the purity of semiconductor materials can reduce or eliminate this noise. Popcorn noise is a low-frequency noise.

6.3.3.3 Thermal (Johnson or Nyquist) Noise

Thermal noise (also called *Johnson* or *Nyquist noise*) occurs in all conducting materials and is caused by the random motion of electrons through a conductor. The thermal noise of a resistor was first observed and explained by Johnson and Nyquist [16,17]. Note that only resistive components generate thermal noise; capacitive and inductive components do not generate thermal noise. The mean square of thermal noise voltage, $V_{n,T}^2$ (in nV), is directly proportional to the resistance of the component, R (in Ω), and can be computed by [18]

$$V_{n,T}^2 = 4k_B T R \Delta f \tag{6.8}$$

where k_B (1.38×10^{-23} J·K^{-1}) is *Boltzmann constant*; T is the temperature (in K) of the circuit; and $\Delta f = f_H - f_L$ (bandwidth in Hz). The *spectral density* of thermal noise voltage signal is a measurement of root mean square noise voltage per square root Hertz (nV/√Hz), described as

$$\frac{V_{n,T}}{\sqrt{\Delta f}} = \sqrt{4k_B T R} \tag{6.9}$$

Thermal noise is characterized as *white noise* or *Gaussian noise*, because it has a uniform energy distribution at all frequencies with a constant power spectral density. Figure 6.1 shows a typical noise density plot of a resistor (√Hz means the square root of Hz). Note that there are two parts in the curve: one is frequency-dependent noise (denoted as $1/f$ *noise*, whose density decreases as frequency f increases); the other is a frequency-independent thermal noise whose density is nearly constant or flat over the broad frequency range (also called *broadband noise*). Typically, the power spectrum of $1/f$ noise falls at a rate of $1/f$. This means the voltage spectrum falls at a rate of $1/\sqrt{f}$.

In general, the *broadband voltage noise* is denoted as $V_{n,BB}$ in volts (RMS), and is computed by

$$V_{n,BB} = V_{BB}\sqrt{BW_n} \tag{6.10}$$

where V_{BB} is the broadband voltage noise density, usually in nV/√Hz; BW_n is the noise bandwidth for a given sensor.

EXAMPLE 6.4

Calculate how much noise is added to a sensor circuit by a 100 kΩ resistor with 1 MHz bandwidth at room temperature (25°C).

SOLUTION

25°C = (273.15 + 25) K = 298.15 K

$$
\begin{aligned}
V_{n,T} &= \sqrt{4k_B T R \Delta f} \\
&= \sqrt{4(1.38 \times 10^{-23}\ \text{J}\cdot\text{K}^{-1})(298.15\ \text{K})(100 \times 10^3\ \Omega)(1 \times 10^6\ \text{Hz})} \\
&= 40.57\ \mu V
\end{aligned}
$$

Thus, a 40.57 μV RMS noise is added to the circuit by this resistor.

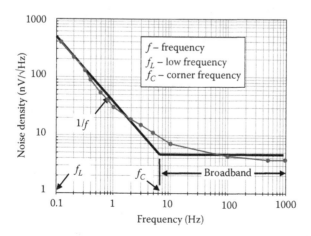

FIGURE 6.1   A typical noise density curve of a resistive component.

6.3.3.4 1/f (Flicker or Pink) Noise

1/f noise is a low-frequency noise with a frequency spectrum such that the power spectral density (PSD) is inversely proportional to the frequence f (i.e., PSD $\propto 1/f$). Both resistors and semiconductors present 1/f noise. This noise was first discovered and successfully measured by J.B. Johnson in 1925 (called *flicker* or *pink noise*) [19]. Thereafter, many scientists, including W. Schottky [20], tried to provide a theoretical explanation. One theory suggests that this noise is produced by a variety of process-dependent mechanisms and can be characterized empirically. Even today, this noise has not been fully understood. 1/f noise is difficult to reduce because it is mainly generated through multiplicative mechanisms and it has fundamentally nonstationary characteristics. The noise reduction methods reported in literature often involve developing a specific component or device. These methods include switched biasing [21], feed forward techniques [22], feedback loops, or the use of different materials and products.

To obtain the RMS of the 1/f noise, $V_{n,1/f}$ (in V), the noise spectral density curve must be integrated over the bandwidth that spans the frequency from f_L to f_C (refer to Figure 6.1), which is

$$V_{n,1/f} = V_{1/f_C}\sqrt{f_C}\sqrt{\int_{f_L}^{f_C}\frac{1}{f}df} = V_{1/f_C}\sqrt{f_C \ln\frac{f_C}{f_L}} \tag{6.11}$$

where V_{1/f_C} (in nV/√Hz or nV $\cdot$ Hz$^{-\frac{1}{2}}$) is *voltage noise density* at f_C—the 1/f corner frequency at which the flicker noise density equals the thermal noise density. f_L is the lowest frequency in the 1/f region (0.1 Hz is typically used). A normalized noise can be defined and calculated by

$$V_{n,\text{norm}} = v_{n,f}\sqrt{f} \tag{6.12}$$

where $V_{n,\text{norm}}$ is the normalized noise at 1 Hz, usually in nanovolts (nV); $v_{n,f}$ is *voltage noise density* at f (in nV/√Hz). f is a frequency in the 1/f region where noise voltage density is known. In some cases, $V_{n,\text{norm}}$ can be read directly from the sensor data sheet.

EXAMPLE 6.5

If the noise density at 0.1 Hz is 20 nV/√Hz, find its normalized noise.

SOLUTION

$$V_{n,\text{norm}} = v_{n,f}\sqrt{f} = (20\ \text{nV}/\sqrt{\text{Hz}})\sqrt{0.1\ \text{Hz}} = 6.32\ \text{nV}$$

EXAMPLE 6.6

In Figure 6.1, find the broadband noise of the resistor for BWn = 1 kHz, and 1/f noise (0.1–6.3 Hz). What is the total noise?

SOLUTIONS

From Figure 6.1, the noise density at 1 kHz is 4.8 (nV/√Hz), thus,
Broadband noise:

$$V_{n,BB} = V_{BB}\sqrt{BW_n} = \left(\frac{4.8\ nV}{\sqrt{Hz}}\right)\sqrt{(1000 - 6.3)\ Hz}$$
$$= 151.31\ nV\,(RMS)$$

The noise density at f_C is 4.8 nV/√Hz, thus,

1/f noise:

$$V_{n,1/f} = V_{1/f_C}\sqrt{f_C\ln\frac{f_C}{f_L}} = \left(4.8\ \frac{nV}{\sqrt{Hz}}\right)\sqrt{(6.3\,Hz)\ln\frac{6.3\,Hz}{0.1\,Hz}} = 24.52\,nV\,(RMS)$$

The total noise is:

$$V_n = \sqrt{V_{n,BB}^2 + V_{n,1/f}^2} = \sqrt{(151.31\,nV)^2 + (24.52\,nV)^2} = 153.28nV\,(RMS)$$

6.3.3.5 Shot Noise

Shot noise (or *Schottky noise*), first reported by Walter Schottky in 1918 [23], is
the time-dependent fluctuations in an electrical signal caused by the discreteness
of its signal carriers. For example, a PN junction in a diode exhibits the discreteness in the form of potential barrier. When the electrons or holes cross the barrier,
the fluctuation occurs and shot noise is produced. Shot noise occurs in all photon
detectors and solid-state devices, such as tunnel junctions, charge-coupled devices
(CCD), and Schottky barrier diodes. Current that flows through a resistor, however,
will not exhibit any fluctuations, because of its continuous flow. Shot noise voltage,
$V_{n,S}$, is proportional to the square root of the average current I and can be characterized by [18]

$$V_{n,S} = \sqrt{2qI\Delta f} \tag{6.13}$$

where q is the electronic charge (1.6×10^{-19} C); Δf is the bandwidth. In contrast to
thermal noise, shot noise cannot be eliminated by lowering the temperature. Instead,
various digital filters such as the wavelet transform filter (WTF), time–frequency
domain filter (TFD), and hybrid gather domain have been proved to reduce shot
noise effectively [24].

6.4 GROUNDING AND SHIELDING TECHNIQUES

To minimize noise and interference, proper grounding and shielding become effective means and necessary steps for many sensor circuits.

6.4.1 GROUNDING AND GROUND LOOPS

Grounding provides safety or signal reference. Safety grounding aims at reducing the voltage differential among the exposed conducting surfaces, while signal reference grounding aims at minimizing the voltage differential among the reference points. Grounding can be classified as *single-point* grounding and *multi-point* grounding. Signal reference grounding should use the single-point grounding to have one connection among all reference points (common ground); while safety grounding should have many connections between exposed conducting surfaces (multi-point grounding). Neither safety nor signal reference ground conducts current (only a signal return path routinely conducts current), thus, a ground is not the return path for a signal. Table 6.7 outlines the functions and applications of groundings and signal return [25].

Ideally, all points of grounding have the same potential. In reality, different points of the same grounding system sit at slightly different potentials. This could be due to the distance between grounding conductors or variations in floor resistance. As a result, a small potential difference exists between grounding points, causing a current flow between two points throughout the system (called *ground loops*). The effects of ground loops depend on the severity of the potential difference between grounding points. Small ground loops inject noise onto a sensor system, while large ground loops can damage sensors and their electronics.

6.4.2 SHIELDING

Shielding is used to suppress or eliminate noise generated from various sources. It is the most efficient and least expensive way to prevent noise coupling through electric fields, magnetic flux, or electromagnetic wave propagation. But it can not reduce conductive coupling. Only filtering can reduce conductive coupling. Filtering will be discussed in Section 6.9.

6.4.2.1 Capacitive Shielding

Since capacitive coupling only occurs when different potentials exist between two conducting objects, if the electrical charge in one electric field is rerouted by a wire,

---

TABLE 6.7
Symbols for Safety or Signal Ground and Return

Symbol	Function	Application
⊥	Safety/power/signal ground	A connection to an electrical ground structure. Used for zero potential reference and electrical shock protection. Normally conducts no current.
/77	Earth or chassis ground	A connection to Earth or the chassis of the circuit. Normally does not conduct current.
▽	Signal return or reference	A conductor that sustains return current for signal path. A reference point from which that signal is measured.

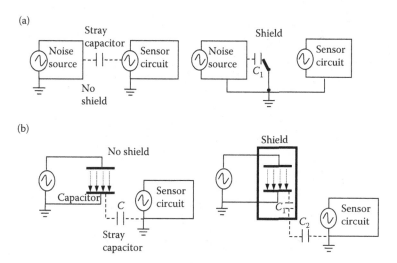

FIGURE 6.2 Capacitive shielding: (a) rerouting method; (b) enclosure method. (Courtesy of Honeywell, Plymouth, Minnesota: 1- and 2-Axis Magnetic Sensors: HMC1021/1022.)

or if the source of the noise is shielded by a metal plate (or box), then the capacitive coupling can be reduced. Figure 6.2 shows how these shielding methods work to reduce coupling [24]. In Figure 6.2a, with the shield connected to a ground (rerouting), the noise currents are shunted around the sensor system through C_1 rather than passing through the sensor. In Figure 6.2b, by introducing a metal box (enclosure), the original stray capacitor C becomes two capacitors, C_1 and C_2 *in series*. Thus, the total capacitance is less than either capacitance alone (recall $1/C_{Total} = 1/C_1 + 1/C_2$).

EXAMPLE 6.7

In Figure 6.2b, the stray capacitance between the capacitor and the sensor circuit is 120 pF without shielding. After adding a shield around the capacitor, the capacitance C_1 between the capacitor and the shield is 50 pF, and the capacitance C_2 between the shield and the sensor circuit is 80 pF. What is the total stray capacitance with shielding?

SOLUTION

$$\frac{1}{C_{Total}} = \frac{1}{C_1} + \frac{1}{C_2} \qquad C_{Total} = \frac{C_1 C_2}{C_1 + C_2} = \frac{(50\,pF)(80\,pF)}{50\,pF + 80\,pF} = 30.77\,pF$$

Thus, the capacitive coupling is reduced from 120 pF to 30.77 pF.

6.4.2.2 Inductive Shielding

Most inductive couplings occur between wires, especially straight wires that encompass significant loop area. Twisted-pair wires can effectively reduce the mutual inductance (see Figure 6.3). Reducing the current change or the loop area also

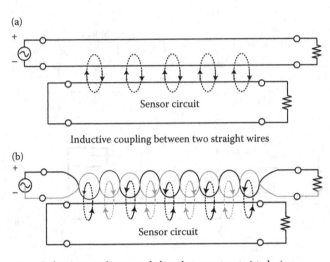

(a)

Inductive coupling between two straight wires

(b)

Inductive coupling canceled out between two twisted wires

FIGURE 6.3 (a) Straight wires creating magnetically coupled loops; (b) Twisted wires eliminating the effective loop coupling.

reduces the inductive coupling. Enclosures can provide magnetic shielding by allowing eddy currents to reflect or absorb interference energy.

6.4.2.3 Electromagnetic Shielding

Electromagnetic coupling can be reduced by adding shielded enclosures (made of conductive materials), implementing proper layout and signal routing, and reducing bandwidth. A shielded conductive enclosure, also known as a *Faraday cage*, can reduce the leak of electromagnetic radiation and block electromagnetic fields. The amount of reduction depends on the material used, its thickness, the size of the shielded volume, and the frequency of the fields of interest.

6.5   DC BRIDGES FOR RESISTANCE MEASUREMENTS

A DC bridge typically uses a constant voltage drive, and it is often used to measure resistance changes in resistive sensors. A DC bridge has a typical Wheatstone bridge configuration. The Wheatstone bridge was originally developed in 1833 by Samuel H. Christie, a mathematics teacher and scientist, to measure and compare wire conductance. It was Charles Wheatstone, a physicist and inventor, who first recognized the value of Christie's circuit and used it extensively in many significant measurements [26]. Over 180 years, the Wheatstone bridge remains the most sensitive and accurate method for precise measurement. To date, the original bridge circuit topology has had many unique modifications that are extensively applied in RCIM sensors. The following sections outline these bridge types and their applications in measuring resistance, capacitance, inductance, and frequency.

A typical bridge consists of four arms connected to form a quadrilateral. Each arm has one or more components, either constant components (resistors, capacitors,

inductors, or their combination) or variable components (e.g., RCIM sensing elements). A source of excitation (DC or AC voltage or current) is supplied across one
of the diagonals and a bridge voltage output is measured across the other diagonal.
It is standard practice to measure bridge output voltages with high impedance (typically >1 MΩ) devices. This section discusses each bridge type and their applications
in RCIM sensor systems.

6.5.1 WHEATSTONE BRIDGE AND ITS BALANCE CONDITION

A Wheatstone bridge is commonly used in resistive sensors (e.g., RTDs, strain gauges,
piezoresistive gauges) to measure resistance change. Figure 6.4 shows a typical
Wheatstone bridge. It consists of an excitation voltage source (V_{ex}) and four resistors
(R_1, R_2, R_3, and R_4). One of the four resistors is the resistive sensor (a variable resistor).
The bridge output V_{out} is the difference between the voltage potential at point B
(V_B) and the voltage potential at point D (V_D). Paths ABC and ADC are two ballast
circuits (i.e., voltage dividers), thus V_{out} is

$$V_{out} = V_B - V_D = \frac{R_2}{R_1 + R_2} V_{ex} - \frac{R_3}{R_3 + R_4} V_{ex} \tag{6.14}$$

When the bridge is balanced (null), the output V_{out} is zero, that is,

$$V_{out} = V_B - V_D = 0 \Rightarrow V_B = V_D \tag{6.15}$$

which results in the *bridge balance condition*:

$$R_1 R_3 = R_2 R_4 \tag{6.16}$$

The above balance condition is independent of voltage excitation V_{ex}, making the
bridge an extremely useful circuit topology. It allows a small change of resistance to
be measured relative to an initial zero value, rather than relative to a large resistance

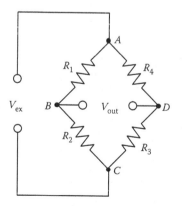

FIGURE 6.4 A Wheatstone bridge structure.

value, thus sensitivity and resolution are significantly increased. All precision measurement starts with this balanced state.

If the resistance values undergo small changes—ΔR_1, ΔR_2, ΔR_3, and ΔR_4 due to a variation in temperature, or excitation line resistance, or their combination, the balance condition will be no longer met. Plugging $R_1 + \Delta R_1$, $R_2 + \Delta R_2$, $R_3 + \Delta R_3$, and $R_4 + \Delta R_4$ into Equations 6.14 and using condition (6.16), then omitting the higher-order terms, the bridge output becomes

$$V_{out} = \frac{R_1 R_3}{(R_1 + R_2)(R_3 + R_4)}\left(-\frac{\Delta R_1}{R_1} + \frac{\Delta R_2}{R_2} - \frac{\Delta R_3}{R_3} + \frac{\Delta R_4}{R_4}\right) V_{ex} \qquad (6.17)$$

One advantage of a Wheatstone bridge is that the temperature effect can be directly eliminated. If R_1 is a strain gauge bonded to a specimen and R_2 is a strain gauge held onto a different specimen with heat sink compound, then R_1 will respond to strain plus temperature, and R_2 will only respond to temperature. Since the bridge subtracts the output of R_1 from that of R_2, the temperature effect is cancelled. The typical full-scale bridge output is 10–100 mV.

EXAMPLE 6.8

Derive the Wheatstone bridge output, V_{out}, in terms of ε_1, ε_2, ε_3, ε_4, GF (Gauge Factor) and V_{ex} for a piezoresistive sensor that has a Wheatstone bridge configuration and contains a strain gauge on each arm: $R_1 = R_2 = R_3 = R_4$, where ε_i ($i = 1, 2, 3$, and 4) is the strain of each gauge.

SOLUTION

From Chapter 2, $GF = (\Delta R/R)/\varepsilon$.
Plug $R_1 = R_2 = R_3 = R_4 = R$, $\Delta R_1/R_1 = \varepsilon_1 \cdot GF$, $\Delta R_2/R_2 = \varepsilon_2 \cdot GF$, $\Delta R_3/R_3 = \varepsilon_3 \cdot GF$, and $\Delta R_4/R_4 = \varepsilon_4 \cdot GF$ into Equation 6.17, resulting in

$$V_{out} = \frac{GF}{4}(-\varepsilon_1 + \varepsilon_2 - \varepsilon_3 + \varepsilon_4) V_{ex} \qquad (6.18)$$

EXAMPLE 6.9

A resistive sensor has a Wheatstone bridge configuration. Its sensing element on each arm has an actual resistance value shown in Figure 6.5 and a nominal resistance of 1 kΩ where $\Delta R = \pm 10$ Ω. If the sensor's sensitivity is 10 mV/V and the excitation voltage is 5 V, find the sensor's output V_{out}.

SOLUTION

There are two ways to find V_{out}:

1. Use Equations 6.14:

$$V_{out} = V_B - V_D = \frac{1010 \ \Omega}{990 \ \Omega + 1010 \ \Omega}(5 \ V) - \frac{990 \ \Omega}{990 \ \Omega + 1010 \ \Omega}(5 \ V) = 0.05 \ V$$

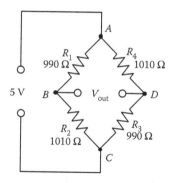

FIGURE 6.5 A resistive sensor's Wheatstone bridge configuration.

2. Use Equation 6.17:

$$V_{out} = \frac{R_1 R_3}{(R_1 + R_2)(R_3 + R_4)} \left(-\frac{\Delta R_1}{R_1} + \frac{\Delta R_2}{R_2} - \frac{\Delta R_3}{R_3} + \frac{\Delta R_4}{R_4} \right) V_{ex}$$

$$= \frac{(990\ \Omega)(990\ \Omega)}{(990\ \Omega + 1010\ \Omega)(990\ \Omega + 1010\ \Omega)}$$

$$\times \left(\begin{array}{c} -\dfrac{990\ \Omega - 1000\ \Omega}{990\ \Omega} + \dfrac{1010\ \Omega - 1000\ \Omega}{1010\ \Omega} \\ -\dfrac{990\ \Omega - 1000\ \Omega}{990\ \Omega} + \dfrac{1010\ \Omega - 1000\ \Omega}{1010\ \Omega} \end{array} \right) (5\,V)$$

$$= 0.049\,V$$

Note that there is a slight difference between the two methods. This is because the second method omits the higher-order terms.

6.5.2 SENSITIVITY OF A WHEATSTONE BRIDGE

Sensitivity of a Wheatstone bridge is defined as the ratio of the maximum expected change in the output voltage to the excitation voltage. Since most of Wheatstone bridges measure from their zero (balance) conditions, thus:

$$\text{Bridge Sensitivity} = \frac{\text{Maximum Change in Output Voltage}}{\text{Excitation Voltage}}$$

$$= \frac{\text{Max}\{V_{out}\} - 0}{V_{ex}} = \frac{\text{Max}\{V_{out}\}}{V_{ex}} \qquad (6.19)$$

Although increasing the excitation voltage increases the output voltage (see Equations 6.14 and 6.17) and improves the sensitivity, it also results in higher power dissipation (recall that power = V_{ex}^2/R) and bigger sensor self-heating errors. On the other hand, lower values of excitation voltage require more gain in the conditioning circuits and also decrease the signal to noise ratio. Typical Wheatstone bridge sensitivities are 1–10 mV/V. In addition, the stability of the excitation voltage or current directly affects the overall accuracy of the bridge output.

EXAMPLE 6.10

If $V_{ex} = 10$ V and the full scale bridge output is 12 mV, find the bridge's sensitivity.

SOLUTION

The full scale bridge output is the maximum bridge output, thus

$$\text{Bridge Sensitivity} = \frac{\text{Max}\{V_{out}\}}{V_{ex}} = \frac{12\,\text{mV}}{10\,\text{V}} = 1.2\,\text{mV/V}$$

Some resistive temperature sensors, such as RTDs, are difficult to measure due to their low resistance (e.g., 100 Ω) that changes only slightly with temperature ($< 0.4\ \Omega \cdot {}^{\circ}\text{C}^{-1}$). To accurately measure these small changes in resistance, special configurations should be used to minimize errors from lead wire resistance. Figure 6.6 shows the three common RTD wirings (the arrow "→" across the resistors means that the resistors are variable or adjustable resistors):

Two-wire RTD connection (Figure 6.6a). The actual resistance that causes V_{out} change is the total resistance of the sensor R_T and the two lead wires ($L_1 + L_2$). If the lead wire resistance is constant, it introduces an offset error and can be easily compensated. However, as temperature varies, the wire resistance also changes, which creates errors in measurement, especially when the wires are long. Thus, two-wire RTD connection is only used with very short lead wires or with a high resistance (e.g., 1 kΩ) sensor.

Three-wire RTD connection (Figure 6.6b). In this connection, L_1 and L_3 carry the measuring current, while L_2 acts as a potential or reference lead only. If the resistances of L_1 and L_3 are perfectly matched, their affects on the temperature

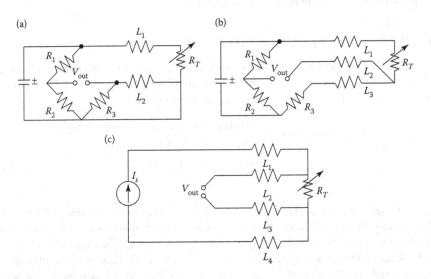

FIGURE 6.6 Three common RTD wire connections: (a) two-wire; (b) three-wire; (c) four-wire connections.

measurement will be cancelled. Thus, L_1 and L_3 can lead up to tens of feet long and usually work well for 100 Ω RTDs.

Four-wire connection (Figure 6.6c). This is the optimum wiring form for RTDs. It removes the error caused by mismatched lead wires resistance. A constant current is passed through L_1 and L_4; while L_2 and L_3 measure the voltage drop across the RTD. With a constant current, the voltage is strictly a function of the resistance R_T and a true measurement is achieved. This connection provides a high degree of accuracy although it is more expensive than two- or three-wire connections.

6.5.3 WHEATSTONE BRIDGE-DRIVEN MEANS

6.5.3.1 Wheatstone Bridge Driven by a Constant Voltage

There are four basic Wheatstone bridge configurations for resistive sensors driven by a constant voltage:

- *Single-element variation* (Figure 6.7a). This configuration is suited for a single sensor (a thermistor, an RTD, or a strain gauge). All the resistances are nominally equal, except the sensor that varies by an amount ΔR. V_{out} is nonlinearly related to ΔR by

$$V_{out} = \frac{\Delta R}{4(R + \Delta R/2)} V_{ex} \qquad (6.20)$$

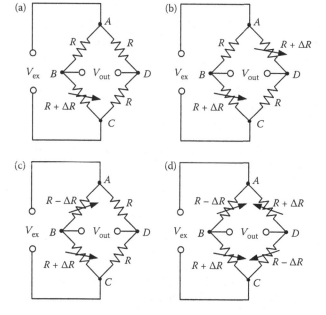

FIGURE 6.7 Four common Wheatstone bridge configurations under a constant voltage drive: (a) single-element; (b) two-diagonal-element; (c) two-adjacent-element; and (d) four-element variations.

- *Two-diagonal-element variation* (Figure 6.7b). This configuration is used for two identical sensors mounted diagonally to each other. For example, strain gauges with their axes in parallel, pressure or flow meters that have one for sensing and the other for reference. In this case, the bridge output is twice that of the single-element varying bridge:

$$V_{out} = \frac{\Delta R}{2(R + \Delta R/2)} V_{ex} \qquad (6.21)$$

- *Two-adjacent-element variation* (Figure 6.7c). This structure contains two adjacently mounted elements, but they vary in opposite directions (e.g., two identical strain gauges: one mounted on the top of a flexing surface, and the other on the bottom). In this case, V_{out} and ΔR are linearly related, and the bridge output is:

$$V_{out} = \frac{\Delta R}{2R} V_{ex} \qquad (6.22)$$

- *Four-element variation* (Figure 6.7d). This bridge produces the largest output for resistance changes and is inherently linear. It becomes an industry-standard configuration for load cells which are constructed from four identical strain gauges. The bridge output is:

$$V_{out} = \frac{\Delta R}{R} V_{ex} \qquad (6.23)$$

6.5.3.2 Wheatstone Bridge Driven by a Constant Current

A Wheatstone bridge can also be driven by a constant current source. A constant current-driven Wheatstone bridge requires an accurate current source since any change in the current will be interpreted as a resistance change. In addition, the drive current must be small to minimize the error caused by power dissipation or self-heating in the resistive sensor. The current drive, though not as popular as the voltage drive, has two advantages:

- The measurement will not be affected by the voltage drop in wire or excitation line resistance. Thus, the bridge can be located remotely from the source of excitation.
- The bridge output is more linear over a large range of ΔR than that of the voltage drive.

Similar to the constant voltage drive, there are four common configurations (see Figure 6.8):

- *Single-element variation* (Figure 6.8a). All the resistances are nominally equal except the sensor's resistance that varies. The output V_{out} is

$$V_{out} = \frac{I_{ex}}{4} \frac{R\Delta R}{(R + \Delta R/4)} \qquad (6.24)$$

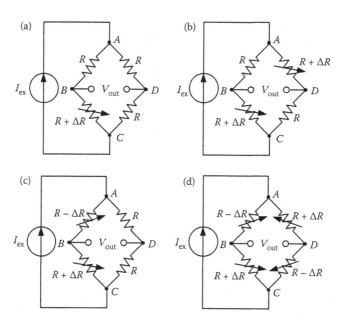

FIGURE 6.8 Four common Wheatstone bridge configurations under a constant current drive: (a) single-element; (b) two-diagonal-element; (c) two-adjacent-element; and (d) four-element variations.

- *Two-diagonal-element variation.* Two identical sensors are mounted diagonally to each other (Figure 6.8b). V_{out} is proportional to ΔR and has a bigger magnitude than that of the single-element varying bridge:

$$V_{out} = \frac{I_{ex}}{2} \Delta R \tag{6.25}$$

- *Two-adjacent-element variation.* Two identical but varying *oppositely* sensors are mounted adjacently to each other (Figure 6.8c). V_{out} is related to ΔR by

$$V_{out} = \frac{I_{ex}}{2} \Delta R \tag{6.26}$$

- *All-element variation* (Figure 6.8d). This bridge configuration has the largest output for the resistance change and is inherently linear.

$$V_{out} = I_{ex} \Delta R \tag{6.27}$$

EXAMPLE 6.11

Compare the performance of a Wheatstone bridge driven by a constant voltage and a constant current with the configuration of: (1) the single-element variation; (2) the two-element variation (diagonally mounted). Assume that the nominal resistance R is 120 Ω, ΔR changes from -100 Ω to $+100$ Ω. The constant voltage source is 5 VDC, and the constant current source is 41.67 mA ($= 5$ V/120 Ω).

SOLUTIONS

1. With the single-element variation, a 5 VDC voltage source yields

$$V_{out} = \frac{\Delta R}{4(R + \Delta R/2)} V_{ex} = \frac{\Delta R}{4(120 + \Delta R/2)}(5) = \frac{\Delta R}{96 + 0.4\Delta R} (\text{Volts})$$

A 41.67 mA current source yields (Note: $V_{ex} = I_{ex}R$):

$$V_{out} = \frac{I_{ex}R}{4} \frac{\Delta R}{(R + \Delta R/4)} = \frac{(41.67 \times 10^{-3})(120)}{4} \frac{\Delta R}{(R + \Delta R/4)} = \frac{\Delta R}{96 + 0.2\Delta R} (\text{Volts})$$

The bridge outputs, as ΔR changes from -100 Ω to $+100$ Ω, are shown in Figure 6.9. It can be seen that the bridge's performance under the constant current excitation is more linear than that under a constant voltage source. Both curves demonstrate good linearity within -20 $\Omega < \Delta R < +20$ Ω.

2. With two-diagonal-element variation, a 5 VDC voltage source produces:

$$V_{out} = \frac{\Delta R}{2(R + \Delta R/2)} V_{ex} = \frac{\Delta R}{2(120 + \Delta R/2)}(5) = \frac{\Delta R}{48 + 0.2\Delta R} (\text{Volts})$$

A 41.67 mA current source produces:

$$V_{out} = \frac{I_{ex}}{2}\Delta R = 0.020835\Delta R \ (\text{Volts})$$

In this case, the bridge output under a constant current drive is linear for the entire measurement range.

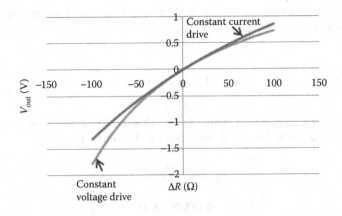

FIGURE 6.9 Bridge performance under a constant current drive vs. a constant voltage drive for a single-element configuration. (From RdF Corporation, Hudson, New Hampshire, USA. With permission.)

6.5.4 Kelvin Bridge

A bridge's excitation line or connecting wire resistances will affect the bridge's output significantly when a sensing element has only a few ohms of resistance. Kelvin bridge (also known as *Kelvin network, double bridge, Thomson bridge*) is a special version of the Wheatstone bridge designed to eliminate the effect of lead and contact resistance and provide accurate measurement of low resistance [27]. Figure 6.10a shows a common Wheatstone bridge. Both R_1 and R_2 are high resistance arms, thus their lead resistances are negligible. R_x (the sensor) and R_4 are low resistance arms, whose lead resistances, R_a and R_b respectively, are not negligible. R_a is added to the unknown R_x resulting in a higher evaluation of R_x. R_b is added to R_4, making the measurement of R_x lower than its actual value.

The Kelvin bridge shown in Figure 6.10b overcomes the lead resistance effect. R_y is a heavy copper yoke of low resistance connected between R_x and R_4. The balance condition of the double bridge is

$$V_{AB} = V_{AmD} \tag{6.28}$$

$$V_{AB} = \frac{R_1}{R_1 + R_2} V_{ex} \tag{6.29}$$

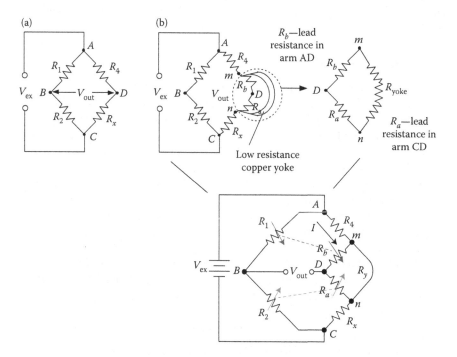

FIGURE 6.10 (a) A standard Wheatstone bridge; (b) a Kelvin double bridge. (From Carr, J.J., *Sensors and Circuits: Sensors, Transducers, and Supporting Circuits for Electronic Instrumentation, Measurement and Control*, PTR Prentice-Hall, Inc., Englewood Cliffs, New Jersey, 1993. With permission.)

while

$$V_{ex} = IR_{AC} = I[R_x + R_4 + (R_a + R_b)//R_y] \tag{6.30}$$

$$V_{AmD} = I\left[R_4 + \frac{R_y}{R_a + R_b + R_y}R_b\right] \tag{6.31}$$

"//" in Equation 6.30 means $(R_a + R_b)$ and R_y are parallely connected. Plugging Equations 6.29, 6.30, and 6.31 into Equation 6.28, yields:

$$R_x = R_4\frac{R_2}{R_1} + \frac{R_bR_y}{R_a + R_b + R_y}\left(\frac{R_2}{R_1} - \frac{R_a}{R_b}\right) \tag{6.32}$$

Thus, the Kelvin bridge eliminates the lead resistance effect on the bridge measurement by introducing two ratio arms R_2/R_1 (main ratio) and R_a/R_b (auxiliary ratio) [28,29]. If we set $R_2/R_1 = R_a/R_b$, the second term on the right-hand side will be zero, the relation reduces to the Wheatstone balance condition. Thus, the Kelvin bridge can remove the lead resistance effect if the two sets of ratio arms have equal resistance ratios.

6.5.5 MEGAOHM BRIDGE

Opposite to Kelvin bridge for low-resistance measurement, a Megaohm bridge is for high-resistance measurement (e.g., an insulator's resistance >1 MΩ). During the measurement, a very high voltage is applied across the insulator, and a microammeter is used to read the current value flowing through the insulator. The insulation resistance is then calculated using Ohm's law. Just like a low-resistance measurement is affected by series lead resistances, a high-resistance measurement is affected by shunt-leakage resistance. Figure 6.11a illustrates how a surface leakage resistance R_S affects the measurement of the high resistance R_X. Here, the actual resistance measured, R_{meas}, is not equal to R_X itself, but the equivalent resistance of R_X and R_S in parallel. In addition, the current I does not just flow through the insulator only (denoted as I_X), but also "leaks" to air (denoted as I_S, $S \rightarrow$ Shunk). By adding a guard ring shown in Figure 6.11b [30], the measuring current I flows through the insulator only and the measured resistance R_{meas} is equal to R_X. Thus, the surface leakage resistance is removed from R_{meas}.

EXAMPLE 6.12

The insulation of a metal-sheath electrical cable is tested using 10,000 V supply and a microammeter. A current of 5 µA is read when the cable is connected without a guard wire, and a current of 1.5 µA is read when the cable is connected with a guard wire. Calculate (1) the resistance of the cable insulation; and (2) the surface leakage resistance.

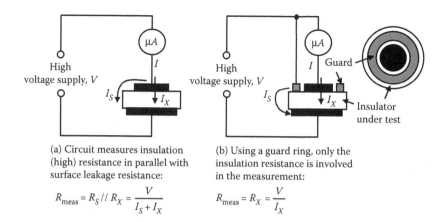

(a) Circuit measures insulation (high) resistance in parallel with surface leakage resistance:

$$R_{meas} = R_S // R_X = \frac{V}{I_S + I_X}$$

(b) Using a guard ring, only the insulation resistance is involved in the measurement:

$$R_{meas} = R_X = \frac{V}{I_X}$$

FIGURE 6.11 High-resistance measurement: (a) without a guard ring; (b) with a guard ring. (From Helfrick, A.D. and Cooper, W.D., *Modern Electronic Instrumentation and Measurement Techniques*, Prentice-Hall, Englewood Cliffs, NJ, 1990; *Type 1644 Megaohm Bridge: User and Service Manual*, IET LABS, Inc., Westbury, NY, 1988. With permission.)

SOLUTIONS

(1) Since the insulation measurement is not affected by the surface leak resistance when using a guard wire, the insulation resistance of the cable can be calculated by

$$R_{meas} = R_X = \frac{V}{I_X} = \frac{10,000 \text{ V}}{1.5 \times 10^{-6} \text{ A}} = 6.67 \times 10^9 \, \Omega$$

(2) $I_S + I_X = 5 \, \mu A$, $I_S = 5 \, \mu A - I_X = 5 \, \mu A - 1.5 \, \mu A = 3.5 \, \mu A$

$$R_S = \frac{V}{I_S} = \frac{10,000 \text{ V}}{3.5 \times 10^{-6} \text{ A}} = 2.86 \times 10^9 \, \Omega$$

A Megaohm bridge using the guard ring principle is shown in Figure 6.12. The surface leak resistance, R_S, is parallel to R_2. Since $R_S \gg R_2$ (because conductivity in air is very low), thus $R_S // R_2 \approx R_2$, resulting in

$$R_X \approx \frac{R_2}{R_1} R_4 \tag{6.33}$$

6.6 AC BRIDGES FOR CAPACITANCE AND INDUCTANCE MEASUREMENTS

An AC bridge has the same structure as a DC bridge, except that:

- The excitation source is an AC voltage or current at a desired frequency.

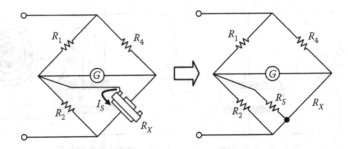

FIGURE 6.12   A Megaohm bridge using a guard ring.

- All four bridge arms are considered as impedance (Z_1, Z_2, Z_3, and Z_4) and consist of either resistors or frequency-dependent components, such as capacitors and inductors, or their combination.
- The detector that measures the bridge output is an AC responding device.

6.6.1   AC Bridge and Its Balance Condition

Figure 6.13 shows a general AC bridge. For the balance condition:

$$V_B = V_D \tag{6.34}$$

or

$$V_A - V_B = V_A - V_D \Rightarrow V_{AB} = V_{AD}$$

that is,

$$I_1 Z_1 = I_2 Z_4$$

Since $I_1 = V/(Z_1 + Z_2)$ and $I_2 = V/(Z_3 + Z_4)$, thus the AC bridge balance condition becomes

$$Z_1 Z_3 = Z_2 Z_4 \tag{6.35}$$

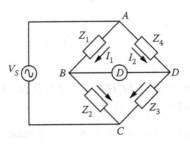

FIGURE 6.13   A general AC bridge.

TABLE 6.8
Component Type Corresponding to a Phase Angle

Phase Angle	Element Type
$<0°$	Capacitor
$=0°$	Resistor
$>0°$	Inductor

TABLE 6.9
Component Type Corresponding to a Complex Number

Real Part	Imaginary Part	Component Type
$\neq 0$	$=0$	Resistor only
$\neq 0$	<0	Capacitor + Resistor
$\neq 0$	>0	Inductor + Resistor
0	<0	Capacitor only
0	>0	Inductor only

Since impedance is a complex number, the above balance condition results in two conditions: equality of the real parts and equality of the imaginary parts, described as follows:

$$\begin{cases} \text{Re}(Z_1 Z_3) = \text{Re}(Z_2 Z_4) \\ \text{Im}(Z_1 Z_3) = \text{Im}(Z_2 Z_4) \end{cases} \tag{6.36}$$

The complex impedance balance condition can also be expressed in polar form:

$$\begin{cases} |Z_1||Z_3| = |Z_2||Z_4| & \text{(Magnitude balance condition)} \\ \angle\theta_1 + \angle\theta_3 = \angle\theta_2 + \angle\theta_4 & \text{(Phase angle balance condition)} \end{cases} \tag{6.37}$$

Table 6.8 lists the possible component type on a bridge arm based on its phase angle value, while Table 6.9 lists the possible component type on a bridge arm based on its complex number.

EXAMPLE 6.13

An AC bridge has the following impedance: $Z_1 = 100 \ \Omega \ \angle 70°$ (inductor); $Z_2 = 300 \ \Omega \ \angle -30°$ (capacitor); $Z_4 = 250 \ \Omega \ \angle 0°$ (resistor). Determine the unknown impedance of Z_3.

SOLUTION

The magnitude balancing requires Z_3 to be

$$|Z_3| = \frac{|Z_2||Z_4|}{|Z_1|} = \frac{(300\ \Omega)(250\ \Omega)}{100\ \Omega} = 750\ \Omega$$

The phase angle balancing requires θ_3 to be

$$\angle\theta_3 = \angle\theta_2 + \angle\theta_4 - \angle\theta_1 = -30° + 0° - 70° = -100°$$

Thus, $Z_3 = 750\ \Omega\ \angle{-100}°$, which is a capacitive element based on Table 6.9.

EXAMPLE 6.14

An AC bridge (see Figure 6.13) is in balance with the following elements: (1) Arm AB: $R = 200\ \Omega$ in series with $L = 15.9$ mH; (2) Arm BC: $R = 450\ \Omega$; (3) Arm AD: $R = 300\ \Omega$ in series with $C = 0.265\ \mu F$. The AC oscillator frequency is 1 kHz. Find the unknown element in Arm CD.

SOLUTION

Arm AB: $Z_1 = R + j\omega L = R + j(2\pi f)L = 200 + j\ 99.85\ (\Omega)$
Arm BC: $Z_2 = R + 0\ j = 450\ (\Omega)$
Arm AD: $Z_4 = R + 1/(j\omega C) = R + (1/j(2\pi f C) = 300 - j\ 600.89\ (\Omega)$

The bridge balance requires:

$$Z_3 = \frac{Z_2 Z_4}{Z_1} = \frac{(450\ \Omega)(300 - j600.89)\ \Omega}{(200 + j99.85)\ \Omega} = 0.01 - j1352.01\ \Omega$$

This indicates that the unknown element in arm CD is a capacitor.

6.6.2   COMPARISON BRIDGE

A comparison bridge measures an unknown capacitance or inductance by comparing it with a known capacitance or inductance. Figure 6.14 shows a comparison bridge for capacitance measurement. Z_1 is a variable resistor; Z_2 is a known capacitor in series with a variable resistor; Z_x is an unknown capacitor in series with an unknown resistor; and Z_4 is a resistor.

Under the bridge balance condition:

$$Z_1 Z_x = Z_2 Z_4 \Rightarrow R_1 \left(R_x + \frac{1}{j\omega C_x} \right) = \left(R_2 + \frac{1}{j\omega C_2} \right) R_4$$

$$R_1 R_x + \frac{R_1}{j\omega C_x} = R_2 R_4 + \frac{R_4}{j\omega C_2}$$

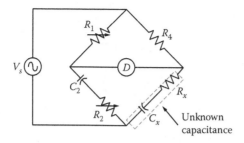

FIGURE 6.14   A comparison bridge for capacitance measurement. (From Helfrick, A.D. and Cooper, W.D., *Modern Electronic Instrumentation and Measurement Techniques*, Prentice-Hall, Englewood Cliffs, NJ, 1990; *Type 1644 Megaohm Bridge: User and Service Manual*, IET LABS, Inc., Westbury, NY, 1988. With permission.)

which results in

$$R_x = \frac{R_2 R_4}{R_1} \tag{6.38}$$

and

$$C_x = \frac{R_1 C_2}{R_4} \tag{6.39}$$

Equations 6.38 and 6.39 indicate that to satisfy both magnitude and phase angle balance conditions, the bridge must contain two variable elements (i.e., R_x and C_x) in its configuration. Figure 6.15 is a comparison bridge for inductance. It is similar to the above capacitance measurement except that the capacitors are replaced by inductors, and Z_1 is fixed and Z_4 becomes variable.

Under the bridge balance condition:

$$Z_1 Z_x = Z_2 Z_4 \Rightarrow R_1(R_x + j\omega L_x) = R_4(R_2 + j\omega L_2)$$

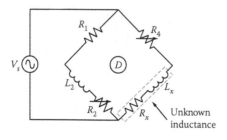

FIGURE 6.15   A comparison bridge for inductance measurement.

which results in

$$R_x = \frac{R_2 R_4}{R_1} \tag{6.40}$$

$$L_x = \frac{R_4 L_2}{R_1} \tag{6.41}$$

6.6.3 Schering Bridge

A Schering bridge in Figure 6.16 is extensively used to measure capacitance of capacitive sensors.

Its balance condition is (// means in parallel)

$$Z_1 Z_x = Z_2 Z_4 \Rightarrow \left(R_1 // \frac{1}{j\omega C_1} \right)\left(R_x + \frac{1}{j\omega C_x} \right) = \frac{1}{j\omega C_2} R_4$$

which results in

$$R_x = \frac{R_4 C_1}{C_2} \tag{6.42}$$

$$C_x = \frac{R_1 C_2}{R_4} \tag{6.43}$$

The *Dissipation Factor* (DF) of a series $R_x C_x$ circuit is defined as

$$DF = \frac{R_x}{1/(\omega C_x)} = R_x C_x \omega \tag{6.44}$$

This factor indicates the quality of a capacitor, that is, how close the phase angle of the capacitor is to the ideal value of 90°. For a Schering bridge:

$$DF = R_x C_x \omega = \frac{R_4 C_1}{C_2} \frac{R_1 C_2}{R_4} \omega = R_1 C_1 \omega \tag{6.45}$$

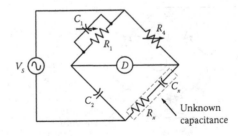

FIGURE 6.16 A Schering bridge for capacitance measurement.

Since R_1 has a fixed value, C_1 thus can be directly adjusted at one particular frequency. The inverse of DF—the ratio of the resistance to the capacitive reactance at a given frequency is called the *quality factor* of the capacitor.

6.6.4 MAXWELL BRIDGE

A Maxwell bridge is used to measure an unknown inductance in terms of a known capacitance. The bridge diagram is shown in Figure 6.17.

Under the bridge balance condition:

$$Z_1 Z_x = Z_2 Z_4 \Rightarrow \frac{R_1}{j\omega R_1 C_1 + 1}(R_x + j\omega L_x) = R_2 R_4$$

which results in

$$R_x = \frac{R_2 R_4}{R_1} \tag{6.46}$$

$$L_x = R_2 R_4 C_1 \tag{6.47}$$

A Maxwell bridge is suitable for a medium Q coil (1–10), but impractical for a high Q coil since R_1 will be very large. (Q is the *quality factor* of an inductor, defined as the ratio of its inductive reactance to its resistance at a given frequency, i.e., $Q = L\omega/R$.)

6.6.5 HAY BRIDGE

A Hay bridge, like a Maxwell bridge, also measures an unknown inductance in terms of a known capacitance, except that C_1 and R_1 are in series as shown in Figure 6.18.

Under the bridge balance condition:

$$Z_1 Z_x = Z_2 Z_4 \Rightarrow \left(R_1 + \frac{1}{j\omega C_1} \right)(R_x + j\omega L_x) = R_2 R_4$$

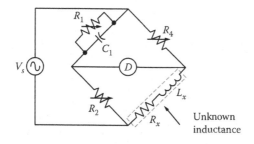

FIGURE 6.17   A Maxwell bridge for inductance measurement.

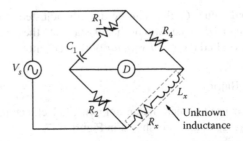

FIGURE 6.18   A Hay bridge for inductance measurement.

which results in

$$\begin{cases} R_x R_1 C_1 = R_2 R_4 C_1 - L_x \\ R_x = R_1 C_1 \omega^2 L_x \end{cases}$$

Solving the above two equations, yields:

$$R_x = \frac{R_1 R_2 R_4 C_1^2 \omega^2}{1 + R_1^2 C_1^2 \omega^2} \tag{6.48}$$

$$L_x = \frac{R_2 R_4 C_1}{1 + R_1^2 C_1^2 \omega^2} \tag{6.49}$$

The Q factor of the unknown impedance $(R_x + j\omega L_x)$ is

$$\frac{L_x \omega}{R_x} = \tan\theta_x = \frac{1}{R_1 C_1 \omega}$$

which is equal to the Q *factor* of the known impedance $[R_1 + 1/(j\omega C_1)]$:

$$\frac{1/(\omega C_1)}{R_1} = \tan\theta_1 = \frac{1}{R_1 C_1 \omega}$$

Rewrite Equation 6.49:

$$L_x = \frac{R_2 R_4 C_1}{1 + (1/Q^2)} \tag{6.50}$$

For a high Q coil (>10), the term $1/Q^2$ can be neglected, thus,

$$L_x \approx R_2 R_4 C_1$$

6.6.6   OWEN BRIDGE

An Owen bridge (see Figure 6.19) measures an unknown inductance by balancing the loads of its four arms that contain capacitors and resistors except for the arm that contains the unknown inductance.

Under the bridge balance condition:

$$Z_1 Z_x = Z_2 Z_4 \Rightarrow \frac{1}{j\omega C_1}(R_x + j\omega L_x) = \left(R_2 + \frac{1}{j\omega C_2} \right) R_4$$

which results in

$$R_x = \frac{C_1}{C_2} R_4 \tag{6.51}$$

$$L_x = R_2 R_4 C_1 \tag{6.52}$$

EXAMPLE 6.15

An Owen bridge reaches a balance point when $C_1 = 0.8\ \mu F$, $C_2 = 0.5\ \mu F$, $R_2 = 400$ Ω, and $R_4 = 450\ \Omega$ under a 1 kHz excitation frequency. Find Z_x.

SOLUTION

$$R_x = \frac{C_1}{C_2} R_4 = \frac{(0.8 \times 10^{-6}\ F)}{(0.5 \times 10^{-6}\ F)}(450\ \Omega) = 720\ \Omega$$

$$L_x = R_2 R_4 C_1 = (400\ \Omega)(450\ \Omega)(0.8 \times 10^{-6}\ F) = 0.144\ H$$

$$Z_x = R_x + j\omega\, L_x = 720\ \Omega + j(2\pi)(1000)(0.144)\ \Omega = 720 + 904.32\, j\ (\Omega)$$

6.6.7   WIEN BRIDGE

A Wien bridge is often used to measure frequency of the voltage source using series RC in one arm and parallel RC in the adjacent arm as shown in Figure 6.20.

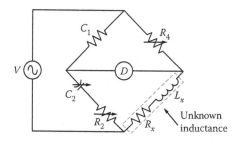

FIGURE 6.19   An Owen bridge for inductance measurement.

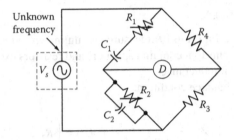

FIGURE 6.20   A Wien bridge for frequency measurement.

Under the bridge balance condition:

$$Z_1 Z_3 = Z_2 Z_4 \Rightarrow \left(R_1 + \frac{1}{j\omega C_1} \right) R_3 = \left(R_2 // \frac{1}{j\omega C_2} \right) R_4$$

which results in

$$\frac{R_1}{R_2} + \frac{C_2}{C_1} = \frac{R_4}{R_3} \tag{6.53}$$

and

$$\omega = \frac{1}{\sqrt{R_1 R_2 C_1 C_2}} \tag{6.54}$$

or

$$f = \frac{1}{2\pi\sqrt{R_1 R_2 C_1 C_2}} \tag{6.55}$$

In most Wien bridges, $R_1 = R_2$ and $C_1 = C_2$, thus Equations 6.53 and 6.55 become

$$R_4 = 2R_3 \quad \text{and} \quad f = \frac{1}{2\pi R_1 C_1}$$

6.6.8   Bridge Selection Considerations

- Selection of voltage or current excitation
- Stability of excitation voltage or current
- Bridge sensitivity
- Full-scale (FS) bridge outputs: 10–100 mV typical
- Precision, low noise amplification/conditioning
- Linearity and linearization

Many techniques are available to linearize bridges, but it is important to distinguish between the linearity of the bridge equation and the linearity of the sensor response to the phenomenon being sensed. For example, if the active element is an RTD, the bridge used to implement the measurement might have perfect linearity; yet the output could still be nonlinear due to the RTD's nonlinearity. Sensor manufacturers employ bridges to improve the nonlinearity in a variety of ways, including keeping the resistive swings in the bridge small, shaping complimentary nonlinear response into the active elements of the bridge, using resistive trims for first-order corrections, and so on.

6.7 RCIM SENSOR OUTPUT CIRCUITS

Sensor outputs are often connected to other circuits for noise rejection, sensor data acquisition, sensor signal conditioning, analog-to-digital conversion (ADC), and so on. Interfacing a sensor to other circuits requires matching the sensor output with the input of the subsequent circuits (e.g., an amplifier input). This section will discuss various sensor output circuits that provide voltage outputs, current outputs, or charge outputs; or convert resistance, capacitance, or inductance outputs into current, voltage, frequency, or time outputs.

6.7.1 VOLTAGE OUTPUT CIRCUITS

Voltage output sensors, such as potentiometers and piezoelectric sensors, produce voltage as their outputs. There are six commonly used sensor voltage output structures as shown in Figure 6.21.

"*Single-ended*" means that one end of the sensor output circuit is grounded (a fixed zero potential, see Figure 6.21a) or is referenced to a common nongrounded ("floating") point (can be any nonzero voltage potential, see Figure 6.21b). If a single-ended sensor "floats" on a voltage source, it becomes a single-ended floating, driven-off-ground sensor (Figure 6.21c). In a single-ended configuration, the sensor

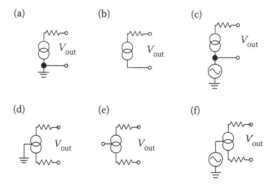

FIGURE 6.21 Commonly used sensor voltage output structures. (a) Single-ended ground, (b) Single-ended floating, (c) Single-ended floating driven-off-ground, (d) Balanced ground, (e) Balanced floating, (f) Balanced driven-off-ground.

has only one output line. Thus, any noise or interference from external fields can be added to this output line. A floating circuit however, can isolate a circuit from interference (e.g., caused by ground loops), but it may have potential safety issues because there is no low impedance path to ground. The term *"balanced"* (or "differential") means the sensor drives output through two *equal* resistance lines. There are balanced ground (Figure 6.21d), balanced floating (Figure 6.21e), and balanced driven-off-ground sensors (Figure 6.21f). A balanced circuit is often used to "cancel out" noise or interference from external fields.

Sensors' voltage outputs are usually sent to the inputs of signal conditioning circuits starting with amplifier circuits. Figure 6.22 shows four common amplifier circuits for taking sensors' outputs [31].

The inverting amplifier in Figure 6.22a provides a resistive isolation, R_1, from the sensor source. Thus, it allows a large range of input voltage (V_{SEN}) and is normally used for high-voltage sensors. Its primary limitation is that it also amplifies the common mode noise. The output of the amplifier is

$$V_{out} = -\frac{R_F}{R_1} V_{SEN} \qquad (6.56)$$

The non-inverting amplifier (Figure 6.22b) presents an impedance R_1 to the sensor and provides a positive gain:

$$\text{Gain} = \frac{V_{out}}{V_{SEN}} = 1 + \frac{R_F}{R_1} \qquad (6.57)$$

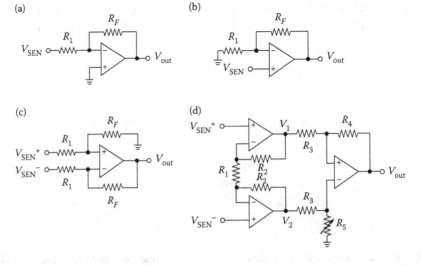

FIGURE 6.22 Four primary types of amplifier circuits: (a) inverting amplifier; (b) non-inverting amplifier; (c) differential amplifier; (d) instrumentation amplifier.

This circuit is often used in high-impedance sensors (e.g., piezoelectric film sensor) [32]. Its main advantages are its simplicity, high input impedance, positive gain, and the low bias current. The disadvantages include its limited input voltage range and its amplification of the common mode noise.

The differential amplifier in Figure 6.22c can be considered to be a combination of an inverting and a non-inverting amplifiers. It amplifies the sensor's output voltage difference:

$$V_{out} = \frac{R_F}{R_1}(V_{SEN^+} - V_{SEN^-}) \tag{6.58}$$

The merits of this amplifier are: it rejects common mode noise, allows large input voltage, and provides resistive isolation from the sensor signal source. A Wheatstone bridge output is often connected to this amplifier.

The instrumentation amplifier in Figure 6.22d is the most popular one. It maximizes the *Common Mode Rejection Ratio* (CMRR) by minimizing the common mode gain [32]. The left two amplifiers form a differential input/differential output amplifier; the right amplifier (a differential amplifier) rejects the common-mode voltage and converts the differential output of the first stage into a single-ended output. The output of each amplifier is [33]

$$V_1 = \left(1 + \frac{R_2}{R_1}\right)V_{SEN^+} - \frac{R_2}{R_1}V_{SEN^-} \tag{6.59}$$

$$V_2 = -\frac{R_2}{R_1}V_{SEN^+} + \left(1 + \frac{R_2}{R_1}\right)V_{SEN^-} \tag{6.60}$$

$$V_{out} = \left[\frac{R_4}{R_3}\left(1 + 2\frac{R_2}{R_1}\right)\right](V_{SEN^+} - V_{SEN^-}) \tag{6.61}$$

For a common mode input, that is, $V_{SEN^+} = V_{SEN^-}$, Equation 6.61 yields a zero output voltage. Hence, the common mode gain is zero, and the CMRR is infinite if $R_5 = R_4$. In reality, the resistances are never an exact match, thus an adjustable resistor is used for R_5.

6.7.2   CURRENT OUTPUT CIRCUITS

Current output sensors, such as DC current sensors, certain humidity sensors, and CMOS temperature sensors, output a current instead of a voltage. In this situation, a resistor (e.g., R_1 in Figure 6.23) is often used to convert sensor current output I_{SEN} into a voltage output. Then, a differential amplifier amplifies the voltage across the

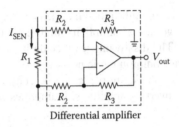

Differential amplifier

FIGURE 6.23 An amplifier circuit for a current output sensor.

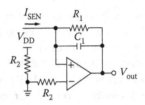

FIGURE 6.24 A transimpedance amplifier circuit for a current output sensor.

resistor R_1. One problem with this method is that the resistor R_1 loads the sensor output. To reduce this loading, the value of resistance R_1 should be chosen much less than R_2. The output V_{out} is expressed by

$$V_{out} = \frac{R_3}{R_2} R_1 I_{SEN} \tag{6.62}$$

A transimpedance amplifier (see Figure 6.24) can also be used to convert a sensor's current output I_{SEN} to a voltage V_{out} [31]. The capacitor C_1 is sometimes needed to stabilize the amplifier when the source has a large capacitance [34]. This provides good impedance buffering of the source and is often used in infrared (IR) smoke detectors and photodetectors. The relationship between V_{out} and I_{SEN} is

$$V_{out} + R_1 C_1 \frac{dV_{out}}{dt} = V_{DD} + R_1 C_1 \frac{dV_{DD}}{dt} - R_1 I_{SEN} \tag{6.63}$$

6.7.3 Charge Output Circuits

Some sensors (e.g., piezoelectric film sensors) measure or detect charge variation. When using this type of sensors, care must be taken since stray capacitance can shunt the sensor output to ground. The capacity of the cable from the sensor to the amplifier and the input impedance of the amplifier are also critical. If the impedance is too small, any small resistance can easily shunt the input sensor signal, resulting in large errors. Special amplifiers called charge amplifiers are available to condition these types of sensor signals. A charge amplifier has very high input impedance, and

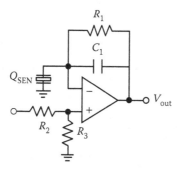

FIGURE 6.25   An amplifier circuit for a charge output sensor.

a capacitor is inserted into the input stage. It is not as sensitive to distributed capacity as a normal operational amplifier (op-amp) would be.

A sensor's charge output can be converted into a voltage using a circuit shown in Figure 6.25. R_1 and C_1 create a high-pass filter (HPF); R_1 also provides a bias path to prevent the inverting input of the op-amp from drifting over time. Any change in charge, Q_{SEN}, will appear almost exclusively across C_1, providing an accurate measurement of sensor output, Q_{SEN}. Thus, the sensor's charge output Q_{SEN} is converted into an output voltage V_{out}. This layout also rejects the common mode noise.

6.7.4   Resistance Output Circuits

Resistive sensors, such as potentiometers, RTDs, thermistors, and magnetoresistive sensors, detect changes in resistance, which is often converted into one of the four quantities: voltage, current, time (RC decay), or oscillator frequency [31]. Among these quantities, the resistance-to-voltage and resistance-to-current conversions are used most.

6.7.4.1   Resistance-to-Voltage Conversion

Resistance-to-voltage conversion can be simply implemented by using a voltage divider (Figure 6.26a) or a constant current source (Figure 6.26b). In Figure 6.26a, a constant voltage source V_{ex} with a resistor R_1 is used to configure a resistive sensor, as a simple voltage divider. In Figure 6.26b, a constant current source flows through the sensor and the voltage across the sensor is measured. Figure 6.26c uses a Wheatstone bridge for resistance-to-voltage conversion, which significantly improves the linearity over the measurement range. The relationship between output voltage V_{out} and resistance R_{SEN} in this circuit is

$$V_{out} = \left(\frac{R_{SEN}}{R_{SEN} + R_{REF}} - 0.5 \right) V_{ex} \qquad (6.64)$$

As mentioned in Section 2.2.2 of Chapter 2, an amplifier circuit shown in Figure 2.3 also provides a linear resistance-to-voltage conversion, and the relationship between the output voltage V_{out} and the sensor resistance R_2 is expressed by Equation 2.5.

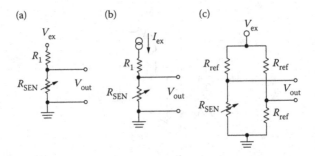

FIGURE 6.26 Resistance-to-voltage conversion by: (a) a voltage divider; (b) a constant current source; (c) a Wheatstone bridge.

Some sensors, such as thermistors, have high resistance and high sensitivity, thus their measurement and signal conditioning circuits are much simpler (no special 3-wire, 4-wire, or Wheatstone bridge connection is necessary). To minimize self-heating, some sensors (e.g., thermistors) are often operated at less than 100 μW.

More accurate resistance-to-current conversion can be achieved by the circuit shown in Figure 6.27. R_{1A}, R_{1B}, R_1, R_2, R_3, and the op-amp form a current source (called *Howland current pump*). C_1 stabilizes this current source and reduces noise. R_4 isolates the sensor from the ground. The voltage on the top of R_4 can detect a sensor [33,34]. This circuit requires accurate resistors, and it is more expensive than the circuits shown in Figures 6.26 and 2.3. However, it provides good linearity in resistance-to-voltage conversion and is often used in thermistors, RTDs, and hot-wire anemometers.

6.7.4.2 Resistance-to-Current Conversion

Resistance-to-current conversion produces a current I_{SEN} that is a function of a sensed resistance R_{SEN}. Figure 6.28 shows such a conversion circuit. I_{SEN} is then sent to other circuits (e.g., an amplifier circuit) for signal conditioning.

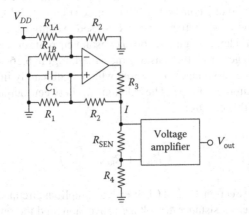

FIGURE 6.27 A more accurate amplifier circuit for a resistance-to-current conversion.

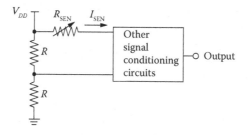

FIGURE 6.28 A resistance-to-current conversion circuit.

6.7.4.3 Resistance-to-Time Conversion (RC Decay)

An RC (resistor–capacitor) circuit shown in Figure 6.29 can be used to measure resistance using time. When switching S to Position 1, the voltage V_C does not immediately reach the applied constant DC voltage V_{DD}. Instead, it takes some time to reach a value close to V_{DD}. Similarly, when switching S back to Position 2, the voltage does not immediately disappear, but it takes some time to reach a value close to zero. Thus, the time it takes for the voltage to reach or decay to a threshold V_C, for example, $0.5\,V_{DD}$, is a measure of the resistance R if C and V_{DD} are fixed. The resistance R can directly relate to the time t by

$$R = \frac{t}{C\ln(V_{DD}/V_C)}\text{(For an } RC \text{ circuit, discharging)} \qquad (6.65)$$

The RC decay method provides an accurate and simple time measurement to determine the resistance value. It can be applied to both resistive and capacitive sensors and to convert resistance or capacitance into time. Its applications include thermistors, capacitive level sensors, and capacitive touch sensors.

EXAMPLE 6.16

Derive Equation 6.65 based on the RC circuit shown in Figure 6.29.

SOLUTION

Applying Kirchhoff's voltage law to the circuit with the switch S in Position 1 results in the *charging circuit equation*:

$$V_{DD} - iR - \frac{Q}{C} = 0$$

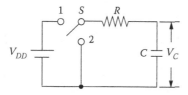

FIGURE 6.29 An RC circuit converting resistance to time.

where Q is the charge of the capacitor; i is the current flowing through the circuit. With the switch S in Position 2, it yields the *discharging circuit equation*:

$$-iR - \frac{Q}{C} = 0$$

Since $i = dQ/dt$, we have $-(dQ/dt)R - (Q/C) = 0$. Integrate this equation and use "$t = 0, Q = CV_{DD}$ and $t = t, Q = CV_C$" as well as "$Q = CV$", resulting in

$$-\int_{CV_{DD}}^{CV_C} \frac{dQ}{Q} = \frac{1}{RC}\int_0^t dt \Rightarrow -\ln Q\Big|_{CV_{DD}}^{CV_C} = \frac{t}{RC} \Rightarrow -[\ln(CV_C) - \ln(CV_{DD})] = \frac{t}{RC}$$

$$\ln(CV_{DD}) - \ln(CV_C) = \frac{t}{RC} \Rightarrow \ln\frac{V_{DD}}{V_C} = \frac{t}{RC} \Rightarrow R = \frac{t}{C\ln(V_{DD}/V_C)}$$

$$V_C = V_{DD}e^{-\frac{t}{RC}} \quad (RC\,\text{circuit, discharging})$$

where RC (often denoted as τ) is the *time constant*.

EXAMPLE 6.17

An *RC* circuit is used to monitor a RTD. If V_{DD} is 5 V, the resistance of the sensor is 100 Ω, and capacitance C is 0.2 μF, what is the half-life of charge decay?

SOLUTION

The half-life of charge indicates $V_C = 1/2V_{DD}$. Applying Equation 6.66, the half-life of charge decay is

$$t = RC\,\ln(V_{DD}/V_C) = (100\ \Omega)(0.2 \times 10^{-6}\ \text{F})\ \ln(2) = 13.86\ \mu s$$

6.7.4.4 Resistance-to-Frequency Conversion: RC Oscillator

A resistance output can be converted into a frequency output using an *RC* oscillator. Figure 6.30a shows a simple and low-cost resistive sensor oscillator circuit (called

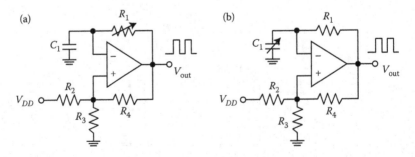

FIGURE 6.30 Relaxation oscillator circuits for: (a) resistive sensors; (b) capacitive sensors.

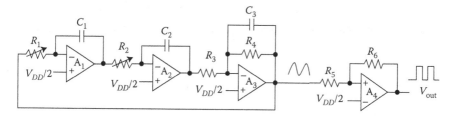

FIGURE 6.31 A schematic of a state variable oscillator for resistive sensors.

relaxation oscillator). Its output is a square wave with a frequency proportional to the change in the sensor resistance [38].

If the capacitor C_1 is a capacitive sensor and R_1 is a fixed resistor (see Figure 6.30b), the same circuit can be used for capacitive sensor measurements, such as capacitive humidity, touch, and level sensors.

For more reliable oscillation startup and higher precision measurement, a state variable *RC* oscillator (see Figure 6.31) can be used [37]. Its output frequency f is proportional to the square root of the product of two resistors R_1 and R_2:

$$f \propto \sqrt{R_1 R_2} \qquad (6.66)$$

Both the relaxation oscillator and the state variable oscillator can be used with resistive sensors and provide a resistance-to-frequency conversion.

6.7.5 Capacitance Output Circuits

A capacitive sensor produces a change in capacitance. This capacitance change can be converted into a voltage change by one of the following methods: *RC decay, integration of current, oscillator frequency,* and *AC bridge*. The AC bridge methods have been discussed in the bridge circuit section.

6.7.5.1 Capacitance-to-Voltage Conversion

In Figure 6.32, the capacitance change is converted into a DC voltage change. The resistance R is grounded to keep the capacitor voltage from drifting outside the amplifier's linear range.

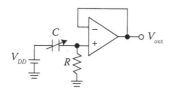

FIGURE 6.32 A capacitance-to-voltage conversion. (From Baxter, L.K., *Capacitive Sensors: Design and Applications*, IEEE Press, New York, 51, 1997.)

6.7.5.2 Capacitance-to-Time Conversion: RC Decay

In the RC decay method indicated in Equation 6.65 and Figure 6.29, the capacitance C relates to time t by

$$C = \frac{t}{R\ln(V_{DD}/V_C)} \quad (RC \text{ circuit, discharging})$$

(6.67)

6.7.5.3 Capacitance-to-Time-to-Voltage Conversion

This method, as shown in Figure 6.33, will integrate a current and measure the elapsed time to reach a preset voltage (V_{REF}). When the switch is closed, the charges on the left plate of capacitor will migrate to the right plate, increasing the output voltage V_{out} of the op-amp. This increased voltage is proportional to the time spent for the migration. The rate of migration depends on the values of V_{REF} and R_1. The comparator at the output trips at a time proportional to C_{SEN} [38].

6.7.5.4 Capacitance-to-Frequency Conversion: RC Oscillator

In the oscillator shown in Figure 6.31, if R_1 and R_2 are fixed, and C_1 and C_2 are replaced by capacitive sensors (e.g., one for the sensor C_{SEN} and the other for the reference C_{REF}), then the oscillator (see Figure 6.34) can be used to accurately measure capacitance. It consists of two integrators (formed by R_1, C_1, and the op-amp A_1; and R_2, C_2, and A_2) and an inverter (R_3, R_4, C_3, and A_3). Each integrator provides a phase shift of 90°, while the inverter provides 180° phase shift. Together, they make a total of 360° phase shift. C_3 in the inverter helps ensure oscillation start-up by providing an additional phase shift. A comparator (A_4) converts the oscillator's sine wave output to a square wave signal. This output frequency is proportional to the square root of the product of two capacitors:

$$f \propto \sqrt{C_1 C_2}$$

(6.68)

This frequency method is reliable and has a low sensitivity to stray capacitance [4]. Absolute quartz pressure sensors and capacitive humidity sensors are examples of capacitive sensors that can use this absolute oscillator.

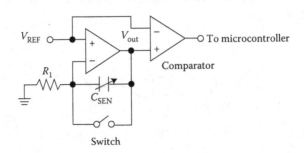

FIGURE 6.33   Capacitive-to-time-to-voltage conversion.

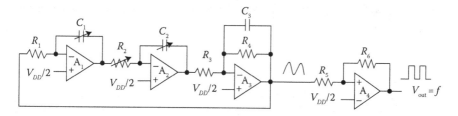

FIGURE 6.34 Schematic of an absolute oscillator for capacitive sensors.

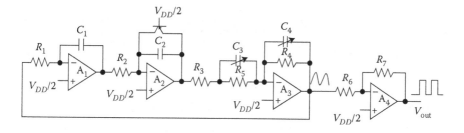

FIGURE 6.35 Schematic of a ratio oscillator to eliminate the dielectric constant variation.

To cancel the effect of a varying dielectric constant due to temperature fluctuation, a ratio oscillator can be used (see Figure 6.35) whose output frequency is proportional to the square root of the ratio of two capacitors:

$$f \propto \sqrt{C_1/C_2} \tag{6.69}$$

This ratio oscillator has two integrators (R_1, C_1, and A_1; R_2, C_2, A_2, and a transistor) and a differentiator (R_3, R_4, R_5, A_3, and sensing capacitors C_3 and C_4). The differentiator provides an 180° phase shift. The comparator (A_4) converts the sine wave output to a frequency f.

Many capacitive sensors, such as oil, air pressure, and acceleration sensors, can use this ratio oscillator to reduce the measurement error due to the variance in the dielectric constant over temperature [5]. For example, in an oil level measurement, two capacitors can be used: one is for sensing, C_{SEN}, and the other is for reference, C_{REF}. The sensing capacitor is placed near the surface of the oil. Thus, it is partially covered by the oil and partially covered by air; while the reference capacitor is completelly immersed in the oil. The oil and/or air serve as the dielectric media. The ratio C_{SEN}/C_{REF} is used to determine the level of the oil.

6.7.6 INDUCTANCE OUTPUT CIRCUITS

Inductance can be measured by many methods, including *impedance, phase shift, bridge, resonance, voltage–current slope* methods. In the impedance method, an inductive sensor is placed in series with a known precision resistor (e.g., a 100 Ω 1% resistor) to form a voltage divider. Then an excitation voltage is applied to the circuit. The voltage potential at the junction between the resistor and inductor can be expressed as

$$V_{out} = \frac{R}{R + j\omega L} V_{ex}$$

If we let $V_{out}/V_{ex} = 1/2$, then,

$$L = \frac{\sqrt{3}R}{2\pi f} \tag{6.70}$$

EXAMPLE 6.18

Derive Equation 6.70.

SOLUTION

The voltage across the resistor can be expressed by

$$\frac{V_{out}}{V_{ex}} = \frac{R}{R + j\omega L}$$

Square the magnitude of both sides of the above equation and let $V_{out}/V_{ex} = 1/2$, and $\omega = 2\pi f$:

$$\left(\frac{1}{2}\right)^2 = \left|\frac{R}{R + j2\pi fL}\right|^2 \Rightarrow \frac{1}{4} = \frac{R^2}{R^2 + (2\pi fL)^2}$$

Thus, $L = \sqrt{3}R/2\pi f$.

One of resonant methods is to place the inductive sensor in parallel with a capacitor of known value to form a tank circuit and then put this tank circuit in series with a resistor. An excitation voltage is then applied, and the output across the tank can be recorded and plotted. The peak value of V_{out} corresponds to the resonant point, and the inductance can be calculated by

$$L = \frac{1}{C\omega^2} \tag{6.71}$$

Figure 6.36 shows this oscillating circuit [39]. Resistor R_{LP} represents the losses in the coil. Its resonant frequency f and Q-factor can be expressed by

$$f = \frac{1}{2\pi\sqrt{LC}} \tag{6.72}$$

$$Q = \frac{R_{LP}}{2\pi fL} \tag{6.73}$$

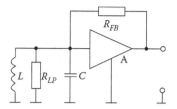

FIGURE 6.36 An oscillator circuit for inductance measurement. (From Hruškovic, M. and Hribik J., Digital capacitance and inductance meter, *Measurement Science Review*, 8, Section 3, No. 3, 61, 2008.)

The voltage–current slope method requires a high-frequency pulse voltage to be placed across the inductive sensor. The duty cycle of the pulse should be below 50%. Then the current is measured and plotted versus voltage. The inductance is determined by the slope of the plot.

6.8 SENSOR COMPENSATION CIRCUITS

To provide reliable and accurate outputs, many sensors use various compensation circuits to adjust their nonlinearity, remove offset or output errors caused by environmental factors such as temperature fluctuation, humidity variation, or normal wear and tear. This section discusses commonly used sensor compensation methods and circuits.

6.8.1 TEMPERATURE COMPENSATION

Temperature compensation is the most important one since almost all RCIM sensors are affected by temperature. Three techniques are used to compensate for temperature variations: passive compensation at the sensor (e.g., the sensor has a built-in temperature insulator), active compensation (e.g., the sensor has a temperature compensation circuit), and digital or software compensation (a microprocessor or a computer is used to correct the sensor outputs based on measured temperature). The first two methods usually require a temperature sensor to implement the compensation. Figure 6.37 shows an anemometer with (Figure 6.37a) and without (Figure 6.37b)

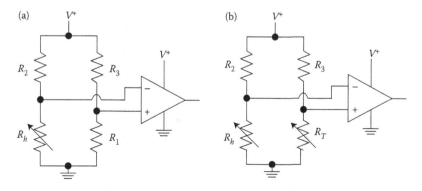

FIGURE 6.37 An anemometer circuit (a) with or (b) without temperature compensation.

temperature compensation [40]. In Figure 6.37b, a temperature sensor R_T is placed in the opposite branch to the heater of the anemometer, R_h, in the Wheatstone bridge. The temperature variation will affect both R_T and R_h, but only their difference affects the sensor output. Thus, the temperature compensation is achieved. Many resistive sensors can use a similar method to reduce temperature-variation caused errors.

6.8.2 NONLINEARITY COMPENSATION

Both digital and analog methods can be used to compensate for nonlinearity of a sensor. The digital linearization often involves a generic equation or a lookup table previously stored in the microprocessor's memory. The generic (linear) equation, for example, $R_T = 100(1 + 0.003917)$ for a RTD, can be used by the microprocessor to calculate the sensor's output (e.g., temperature T) based on the measured value (e.g., resistance $R_T = 139.11\ \Omega$), resulting in $T = 100.03°C$; while a lookup table contains a number of pre-calculated outputs (e.g., temperature values) for various possible measured values (e.g., resistance values). Once a measured value (e.g., resistance) is obtained, the microprocessor directly reads the corresponding linearized output (e.g., temperature) from the table. Thus, for digital compensation, a microprocessor is required to perform the calculations and interpolation (if the lookup table is used). Figure 6.38 shows an analog approach, a physical circuit, to perform nonlinearity compensation of a PT100 RTD. Without any linearization, the output of a sensor will deviate from its linear model about −1.5–0.5°C within −50–125°C operating range; while using this circuit model, the deviation from its linear model (the linearity error) can be reduced to a maximum of 0.05°C within the same operating range.

6.8.3 OFFSET ERROR COMPENSATION

Many sensors suffer from offset errors. These offsets must be removed to obtain an accurate measurement. There are several basic circuit designs for offset

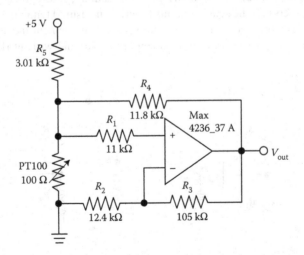

FIGURE 6.38 A PT100 RTD nonlinearity compensation circuit.

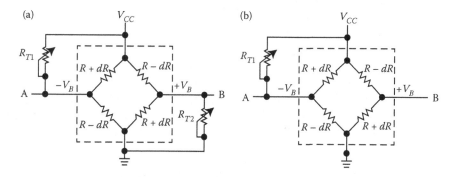

FIGURE 6.39   Offset error compensation circuits: (a) symmetrical offset trim; (b) asymmetrical offset trim.

compensation. Taking a resistive pressure sensor as an example, two methods can be used to remove the offset errors: symmetrical offset trim and asymmetrical offset trim [41]. In the symmetrical offset trim method (see Figure 6.39a), the two trim resistors (potentiometers) R_{T1} and R_{T2} are required to move the output signal level on each side to $V_{CC/2}$ for the lowest pressure point. In the asymmetrical offset trim method shown in Figure 6.39b, only one trim resistor R_{T1} is used to move the output level of only one side to the same voltage level as the other side.

6.9   SENSOR SIGNAL CONDITIONING, PASSIVE AND ACTIVE FILTERS

Most sensors' raw output signals are not ready for immediate use. Instead, these sensors' outputs require signal conditioning that can remove or attenuate unwanted components (e.g., noise) in the sensor signals and increase the sensors' signal levels or magnitudes. A signal conditioning may involve one or more of the following processes: amplification, frequency-selective filtering, mathematical operations (e.g., differentiation or integration), noise attenuation, or a simple DC-level translation. Among them, op-amps and various passive or active filters that consist of various electronic components such as resistors, capacitors, inductors, or other op-amps are the main components of a signal conditioning. Filters are essential to sensor circuits. Without proper filtering, a sensor's signal cannot be interpreted correctly, causing a false alarm. This section will discuss various commonly used filters for sensor signal conditioning.

6.9.1   FILTERING

In sensor circuits, the conductive noise or coupling can be reduced or eliminated by two methods: (1) frequency-selective filters and (2) common or differential modes. Frequency-selective filters include low-pass filters (LPF) that pass the low-frequency components and reject the high-frequency components of a signal; high-pass filters (HPF) that pass the high-frequency components and reject the low-frequency components of a signal; band-pass filters (BPF) that allow only certain range of frequency components to pass; and notch filters (NF) that reject specific frequency components

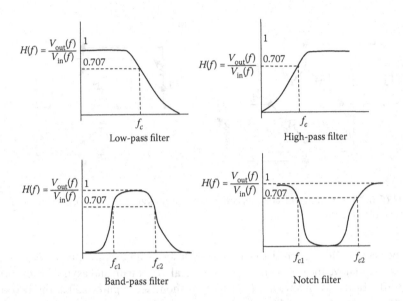

FIGURE 6.40 Frequency-selective filters used to remove conductive noise.

as shown in Figure 6.40. A transfer function, defined as the ratio of the output signal $V_{out}(f)$ to the input signal $V_{in}(f)$, is used to describe the characteristics of these filters:

$$\text{Transfer function: } H(f) = \frac{V_{out}(f)}{V_{in}(f)} \tag{6.74}$$

In Figure 6.40, the *corner* or *cutoff* frequency, f_c, is the frequency at which the transfer function has an amplitude of 0.707. One or more filters may be used in a signal conditioning circuit, depending on the frequency range of the noise in a signal.

6.9.2 CHARACTERISTICS OF A FILTER

A filter is an electrical network that alters the amplitude and/or phase characteristics of a signal [42] and removes unwanted components in the signal. Strictly speaking, a filter cannot remove or alter the component frequencies of that signal, but attenuate or change the *relative amplitudes* of the various frequency components and/or their *phase relationships*, so that the unwanted components are not dominant or noticeable in the output/final signal, while the real sensor signal is emphasized.

The characteristics of a filter can also be described by its *transfer function* in Laplace domain, $H(s)$, defined as the ratio of the Laplace transform of its output, $V_{out}(s)$, and the Laplace transform of its input, $V_{in}(s)$, i.e.:

$$H(s) = \frac{V_{out}(s)}{V_{in}(s)} \tag{6.75}$$

where s is the Laplace variable. The *order* of a filter is the highest power of the variable s in the denominator of its transfer function and it is usually equal to the total number of capacitors and inductors in the filter. Higher-order filters provide more effective discrimination between signals at different frequencies, but they are more expensive due to having more components (more noise resources) and requiring the more complicated design. If s is replaced by $j\omega$ (j is the imaginary unit, which satisfies the equation $j^2 = -1$; ω is the angular frequency, $\omega = 2\pi f$), the transfer function in Laplace domain becomes the transfer function in frequency domain:

$$H(j\omega) = \frac{V_{out}(j\omega)}{V_{in}(j\omega)} \tag{6.76}$$

The *magnitude* or *gain* of the above transfer function is a function of frequency ω, and it indicates the filter's performance over a broad frequency range. The *phase angle* of this transfer function gives the amount of phase shift at each frequency. The phase characteristics of a filter are important, especially when the time relationships between signal components at different frequencies are critical.

EXAMPLE 6.19

Derive the transfer function of the *RLC* filter shown in Figure 6.41.

Solution

The equivalent impedance of Z_R, Z_L, and Z_C is: $Z_R = R$, $Z_L = j\omega L = sL$, $Z_C = 1/(j\omega C) = 1/(sC)$, respectively.

$$Z_L // Z_C = \frac{Z_L Z_C}{Z_L + Z_C} = \frac{(sL)\left(\dfrac{1}{sC}\right)}{sL + \dfrac{1}{sC}} = \frac{Ls}{LCs^2 + 1}$$

Applying the voltage divider equation yields (// means a parallel connection):

$$V_{out} = \frac{Z_L // Z_C}{Z_R + Z_L // Z_C} V_{in} = \frac{Ls/(LCs^2 + 1)}{R + Ls/(LCs^2 + 1)} V_{in} = \frac{Ls}{RLCs^2 + Ls + R} V_{in}$$

Thus, the transfer function is

$$H(s) = \frac{V_{out}}{V_{in}} = \frac{Ls}{RLCs^2 + Ls + R}$$

This is a second-order system.

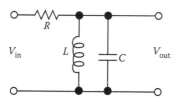

FIGURE 6.41 An RLC filter.

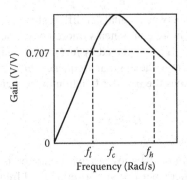

FIGURE 6.42 Definition of the Q value of a filter.

The *cutoff frequency* of a filter is defined as the frequency at which the gain or magnitude of a filter's transfer function is equal to 0.707 or −3 dB in the logarithmic scale:

$$20 \log(0.707) = -3 \text{ dB}$$

This is because at the cutoff frequency, the power of a filter's output signal is dissipated to the half of its input signal:

$$\frac{P_{\text{out}}}{P_{\text{in}}} = \frac{1}{2} \Rightarrow \frac{P_{\text{out}}}{P_{\text{in}}} = \frac{V_{\text{out}}^2}{V_{\text{in}}^2} = \frac{1}{2} \Rightarrow \frac{V_{\text{out}}}{V_{\text{in}}} = \frac{1}{\sqrt{2}} = 0.707$$

Q value of a filter indicates the "sharpness" of the amplitude response of the filter, defined as

$$Q = \frac{f_c}{f_h - f_l} \tag{6.77}$$

where f_l is the lower frequency at −3 dB or 0.707; f_h is the higher frequency at −3 dB or 0.707; and f_{ctr} is the center frequency (see Figure 6.42). Their relationship is

$$f_{\text{ctr}} = \sqrt{f_l f_h} \tag{6.78}$$

6.9.3 Passive Filters

In passive filters, all components are *passive* electrical or electronics components (e.g., resistors, capacitors, and inductors). These components do not require additional power supplies to operate. Thus, a passive filter is the simplest implementation of a filter for a given transfer function. Figure 6.43 shows the second-order low-pass, high-pass, band-pass, and notch filters.

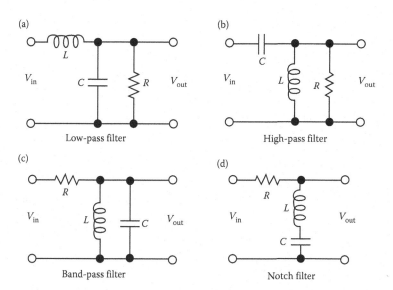

FIGURE 6.43 Examples of second-order passive filters: (a) low-pass filter; (b) high-pass filter; (c) band-pass filter; (d) notch filter.

Passive filters work well at very high frequencies because they are not restricted by the bandwidth limitations that active filters have. Passive filters are often used in applications that involve larger current or voltage levels than can be handled by active filters. Passive filters also generate less noise compared to active filters. The noise that passive filters produce is simply the thermal noise from their resistive components. The disadvantages of passive filters include: (1) no signal amplification function; (2) more expensive to achieve high accuracy (e.g., 1% or 2%) and small size due to using inductors; and (4) difficult to design complex or higher than second-order passive filters.

EXAMPLE 6.20

(1) Derive the transfer function of the 2nd-order high-pass filter shown in Figures 6.43b. (2) Draw the bode diagram of the filter if $R = 1$ kΩ, $C = 0.1$ µF, and $L = 1$ mH. (3) In Figures 6.43b, if the inductor is removed, the original 2nd-order high-pass filter will become a simple 1st-order high-pass filter. Derive the transfer function of this resultant 1st-order high-pass RC filter, and then (4) draw its bode diagram given $R = 1$ kΩ, $C = 0.1$ µF.

Solutions

(1) The equivalent impedance of Z_R, Z_L, and Z_C is:
$Z_R = R$, $Z_L = j\omega L = sL$, $Z_C = 1/(j\omega C) = 1/(sC)$.
Z_R and Z_L are parallel:

$$Z_R // Z_L = \frac{Z_R Z_L}{Z_R + Z_L} = \frac{R(sL)}{R + sL} = \frac{RLs}{R + Ls}$$

Applying voltage divider yields:

$$V_{out} = \frac{Z_R/\!/Z_L}{Z_C + Z_R/\!/Z_L}V_{in} = \frac{\dfrac{RLs}{R+Ls}}{\dfrac{1}{sC} + \dfrac{RLs}{R+Ls}}V_{in} = \frac{RLCs^2}{RLCs^2 + Ls + R}V_{in}$$

Thus, the transfer function is:

$$H(s) = \frac{V_{out}}{V_{in}} = \frac{RLCs^2}{RLCs^2 + Ls + R}$$

This is a 2nd order system.

(2) Plug in $R = 1\ k\Omega$, $C = 0.1\ \mu F$, and $L = 1\ mH$, the transfer function becomes:

$$H(s) = \frac{V_{out}}{V_{in}} = \frac{RLCs^2}{RLCs^2 + Ls + R} = \frac{(1000)(1\times 10^{-3})(0.1\times 10^{-6})s^2}{(1000)(1\times 10^{-3})(0.1\times 10^{-6})s^2 + 0.001s + 1000}$$

$$= \frac{0.1\times 10^{-6}s^2}{0.1\times 10^{-6}s^2 + 0.001s + 1000}$$

Use MATLAB with commands:

NUM=[0.0000001 0 0];	% if the numerator is $(0.01\times 10^{-6}s^2 + 0s + 0)$
DEN=[0.0000001 0.001 1000];	% if the denominator is $(0.01\times 10^{-6}s^2 + 0.0001s + 1)$
SYS=TF(NUM,DEN);	% define the transfer function
bode(SYS)	% draw the bode plot
grid;	% put grid in the magnitude plot by click the plot first
grid;	% put grid in the phase angle plot by click the plot first

yielding a bode diagram below:

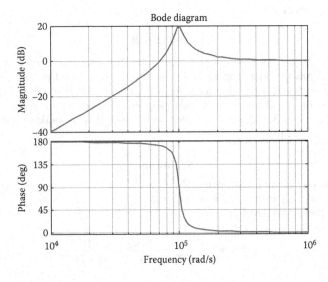

Bode diagram

(3) If L is removed from Figures 6.43b,

$$V_{out} = \frac{Z_R}{Z_C + Z_R} V_{in} = \frac{R}{\dfrac{1}{sC} + R} V_{in} = \frac{RCs}{RCs + 1} V_{in}$$

$$H(s) = \frac{V_{out}}{V_{in}} = \frac{RCs}{RCs + 1}$$

(4) Plug in $R = 1\ k\Omega$ and $C = 0.1\ \mu F$ into the transfer function in (3), yields

$$H(s) = \frac{V_{out}}{V_{in}} = \frac{RCs}{RCs + 1} = \frac{0.0001s}{0.0001s + 1}$$

Use MATLAB with commands:

```
NUM=[0.0001 0];        % for the numerator of (0.0001s + 0)
DEN=[0.0001 1];        % for the denominator of (0.0001s + 0s + 1)
SYS=TF(NUM,DEN);       % define the transfer function
bode(SYS)              % draw the bode plot
grid;                  % put grid in the magnitude plot by click the plot first
grid;                  % put grid in the phase angle plot by click the plot first
```

yielding a Bode diagram below:

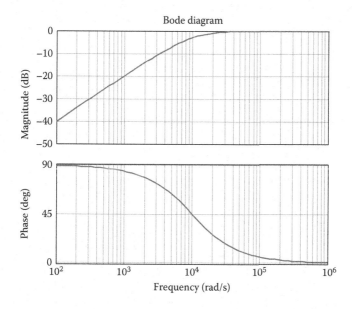

6.9.4 ACTIVE FILTERS

Active filters use active components, especially op-amps, together with resistors and capacitors (note: no inductors) in their designs. Active filters have very high input impedance and very low output impedance, which is ideal for sensor signal conditioning. Theoretically, active filters can provide an infinite gain; in reality, only a limited gain can be provided. Active filters are easier to design than passive filters, and are very accurate even though low-tolerance resistors and capacitors are used. Active filters can also amplify a sensor signal since they use op-amps. Performance of active filters at high frequencies is limited by the gain-bandwidth product of the amplifying components. Active filters generate noise due to the amplifying circuitry. Using low-noise amplifiers can minimize this noise.

Figure 6.44 shows commonly used basic active filters. By cascading two or more of these circuits, higher-order filters can be obtained. Figure 6.44a is a second-order LPF (this configuration is called *Sallen–Key* topology). It has excellent gain accuracy and is often used as a building block for higher-order filters. A LPF can filter out high-frequency noise from sensor signals. Interchanging resistors and capacitors in the circuit will result in a HPF as shown in Figure 6.44b. This filter can reject unwanted low-frequency component (e.g., DC offset). In Figure 6.44a, if we move C_2 from the ground path to the negative terminal side (marked as "−"), and then exchange C_2 with R_2, a second-order band-pass filter is produced (see Figure 6.44c). This filter provides a sharp peak at the frequencies of interest and allows only these frequency signals to pass. Figure 6.44d is a third-order notch filter in twin-T configuration. It provides a low impedance for all frequencies of interest and a high impedance for the frequency component to be rejected, determined by the values of resistors and capacitors chosen. These active filters are provided with various electronic packages or chips that contain one or more filters inside.

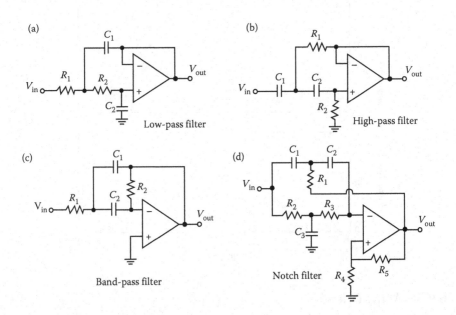

FIGURE 6.44 Examples of various active filters. (a) low-pass filter; (b) high-pass filter; (c) band-pass filter; (d) notch filter.

6.9.5   A DESIGN EXAMPLE OF SENSOR SIGNAL CONDITIONING CIRCUITS

In this section, an electrocardiogram (ECG) signal from a human subject (see Figure 6.45) is used to demonstrate how to design a sensor signal conditioning circuit. A typical ECG signal obtained using the Ag–AgCl electrodes has an amplitude of 0.1–10 mV and frequency of 0.01–250 Hz. Signals like these are very vulnerable to noise. Any movement during the measurement, such as respiration, motion artifacts, muscle contraction, or power line interference, RFI, and EMI, could add noise to the ECG signal. That is why it is hardly seen the ECG waves from Figure 6.45. The signal conditioning is to remove/attenuate various noise from a sensor's raw signal, and it usually involves the following procedures.

6.9.5.1   Amplification Circuit

The signal conditioning starts with amplification. An *Analog Devices'* AD8220 instrumentation amplifier is chosen for amplification due to its wide operating range, low input bias current (10 pA), high CMRR (common-mode rejection ratio), and adjustable gain. The gain G is set at 19 to avoid output voltage saturation. Thus, a typical 1.0 mV ECG signal is amplified to 19.0 mV. The gain resistor R_g is determined using the formula in the *Analog Devices' Designer's Guide* [43]:

$$R_g = \frac{49.4 \text{ k}\Omega}{G - 1} = \frac{49.4 \text{ k}\Omega}{19 - 1} = 2.74 \text{ k}\Omega \tag{6.79}$$

6.9.5.2   High-Pass Filter

After the above instrumentation amplifier, the ECG signal is sent to a non-inverting active HPF to eliminate its DC offset (a low-frequency component) and further amplify the ECG signal. The filter circuit (see Figure 6.46) consists of an MCP6271 op-amp, an RC filter (the capacitor Z_2 and the resistor Z_4), and the gain resistors Z_{10} and Z_{11} arranged in a Sallen–Key configuration. This is a first-order filter.

Since an ECG signal has a frequency range of 0.01–100 Hz, a cutoff frequency f_c of 0.033 Hz is chosen to ensure the useful ECG signals pass through the HPF. Z_2 was set to 6.8 μF for a faster response. The resistor Z_4 therefore should have a value of

$$Z_4 = \frac{1}{2\pi Z_2 f_c} = \frac{1}{2\pi(6.8 \times 10^{-6}\text{F})(0.033 \text{ Hz})} = 710 \text{ k}\Omega \tag{6.80}$$

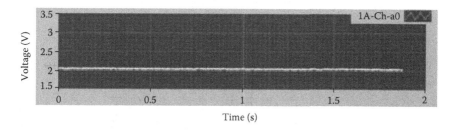

FIGURE 6.45   A typical raw ECG signal directly obtained from a human body.

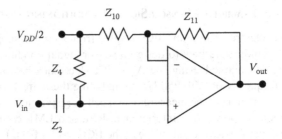

FIGURE 6.46 Circuit diagram of a first-order HPF with the Sallen–Key configuration.

In addition, Z_{10} and Z_{11} are set to 806 Ω and 13 kΩ, respectively, which results in an HPF gain G_{HPF} of

$$G_{HPF} = 1 + \frac{Z_{11}}{Z_{10}} = 1 + \frac{13000 \text{ k}\Omega}{806 \text{ }\Omega} = 17.13 \tag{6.81}$$

6.9.5.3 Low-Pass Filter

Following the HPF is the LPF to remove the high-frequency noise components from the ECG signal. A 160 Hz cutoff frequency is chosen for the LPF based on the normal ECG frequency range (0.01–100 Hz) as well as the American Heart Association (AHA) guidelines for minimum ECG frequency bandwidth—AHA recommends a minimum bandwidth of 150 Hz for children between the ages of 12 and 16 years; and a minimum bandwidth of 125 Hz for adults [44]. The LPF is a fifth-order active Bessel filter, formed by three active filters cascaded together (see Figure 6.47). The first filter is an active first-order LPF that has a real pole in its transfer function. Both the second and the third filters use second-order Sallen–Key topologies, and each of them has a pair of complex conjugate poles in its transfer function. All three filters use unity gain, thus less resistors in the circuit and less thermal noise caused by resistors. This LPF has excellent transient and linear phase responses [45].

The resistors and capacitors in the LPF are determined as follows, considering Filter 2 first:

$$f_c = \frac{1}{2\pi\sqrt{R_{21}R_{22}C_{21}C_{22}}} = 160 \text{ Hz} \tag{6.82}$$

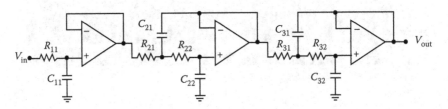

FIGURE 6.47 Circuit diagram of the fifth-order active Bessel LPF.

Let m be the ratio of R_{21} and R_{22}, n be the ratio of C_{21} and C_{22}, that is,

$$\begin{cases} R_{21} = mR \\ R_{22} = R \end{cases} \begin{cases} C_{21} = nC \\ C_{22} = C \end{cases}$$

Let Q-factor = 1 (for better filter performance at cutoff frequency without significant overshoot), thus,

$$Q = \frac{\sqrt{R_{21} R_{22} C_{21} C_{22}}}{C_{22}(R_{21} + R_{22})} = \frac{\sqrt{mn}}{m + 1} = 1 \tag{6.83}$$

Choose a value for m, and determine n based on the above Q setting:
If $m = 1$, then $n = 4$ from Equation 6.83.
Select C value and calculate R based on f_c:
C should not be too small since a low capacitor value can result in significant errors due to parasitic capacitance. Chose $C = 0.068 \ \mu F$, thus,

$$\begin{cases} C_{21} = nC = 4 \times (0.068 \ \mu F) = 0.272 \ \mu F \\ C_{22} = C = 0.068 \ \mu F \end{cases}$$

From Equation 6.82,

$$R = \frac{1}{2\pi C f_c \sqrt{mn}} = \frac{1}{2\pi(0.068 \times 10^{-6} \ F)(160 \, Hz)\sqrt{(1)(4)}} = 7.318 \ k\Omega$$

thus,

$$\begin{cases} R_{21} = mR = (1)(7.318) = 7.318 \ k\Omega \\ R_{22} = R = 7.318 \ k\Omega \end{cases}$$

Similarly, the resistors and capacitors in Filter 1 can be determined by letting $f_c = 160$ Hz, $Q = 1$, $n = 4.5$, $m = 2$, $C = 0.068 \ \mu F$, and calculating $C_{11} = 0.068 \ \mu F$, $R_{11} = 14.628 \ k\Omega$; Filter 3 can be designed by choosing $C_{31} = 0.306 \ \mu F$, $C_{32} = 0.068 \ \mu F$, $R_{31} = 9.752 \ k\Omega$, and $R_{32} = 4.876 \ k\Omega$. All op-amps in the LPF are MCP6271. This CMOS chip has a 2 MHz *Gain Bandwidth Product* (GBWP) and a 65° phase margin. It also supports rail-to-rail input and output swing [46].

6.9.5.4 Notch/Band-Reject Filter

A *notch* or *band-reject* (*band-stop*) filter can suppress certain frequency or range of frequencies in a signal. The 60 Hz power line interference in ECG signal can be rejected using a notch filter, thus the cutoff frequency for the notch filter is $f_n = 60$ Hz. This requires a high-quality Q factor to achieve the steeper roll-off. A Twin-T notch filter is one of a few RC networks capable of providing a deep notch at a particular frequency. Another advantage of the Twin-T configuration is that the Q factor can be altered via the inner gain G_n without modifying the notch frequency f_n. The two T-shaped RC filters combined with a MCP6271 op-amp form an active notch filter (see Figure 6.48). A Q factor of 2.5 is chosen, even though a Q factor of 50 or more is achievable. The op-amp provides low output impedance and high input impedance, making it possible to use large resistance values in the "T" filters and only small capacitance values are required, even at low frequencies. A +5 V power supply is provided to the MCP6271 (similar to V_{DD} in an LPF circuit). All grounds are connected to a 2.5 V reference (similar to $V_{DD}/2$ in an LPF circuit).

With $f_n = 60$ Hz and $C = 0.47$ μF, the resistor value in the filter is

$$R = \frac{1}{2\pi f_n C} = \frac{1}{2\pi(60\,\text{Hz})(0.47 \times 10^{-6}\,\text{F})} = 5.647\,\text{k}\Omega$$

To achieve a higher Q factor of approximately 2.5, the resistance values for $R_1 = 1$ kΩ and $R_2 = 800$ Ω are chosen, thus

$$Q = \frac{R_1}{2(R_1 - R_2)} = \frac{1000\,\Omega}{2(1000\,\Omega - 800\,\Omega)} = 2.5$$

The resulting filter gain is

$$G_n = 1 + \frac{R_2}{R_1} = 1 + \frac{800}{1000} = 1.8$$

The block diagram of the overall ECG signal conditioning circuit is shown in Figure 6.49. The high-pass, low-pass, and notch filters are connected in cascading

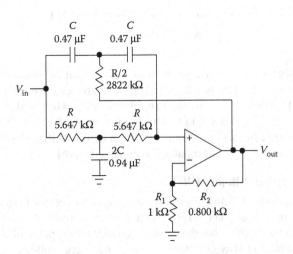

FIGURE 6.48   Circuit diagram of a Twin-T notch filter.

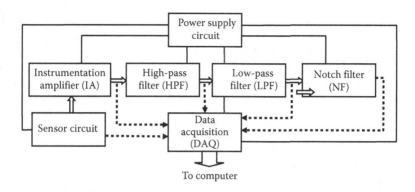

FIGURE 6.49 Block diagram of the entire signal conditioning circuits.

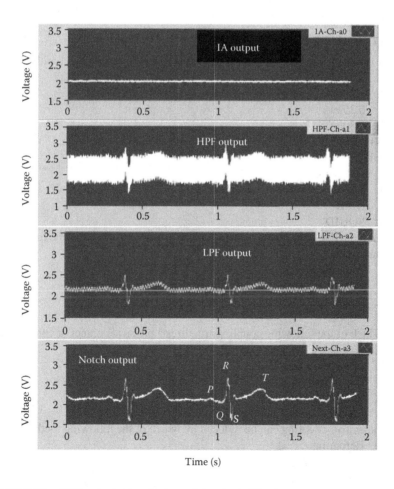

FIGURE 6.50 ECG outputs at each stage of signal conditioning.

series. The ±5 V power supply circuit is located between the sensor circuit and the USB6009 data acquisition (DAQ) device.

This signal conditioning circuit is tested on a real ECG signal from a human subject. The ECG output at each filter is acquired by the USB6009 DAQ system and the result is displayed in Figure 6.50 using LabVIEW. It can be seen that in the initial amplification (using an instrumentation amplifier that amplifies the 1 mV ECG signal to 19 mV), no significant ECG features are observed. The output of the HPF starts to show the peaks of the ECG signal, but there is high-frequency noise riding onto it. This high-frequency noise is removed by the LPF, but the power line interference is still visible. The consecutive notch filter is able to remove the power line noise, making the base signal much clearer. The P, Q, R, S, T peaks are distinguishable. This final output of the ECG waveform indicates that the designed signal conditioning circuit is a very powerful tool to enhance the ECG sensor signal and attenuate noises.

EXERCISES

Part I: Concept Problems

1. A typical capacitive proximity sensor can detect a metal object within:
   A. 0.02 m
   B. 0.20 m
   C. 0.25 m
   D. 0.30 m
2. Which of following sensors/devices can detect an extremely small magnetic field?
   A. Hall sensor
   B. AMR
   C. GMR
   D. SQUID
3. An inductive or magnetic sensor is more sensitive to which of the following materials?
   A. Aluminum
   B. Brass
   C. Steel
   D. Copper
4. Which of the following couplings requires a physical contact or connection between two elements or circuits?
   A. Conductive coupling
   B. Capacitive coupling
   C. Inductive coupling
   D. Electromagnetic coupling
5. Capacitive coupling can be reduced by which of the following methods?
   A. Rerouting
   B. Twisted wires
   C. Filtering
   D. *Faraday* cage

6. With which of the following frequencies, a signal will create the largest parasitic capacitance when it passes through one of two closely placed circuits?
   A. 1 kHz
   B. 10 kHz signal
   C. 100 kHz signal
   D. 1 MHz signal

7. Under which of the following frequency ranges, do cables act as antennas and emit or receive electromagnetic waves?
   A. $f \leq 20$ MHz
   B. 20 MHz $< f \leq 200$ MHz
   C. $f > 200$ MHz
   D. $f > 500$ MHz

8. Under which of the following frequency ranges, does a PCB trace in a circuit acts as an antenna and emits electromagnetic waves?
   A. $f \leq 20$ MHz
   B. 20 MHz $< f \leq 200$ MHz
   C. $f > 200$ MHz
   D. $f > 1000$ MHz

9. To reduce inductive coupling, the area of a current loop should be
   A. As small as possible
   B. As large as possible
   C. Proportional to the magnitude of the current flowing in it
   D. Inversely proportional to the magnitude of the current flowing in it

10. Which of the following noises is also called "broadband noise"?
    A. White
    B. Crosstalk
    C. Thermal
    D. Schottky

11. Which of the following noises can be minimized by decreasing the circuit density?
    A. Popcorn
    B. Crosstalk
    C. Thermal
    D. $1/f$

12. Which of the following noises can be eliminated by reducing temperature?
    A. Popcorn
    B. Shot
    C. Thermal
    D. $1/f$

13. Which of the following noises can be reduced by increasing a material's purity?
    A. Popcorn
    B. Crosstalk
    C. Thermal
    D. Shot

14. Match each of the following formula to its corresponding noise type:

$$1/f \qquad\qquad V_n = \sqrt{2qI\Delta f}$$

$$\text{Shot} \qquad\qquad V_n = 2\sqrt{k_B TR\Delta f}$$

$$\text{Broadband} \qquad\qquad V_n = V_{BB}\sqrt{BW_n}$$

$$\text{Thermal} \qquad\qquad V_n = V_{nw}\sqrt{f_C \ln\frac{f_C}{f_L}}$$

15. Which of the following bridges is unbalanced?

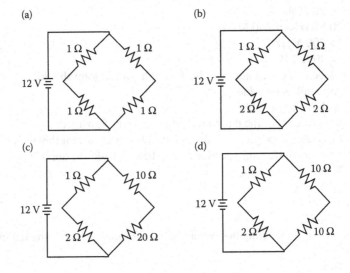

16. Which of the following bridges is a DC bridge?
    A. Hay bridge
    B. Wien bridge
    C. Kelvin bridge
    D. Schering bridge
17. Which of the following bridges is used to measure an unknown capacitance?
    A. Maxwell bridge
    B. Wien bridge
    C. Owen bridge
    D. Schering bridge
18. Why does a Wheatstone bridge need to be balanced first?
19. In the following sensor output circuits, which one(s) can cancel out an inter-
    ference with an external field (e.g., an audio signal or a power-line source)?

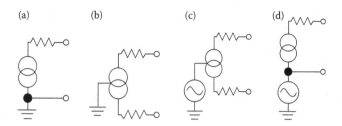

(a) (b) (c) (d)

20. The main advantage of the non-inverting gain amplifier shown in Figure 6.22b are:
    A. Simplicity
    B. High input impedance
    C. Low bias current
    D. Broad input voltage range
21. Explain the features and functions of the instrumentation amplifier shown in Figure 6.22d.
22. Which of the following circuits is a transimpedance amplifier?

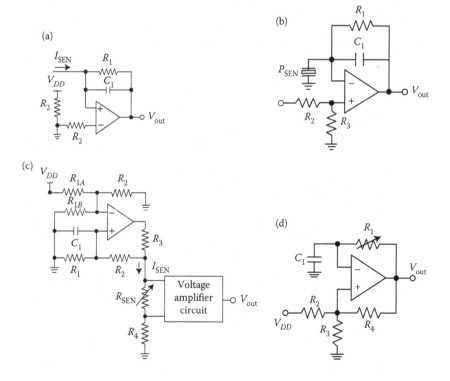

23. Which of the following signals can NOT be converted into time by an RC decay circuit?

A. Resistance
B. Capacitance
C. Voltage
D. Frequency

24. The *oscillator frequency* method can accurately measure the resistance of resistive sensors or capacitance of capacitive sensors with the output of:
    A. Voltage
    B. Current
    C. Charge
    D. Time

25. *Integration of Current* method is used to measure
    A. Current
    B. Elapsed time
    C. Frequency
    D. Voltage

Part II: Calculation Problems

26. A stray capacitance between a capacitor and a sensor circuit is $C = 140$ pF (see Figure 6.2b). To reduce the stray capacitance effect in the sensor circuit, a metal mesh shield is put around the capacitor. As a result, the capacitance C_1 between the capacitor and the shield is 45 pF, and the capacitance C_2 between the shield and the sensor circuit is 70 pF. How much the stray capacitance is reduced with the metal mesh shielding?

27. Every component in a circuit, especially an active component, adds noise to the circuit. If a circuit contains a resistor, a capacitor, and an amplifier, and their contributed noises are $V_{n,R} = 20$ μV, $V_{n,C} = 35$ μV, and $V_{n,A} = 120$ μV, respectively. What is the total noise in the circuit?

28. Compare the Johnson noise generated by a 1 kΩ resistor and 1 MΩ resistor at room temperature (25°C) in RMS (root-mean-squared) value measured across a 10 kHz bandwidth.

29. Figure 6.51 is the voltage noise density graph of a piezoresistive pressure sensor. Calculate its total noise.

30. A sensor's amplifier circuit is affected by a nearby digital circuit when the digital circuit switches its logic levels between 0.5 V (logic ZERO) and 3.6 V (logic ONE) within 12 ns and its current changes from 0 to 8 mA within 100 ns. What type of interference is the sensor circuit experiencing?

31. A magnetoresistive sensor has several noises: thermal noise: 38.94 nV, Johnson noise: 27.88 nV, shot noise: 17.65 nV. What is the sensor's total noise?

32. A multi-plate capacitive sensor has a circuit diagram as shown in Figure 6.52. If $C_0 = 100$ fF, $\Delta C/\Delta x = 10$ aF/Å, $\Delta x = 0.003$Å, and $V_0 = 5$ V, find: V_x.

33. The maximum trace length in a sensor circuit is 3 mm. If a nearby signal line is carrying a clock signal with a wavelength of 0.01 m, what is the maximum frequency of the clock signal allowed in the signal line, so that

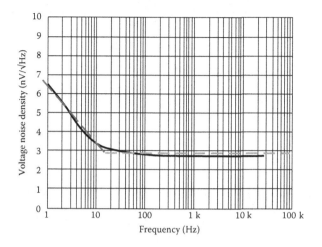

FIGURE 6.51 Voltage noise density graph of a piezoresistive pressure sensor.

it has no electromagnetic interference with the sensor circuit and the trace will not act as an antenna?

34. Derive Equation 6.32 with details for Kelvin double bridge (refer to Figure 6.10b).
35. Derive an expression for R_x in a Megaohm bridge (refer to Figure 6.12).
36. A temperature sensor, R_x, is placed in a Wheatstone bridge (see Figure 6.53). The ratio R_4/R_1 is 1/100. At the first bridge balance, R_2 is adjusted to 1000.5 Ω. The value of R_x is then changed due to the temperature change, therefore, R_2 is adjusted again to 1002.7 for the new balance. Find the change of R_x due to the temperature change.
37. Murray loop test (see Figure 6.54) is used to locate ground fault in a telephone system. The total resistance R $(= R_1 + R_2)$ is measured by a Wheatstone bridge, and its value is 255 Ω. The conditions for Murray loop test are: $R_3 = 900$ Ω and $R_4 = 510$ Ω. Find the location of the fault in meter, if the length per Ohm is 34.89 m (Hint: a bridge balance condition:

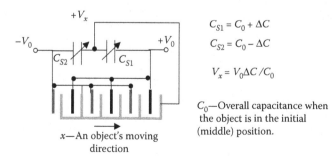

$$C_{S1} = C_0 + \Delta C$$
$$C_{S2} = C_0 - \Delta C$$

$$V_x = V_0 \Delta C / C_0$$

C_0—Overall capacitance when the object is in the initial (middle) position.

FIGURE 6.52 A circuit diagram of a multi-plate capacitive sensor.

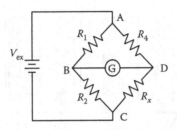

FIGURE 6.53   A Wheatstone bridge circuit for temperature measurement.

$R_1R_4 = R_2R_3$. After finding R_1 and R_2, calculate $R_1 \times 34.89$ for X_1 and $R_2 \times 34.89$ for X_2.)

38. An AC bridge (see Figure 6.13) is in balance with the following constants: Arm AB, $R = 100\ \Omega$ in series with $C = 0.5\ \mu F$; Arm AD, $R = 250\ \Omega$ in series with $L = 1.6\ mH$; Arm BC, $R = 400\ \Omega$. The oscillator frequency is 1.2 kHz. Find the constants of Arm CD.

39. A capacitive pressure sensor has a bridge circuit shown in Figure 6.55. The sensor is originally in balance with the capacitance, C_0, and $C_0 = C_1 = C_2 = 12\ pF$. When a pressure is applied, the top plate moves down by $\Delta x = 0.007 \mathring{A}$ and C_1 becomes $C_0 + \Delta C$, while C_2 becomes $C_0 - \Delta C$, where $\Delta C = 8.5\ \Delta x$ aF/$\mathring{A}$. (1) What is the equivalent capacitance of the two capacitors after the pressure is applied? (2) Find $V_x (= V_1\ \Delta C/C_0)$ if the magnitude of $V_1 = 0.5\ V$; (3) if $V_1 = 0.5\ V$, $N_1 = 10$ turns, $N_2 = 100$ turns, find V_2. (Hint: $V_1/V_2 = N_1/N_2$, a-alto, aF = 10^{-18} F.)

40. A student used a Wien bridge to measure the unknown frequency of a voltage source. He started with measuring a known 60 Hz voltage source by selecting $C_1 = 1.5\ \mu F$, $C_2 = 2\ \mu F$, and setting R_2 to 1 kΩ. What is the value of R_1 that he had adjusted to in order to achieve the bridge's balance condition? He then used the same bridge to measure the unknown frequency of the voltage source. He found that he has to adjust R_1 to 587 Ω to make the bridge balance again. What is the unknown frequency that he measured?

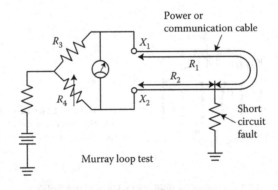

FIGURE 6.54   A Murray loop test circuit.

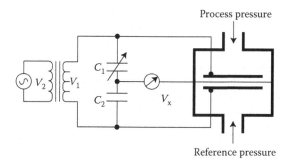

Process pressure

Reference pressure

FIGURE 6.55   A capacitive pressure sensor.

41. In Figure 6.22a and c, if a noise ΔV_{SEN} is added to each sensor output (V_{SEN}, V_{SEN^+}, or V_{SEN^-}), prove that the circuit's output, V_{out}, in (c) will not be affected by ΔV_{SEN}.

42. Derive Equation 6.62 based on the circuit shown in Figure 6.23.

43. An RC circuit is used to monitor a thermistor. If V_{DD} is 5 V, the resistance of the sensor is 2 kΩ, and capacitance C is 10 pF, find the half-life of charge decay.

44. In Figure 6.43c and d, if the inductors are removed, both the original 2nd-order band-pass filter (Figure 6.43c) and the 2nd-order notch filter (Figure 6.43d) will become a simple 1st-order low-pass filter. (1) Derive the transfer function of the filter after the inductor is removed from Figures 6.43c and 6.43d. (2) Given $R = 1$ kΩ, $C = 0.1$ μF, draw a bode diagram of the resultant low-pass RC filter using MATLAB.

45. (1) Derive the transfer function of the 2nd-order low-pass filter shown in Figure 6.43a. (2) Draw the bode diagram of this low-pass filter knowing $R = 1$ kΩ, $C = 0.1$ μF, and $L = 0.1$ mH, and compare this filter with the 1st-order low-pass filter in Problem 44.

REFERENCES

1. Andreev, S.J. and Koprinarova, P.D., Magnetic properties of thin film AMR sensor structures implemented by magnetization after annealing, *J. Optoelectronics Adv. Materials,* 7(1), 317–320, 2005.
2. Kester, W., Bridge Circuit, Section 2. Seminars, Analog Devices, Inc., page 1.
3. Introduction to NVE GMR Sensors, NVE Corporation, pp. 13–14.
4. Baxter, L.K., *Capacitive Sensors: Design and Applications*, IEEE Press, New York, 1997, 51.
5. Lepkowski, J., Designing operational amplifier oscillator circuits for sensor applications, AN866, Microchip Technology Inc., DS00866, 2003.
6. *Inductive Proximity Sensors*, Allen-Bradley. Available at: www.ab.com/catalogs.
7. Caruso, M.J. and Withanawasam, L.S., Vehicle detection and compass application using AMR magnetic sensors, 1999. Available at:http://masters.donntu.edu.ua/2007/kita/gerus/library/amr.pdf
8. Magnetic (AMR) sensor AS series, Murata Manufacturing Co., Ltd., 2008.

9. Fowler, K.R., *Electronic Instrument Design: Architecting for the Life Cycle*, Oxford University Press Inc., New York, 1996, Chap. 2.

10. Carr, J.J., *Sensors and Circuits: Sensors, Transducers, and Supporting Circuits for Electronic Instrumentation, Measurement and Control*, PTR Prentice-Hall, Inc., Englewood Cliffs, New Jersey, 1993.

11. Federal Standard 1037C - August 7, 1996: Glossary of Telecommunication Terms.

12. Kaiser, K.L., *Electromagnetic Shielding*, Taylor & Francis Group, Boca Raton, Florida, USA, Chap. 2, 2006.

13. Weber, I.J. and Mclean, D.I., *Electrical Measurement Systems for Biological and Physical Scientists*, Addison-Wesley Publishing Company, Philippines, Chap. 4, 1975.

14. Doren, V.T., Grounding and shielding electronic systems, Notes from video seminar on the NTU Satellite Network, November 11 and 12.

15. *Noise, Cross-talk, Jitter, Skew, and EMI Backplane Designer's Guide*, Fairchild Semiconductor Corporation, March 2002. pp. 6–7.

16. Johnson, J.B., *Phys. Rev.* 32, 97, 1928.

17. Nyquist, H., *Phys. Rev.* 32, 110, 1928.

18. Vassos, B.H. and Ewing, G.W., *Analog and Digital Electronics for Scientists*, 3rd ed., John Wiley & Sons, Inc., New York, pp. 23–24, 1985.

19. Milotti, E., *1/f* noise: A pedagogical review. Available at:http://arxiv.org/ftp/physics/papers/0204/0204033.pdf.

20. W. Schottky, *Phys. Rev.* 28, 74, 1926.

21. Klumperink, E. et al., Reduction of 1/f noise by switched biasing: An overview, *IEEE Electron Device Letters*, 21, 43, 2002.

22. Römisch, S. and Ascarrunz, F., Improved characterization of feed-forward noise cancellation scheme for microwave amplifiers, *Microwave Symposium Digest, IEEE MTT-S International*, 1181, June 2004.

23. Jong, M.D., Sub-Poissonian shot noise, *Physics World*, August 1996, p. 22.

24. Stein, J.A. and Langston, T., A review of some powerful noise elimination techniques for land processing, *EAGE 69th Conf. & Exhibition*, London, June 11–14, 2007.

25. Carr, J.J., *Elements of Electronic Instrumentation and Measurement*, 2nd ed., Prentice-Hall, Englewood Cliffs, NJ, Chap. 15, 1986.

26. Basic Bridge Circuits, Application Note, AN117, Dataforth Corp., Tucson, Arizona.

27. *McGraw-Hill Dictionary of Scientific & Technical Terms*, 6th ed., McGraw-Hill Companies, Inc., New York, 2003.

28. Helfrick, A.D. and Cooper, W.D., *Modern Electronic Instrumentation and Measurement Techniques*, Prentice-Hall, Englewood Cliffs, NJ, 1990.

29. Bell, D.A., *Electronic Instrumentation and Measurements*, 2nd ed., Prentice-Hall, Englewood Cliffs, NJ, 1994.

30. *Type 1644 Megohm Bridge: User and Service Manual*, IET LABS, Inc., Westbury, NY, 1988.

31. Blake, K., Analog sensor conditioning circuits—An overview, Microchip AN990. Microchip Technology Inc., 2005.

32. Tandeske, D., *Pressure Sensors: Selection and Application*, 1st ed., Marcel Dekker, Inc., New York, pp. 155, 1990.

33. Alciatore, D.G. and Histand, M.B., *Introduction to Mechatronics and Measurement Systems*, 3rd ed., McGraw-Hill, New York, pp. 158, 2007.

34. Blake, K., and Bible, S. Amplifying high-impedance sensors—photodiode example, AN951, Microchip Technology Inc., Chandle, Arizona, USA, DS00951, 2004.

35. Baker, B., Using single supply operational amplifiers in embedded systems, AN682, Microchip Technology Inc., Chandle, Arizona, USA, DS00682, 2000.

36. Baker, B., Precision temperature-sensing with RTD circuits, AN687, Microchip Technology Inc., Chandle, Arizona, USA, DS00687, 2003.
37. Haile, E. and Lepkowski, J., Oscillator circuits for RTD temperature sensors, AN895, Microchip Technology Inc., Chandle, Arizona, USA, DS00895, 2004.
38. Richey, R., Resistance and capacitance meter using a PIC16C622, AN611, Microchip Technology Inc., Chandle, Arizona, USA, DS00611, 1997.
39. Hruškovic, M. and Hribik J., Digital capacitance and inductance meter, *Measurement Science Review*, 8, Section 3, No. 3, 61, 2008.
40. Sosna, C., Buchner, R., and Lang, W., A temperature compensation circuit for thermal flow sensors operated in constant-temperature-difference mode, *IEEE Trans. Instr. Meas.*, 59, 1715, 2010.
41. Low-power signal conditioning for a pressure sensor, Application Report, SLAA034, Texas Instruments Inc., Dallas, Texas, USA, p. 13, 1998.
42. Lacanette, K., A basic introduction to filters—active, passive, and switched capacitor, National Semiconductor Application Note 779, 1991.
43. Count, L. and Kitchin, C., *A Designer's Guide to Instrumentation Amplifiers*, 2nd ed., Analog Devices Inc., Norwood, Massachusetts, USA, 2004.
44. Rijnbeek, P., Kors, J., and Witsenbur, M., Minimum bandwidth requirements for recording of pediatric electrocardiograms, *Circulation 2001*. American Heart Association, No. 104, ISSN: 1524-4539, pp. 3087–3090, 2001.
45. Karki, J., Active low pass filter design, *Application Report*, Texas Instrument Inc., Dallas, Texas, USA, September 2002.
46. MCP6271/2/3/4/5, Microchip Technology Inc., Chandler, Arizona, USA, 2004.

Index

A

Absolute accuracy, 7
Absolute probe, 188
AC, *see* Alternating current (AC)
AC bridges, 327; *see also* DC bridge; RCIM
    sensor circuitry
  and balance condition, 328–330
  bridge selection considerations, 336–337
  comparison bridge, 330–332
  example, 329–330
  Hay bridge, 333, 334
  impedance, 329
  Maxwell bridge, 333
  Owen bridge, 335
  Schering bridge, 332–333
  Wien bridge, 335, 336
Accelerometer, 123–124
  ADXL-50, 1
  capacitive, 117, 119, 120
  piezoelectric, 132–133
  piezoresistive, 74–75
Accuracy, 7
Active filters, 357–358; *see also* Passive filters
Active line, 310
Active magnetic bearing (AMB), 138
Active sensors, 99
Acupuncture points, resistance measurement at,
  90–91
Adjustable inductor, 155
Aggressor line, *see* Active line
AHA, *see* American Heart Association
  (AHA)
All-fiber sensor, 281
Alternating current (AC), 178
Aluminum
  electrical resistivity, 167
  minimum thickness, 166
  physical properties, 180
  relative permeability, 156
  temperature coefficients, 34, 167
AMB, *see* Active magnetic bearing (AMB)
American Heart Association (AHA), 360
Ampere's law, 162–163
Amplification circuit, 359
Amplifier circuits
  for charge output sensor, 341
  for current output sensor, 340
  low-impedance, 118
  for resistance output sensor, 342

  transimpedance, 340
  types, 338
Anisotropic magnetoresistive sensor (AMR
  sensor), 223; *see also* Giant
  magnetoresistive sensor (GMR sensor)
  and barber-pole structure, 246–249
  effect, 240, 241, 242–243
  materials, 239
  MR-type potentiometer circuit, 248
  permalloy, 241, 242
  R–H characteristics, 247
  sensor materials and circuit configurations,
    249, 250
Application-specific integrated circuit (ASIC), 99
Area-variation-based sensors, 120
  accelerometer, 123–124
  characteristics, 120
  example, 121–122
  position sensor, 124–125
  pressure sensor, 122
  sensing principle, 120
  sensor design, 120–121
ASIC, *see* Application-specific integrated
  circuit (ASIC)

B

Balanced wheatstone bridge circuit, 212, 213
Ballistic magnetoresistance effect (BMR effect),
  240, 245, 246
Band-pass filters (BPF), 351
Band-reject filter, *see* Notch reject filter
Bandwidth, 16–18
  measurement, 285
  noise, 311
Barber poles, 246
  AMR sensor, 248
  sensor, 247
BARC III, *see* Bead array counter (BARC III)
Barkhausen sensors, 223, 270; *see also* Wiegand
  sensors
  Barkhausen effect, 270, 271
  impact toughness tester, 272
  operating principle, 271
  stress sensor, 272, 273
Bead-type thermistor, 46, 48
Bead array counter (BARC III), 254
Beam-type sensors, 72
Beta tolerance, 46
Bias, *see* Offset

Bioresistance/bioimpedance sensors
   designs and applications, 88–91
   materials and characteristics, 87–88
   modeling, 86–87
   sensing principles, 85
   types, 85–86
Bioresistive sensors, 25, 85
Biot–Savart law, 162
Blood pressure sensor
   hysteresis, 65
   piezoresistive, 76
   strain-gauge-type, 64
BMR effect, *see* Ballistic magnetoresistance
     effect (BMR effect)
Body composition monitor, 90
BPF, *see* Band-pass filters (BPF)
Broadband noise, *see* Frequency-independent
     thermal noise
broadband voltage noise, 311
Bulk optic sensor, 281
Burst noise, *see* Popcorn noise

C

Cadmium selenide sensor (CdSe sensor), 56
   relative sensitivity *vs.* wavelength curves, 93
Cadmium sulfide sensor (CdS sensor), 49, 56
   ISL2902 CdS photoresistor datasheet, 54
   PGM 1200 CdS photoresistor, 56
   relative sensitivity *vs.* wavelength curves, 93
Calibration, 14
   error, 5, 16
   one-point, 14–15
   two-point, 15–16
Capacitance, 100, 303; *see also* AC bridges;
     Inductance
   calculations for configurations, 102–103
   in capacitive pressure sensor, 118
   change, 123, 141
   charge storage, 100
   of flat plate capacitor, 120
   function of geometry, 101
   values of capacitive sensors, 304
Capacitance output circuits, 345; *see also*
     Resistance output circuits
   capacitance-to-frequency conversion,
     346, 347
   capacitance-to-time-to-voltage conversion,
     346
   capacitance-to-time conversion, 346
   capacitance-to-voltage conversion, 345
Capacitive
   accelerometer, 117, 119–120, 123
   coupling, 306–307
   defect detection, 117
   distance, 117–118
   finger print detector, 146

level sensor, 130–131
pressure sensor, 118–119
proximity sensors, 127
shielding, 314, 315
tactile sensor, 134
touchscreen, 146–147
touch sensing, 133
Capacitive sensors, 99; *see also* Parallel-plate
     capacitive sensors
   arrays, 144–147
   CCS, 135–143
   Coulomb's law, 106–107
   excitation frequencies, 111–112
   Gauss's law, 107–109
   parallel-plate capacitor, 106
   parallel capacitor, 104–105
   piezoelectric effect, 110–111
   spherical capacitor-based sensors, 143–144
   spherical capacitor, 105
Capacitor, 100
   electronic component, 100
   parallel-plate, 106
   parallel, 104–105
   in series, 101
   spherical, 105
   types, 101
Carbon nanotube chemoresistor, 82
Carrier concentration, 49, 50, 227
   accepted values of intrinsic, 227
   temperature dependence of intrinsic, 52
CCD, *see* Charge-coupled devices (CCD)
CCS, *see* Cylindrical capacitive sensor (CCS)
Cell adhesion strength characterization, 133–134
Charge-coupled devices (CCD), 295, 313
Charge amplifiers, 340, 341
Charge carrier density, 225, 227
Charge output circuits, 340–341
Chemoresistive sensors, 25, 77
   chemoresistive effect, 77–78
   design, 81–83
   e-nose, 84–85
   gas sensor, 79
   groundwater monitoring system, 84
   hygristor, 84
   MMOS characteristics, 79–80
   organic material sensors characteristics,
     80–81
   polymer characteristics, 80–81
Chemoresistors, *see* Chemoresistive sensors
CHS, *see* Cylindrical Hall sensor (CHS)
CIP, *see* Current-in-plane (CIP)
CMOS, *see* Complementary metal-oxide
     semiconductors (CMOS)
CMR effect, *see* Colossal magnetoresistance
     effect (CMR effect)
CMRR, *see* Common mode rejection ratio
     (CMRR)

Coating thickness detector, 144
Coaxial capacitor touch sensor, 137
Coercive force, 172
Colossal magnetoresistance effect (CMR effect),
        240, 246
Colpitts oscillator, 212, 213
Common mode rejection ratio (CMRR),
        339, 359
Comparison bridge, 330–332
Complementary metal-oxide semiconductors
        (CMOS), 2
    capacitive sensor, 133
    CMOS-compatible vertical Hall angle
        sensor, 230
    coils, 179
Computed tomography (CT), 267
Conducting polymers, 80–81
Conductive coupling, 306
Conductive liquid level measurement, 139–140
Conductometric sensors, 77
Constantan
    electrical resistivity, 167
    gauge factor, 65
    temperature coefficients, 167
    ultimate elongation, 65
Copper (Cu), 180
    characteristics, 40
    electrical resistivity, 167
    gauge factor, 65
    Lorenz numbers, 40
    loss, 174
    minimum thickness, 166
    physical properties, 180
    relative permeability, 156
    temperature coefficients, 167
    temperature coefficients, 34
    ultimate elongation, 65
Core loss, 174
Cosine error, 169
Cotton–Mouton effect, 277, 278
Coulomb's law, 106–107
CPP, see Current-perpendicular-to-plane (CPP)
Creep, 66, 68
Crosstalk, 308, 310
CT, see Computed tomography (CT)
Currency counter, 254, 255
Current-in-plane (CIP), 251
Current-perpendicular-to-plane (CPP), 251
Current output circuits, 339, 340
Cylindrical capacitive sensor (CCS), 135
    on dielectric constant variation, 141–143
    on dielectric media movement, 140–141
    on electrode movement, 136–140
    principles of, 136
Cylindrical configurations, 233, 234
Cylindrical core, 185
Cylindrical Hall sensor (CHS), 233

D

Data acquisition (DAQ), 364
DC, see Direct current (DC)
DC bridge, 316; see also RCIM sensor circuitry
    Kelvin bridge, 325–326
    Megaohm bridge, 326, 327
    Wheatstone bridge-driven means, 321–324
    Wheatstone bridge and balance condition,
        317–319
DC component, see Offset
DC offset, see Offset
Densities of states (DOS), 245
Deoxyribonucleic acid (DNA), 134
Detecting probes, see Capacitive sensors
DF, see Dissipation Factor (DF)
Diagnostic ratio, 308
Dielectric-constant-variation-based sensors, 125
    angular speed sensor, 131–132
    applications, 127
    characteristics, 125–126
    chemical sensing, 129
    design, 127
    droplet detector, 131
    level measurement, 129–130
    pressure measurement, 128–129
    sensing principle, 125–126
    water level detection, 128
Differential probe, 188
Differential synchros, 203
Dipole, 110
Direct current (DC), 293, 304
    magnetometer, 208, 209
    SQUID, 290–291
Dissipation constant, 46
Dissipation Factor (DF), 332
DNA, see Deoxyribonucleic acid (DNA)
Door security system, 236–237
DOS, see Densities of states (DOS)
Double bridge, see Kelvin bridge
Driver-pickup probe, see Reflection probe
Dual-plate capacitive sensor design, 115
Dual Hall sensors, 234

E

e-nose, see electronic nose (e-nose)
E-shaped core, 186
Easy axis, 242
ECG signal, see Electrocardiogram signal (ECG
        signal)
ECU, see Electronic control unit (ECU)
Eddy current, 164, 165
    force sensor, 191
    loss, 174
    sensor, 184
EEG, see Electroencephalogram (EEG)

Electret effect, 129
Electrical resistance, 25
Electric field, 100, 104
  analogies, 168
  disturbance, 147
  Gauss's law for, 107–109
Electric model, 87
Electrocardiogram signal (ECG signal), 359
Electrode-property-variation-based sensors, 132
  capacitive DNA biosensors, 134
  capacitive tactile sensor, 134
  capacitive touch sensing, 133
  cell adhesion strength characterization,
      133–134
  characteristics, 132–133
  electropolymerized capacitive chemical
      sensor, 135
  materials and design, 133
  piezoelectric pressure sensor, 134
  sensing principle, 132–133
Electrode movement
  capacitive sensors, 136
  capacitive touch transducer, 136
  characteristics, 136
  concentric capacitive displacement sensor,
      136–138
  conductive liquid level measurement,
      139–140
  design and applications, 136
  radial displacement measurement, 138–139
  sensing principle, 136
Electroencephalogram (EEG), 295
Electromagnetic
  coupling, 308
  shielding, 316
Electromagnetic interference (EMI), 181, 306
Electromotive force (EMF), 161
Electronic control unit (ECU), 31
electronic nose (e-nose), 84–85
Electropolymerization, 135
Electropolymerized capacitive chemical
    sensor, 135
Electrostatic discharge (ESD), 146
EMF, see Electromotive force (EMF)
EMI, see Electromagnetic interference (EMI)
ESD, see Electrostatic discharge (ESD)

F

Faradaic processes, 86
Faraday cage, see Shielded conductive enclosure
Faraday current sensor, 286
Faraday effect, 276, 277
Faraday rotation, see Faraday effect
Faraday's disk, 161, 162
Faraday's law of electromagnetic induction, 161
Fast Fourier transform (FFT), 271

FBG sensor, see Fiber Bragg grating sensor
    (FBG sensor)
FEA, see Finite element analysis (FEA)
Ferrite cores, 170
Ferromagnetic core design, 185, 186
Ferromagnetic materials (FM materials), 223, 241
Fetal magneto-cardiography (FMCG), 295
FFT, see Fast Fourier transform (FFT)
Fiber-Terfenol-D hybrid sensor, 287
Fiber Bragg grating sensor (FBG sensor), 281
Figure of merit (FOM), 285, 294
Filtering, 351–352
Finite element analysis (FEA), 165
Flat-plate capacitive sensors, see Parallel-plate
    capacitive sensors
Flicker noise, see 1/f noise
Flow rate meter, 237
Fluxgate sensors, 197–198
  core types of, 206
  DC magnetometer, 208, 209
  design and application of, 206
  features of, 202–203
  PCB planar fluxgate, 209
FMCG, see Fetal magneto-cardiography
    (FMCG)
FM materials, see Ferromagnetic materials
    (FM materials)
FOM, see Figure of merit (FOM)
Force sensor
  A201, 16
  eddy-current, 191
  LPM 560 Micro, 76
  magnetostrictive, 259, 262
  piezoresistive, 76
Foucault current, see Eddy current
Frequency-independent thermal noise, 311
Frequency-selective filters, 351–352
FS, see Full scale (FS); Full span (FS)
FSO, see Full scale output (FSO); Full-span
    output (FSO)
Full-span output (FSO), 6
Full scale (FS), 5, 66
Full scale output (FSO), 176
Full span (FS), 5

G

Gain Bandwidth Product (GBWP), 361
Gallium orthophosphate (GaPO$_4$), 110
Gauge factor (GF), 59, 60, 64
  for materials, 65
  range of strain gauges, 65
  semiconductor, 68
  TCS and, 66
  temperature curve, 67
Gauge resistance, 64
Gauge spring element, 70

Gauss's law, 107–109
GBWP, *see* Gain Bandwidth Product (GBWP)
Geophysical fluid flow cell (GFFC), 143–144
GF, *see* Gauge factor (GF)
GFFC, *see* Geophysical fluid flow cell (GFFC)
Giant magnetoresistive sensor (GMR sensor),
    223, 246
  effect, 239, 240, 243–244
  magnetic recording system, 254
  measuring PCB trace current, 255
  microscopic illustration, 251
  and multilayer structure, 250–252
  read sensor, 254
  spin valve, 253
GMR sensor, *see* Giant magnetoresistive sensor
    (GMR sensor)
Gold (Au)
  electrical resistivity, 167
  minimum thickness, 166
  temperature coefficients, 167
  temperature coefficients, 34
Grounding, 314
Ground loops, 314

H

Hall effect, 223, 224
  example, 225
  Hall coefficient, 225
  in metals, 226–227
  in semiconductors, 227–228
Hall sensors, 223, 224; *see also* Magnetic sensors
  cylindrical configurations, 233, 234
  door security system, 236–237
  flow rate meter, 237
  Hall current sensor, 235, 236
  Hall position sensor, 234, 235
  input and output resistance, 230
  magnetic system, 238
  motor control, 237
  multiaxis configurations, 234
  noise, 230, 231
  nonlinearity, 230
  ohmic offset, 230
  operating principle, 228
  sensitivity, 230
  SIEMENS KSY-46, 231
  temperature coefficient, 230
  transfer function, 229
  vertical configurations, 232, 233
Hall voltage, 224
Hay bridge, 333, 334
Heat capacity, 46
High-pass filter (HPF), 341, 351, 359–360
High temperature SQUID (HTS), 292
Homopolar generator, *see* Faraday's disk
Howland current pump, 342

HPF, *see* High-pass filter (HPF)
HTS, *see* High temperature SQUID (HTS)
Hybrid probe, 189
Hybrid sensor, 258, 281; *see also* Fiber-
    Terfenol-D hybrid sensor
Hysteresis, 8, 186
  of blood pressure sensor, 65
  error, 8
  loss, 174
  magnetic hysteresis loop, 171–172
  of strain gauge, 64
  switching, 169

I

ID, *see* Inside diameter (ID)
IEC, *see* International Electrotechnical
    Commission (IEC)
Imaging sensor, piezoresistive, 76
Impact toughness tester, 272
Indium antimonide (InSb), 56
Indium tin oxide (ITO), 133
Inductance, 87, 155–159, 185
  factor, 186
  mutual, 195
  output circuits, 347–349
Inductive
  air coil sensors, 179–184
  coupling, 307
  displacement sensor, 190
  proximity sensor, 190
  shielding, 315, 316
  turbine engine speed sensor, 210
Inductive air coil sensors, 178, 179
  applications, 183–184
  coil diameter, 181
  EMI, 181–182
  example, 182–183
  features of, 180, 181
  with ferromagnetic cores, 178
  target thickness, 181
  temperature stability, 181
  types of, 179, 180
Inductive sensors, 153; *see also* Magnetic
    sensors
  characteristics of, 184, 185
  coil materials, 169
  core materials, 169–171
  Eddy-current force sensor, 191
  with ferromagnetic cores, 184
  frequency response, 175–176
  housing, cable, and target materials, 172, 173
  inductance, 155–159
  inductors, 154, 155
  magnetic field, 155–159
  number of turns, 175
  operating principles, 178–179

Inductive sensors (*Continued*)
  oscillator and signal processing circuits,
    211–214
  parameters, 170
  physical laws and effects, 159–168
  power losses in, 173–174
  Q factor, 175
  sensor design, 185–189
  stability, 176
  terminologies, 168–169
  thread detector, 191
  types, 176–178
  WayCon's TX series eddy-current
    sensors, 185
Inductor, 153, 154, 169, 216
  circuit symbols, 155
  slug-tuned, 155
  types, 154, 155
  variable, 191
Indwelling electrode, 88
Input resistance, 230
Inside diameter (ID), 187
Interferometric technique, 283, 284
International Electrotechnical Commission
    (IEC), 169
International Practical Temperature Scale
    (IPTS), 41
IPTS, *see* International Practical Temperature
    Scale (IPTS)
Iron (Fe)
  electrical resistivity, 167
  temperature coefficients, 167
  temperature coefficients, 34
ITO, *see* Indium tin oxide (ITO)

J

Johnson noise, *see* Thermal—noise
Josephson effect, 288
  AC, 290
  DC, 289, 290
Joule effects, 256, 257

K

KCl-filled microelectrode, 89–90
Kelvin bridge, 325–326
Kelvin network, *see* Kelvin bridge
KP125 pressure sensor, 3

L

Larmor equation, 266
Lattice scattering, 50
Lead (Pb)
  electrical resistivity, 167
  Lorenz numbers, 40

  minimum thickness, 166
  temperature coefficients, 167
Lead selenide (PbSe), 56
Lead sulfide (PbS), 56
Light-dependent resistor (LDR), *see* Photoresistor
Linear potentiometer, 29–30
Linear variable differential transformers
    (LVDT), 178, 193
  features, 201–202
  plant growth monitor, 205–206
  pressure sensor, 205
  sensing principles, 195–197
Liquid electrolyte, 82
Lithium battery, 295
Lithium niobate ($LiNbO_3$), 110
Lithium tantalite ($LiTaO_3$), 110
Long-term stability, 176
Lorentz force, 159, 160, 240, 269
Low-pass filters (LPF), 351, 360–361
Low-temperature SQUID (LTS), 292
LPF, *see* Low-pass filters (LPF)
LTS, *see* Low-temperature SQUID (LTS)
LVDT, *see* Linear variable differential
    transformers (LVDT)

M

mA, *see* milliampere (mA)
Magnetic
  hysteresis loop, 171, 172
  moments, 266
  recording system, 254
  sample/film sensor, 281
  system, 238
Magnetic core, 184
  $B–H$ characteristics, 170, 171
  magnetic hysteresis loop, 171, 172
Magnetic field, 155–159, 184
  analogies, 168
  detection ranges, 224
Magnetic hysteresis curve, *see* Magnetization
    curve
Magnetic random-access memory (MRAM),
    245, 246
Magnetic resonance imaging (MRI), 156,
    267–268, 270
  sensor, 223
Magnetic resonance sensors
  MRI devices, 270
  operating principles, 268
  resonant magnetic field sensors, 268–269
Magnetic sensors, 223
  Barkhausen sensors, 270–273
  magnetic field detection ranges, 224
  magnetostrictive magnetoelastic sensors,
    256–266
  MRI, 267–268

nuclear magnetic resonance, 266–267
SQUID, 287–295
Wiegand sensors, 273–276
Magnetization curve, 171
Magneto-encephalography (MEG), 295
Magneto-optical effects, 276
   Cotton–Mouton effect, 277, 278
   Faraday effect, 276, 277
   Malus's law, 278–279
   MOKE, 279–280
   Voigt effect, 277, 278
Magneto-optical Kerr effect (MOKE), 223, 279, 280, 286, 287
Magneto-optical sensors, 223, 276; *see also* Magnetic sensors
   Faraday current sensor, 286
   fiber-Terfenol-D hybrid sensor, 287
   magneto-optical effects, 276–280
   materials and characteristics, 284–285
   MO disk reader, 286
   operating principles, 281–284
Magnetomotive force (MMF), 163, 164
Magnetoresistance effect (MR effect), 223, 239
   AMR effect, 241–243
   biochips, 254
   BMR effect, 245, 246
   CMR effect, 240, 246
   comparison, 247
   currency counter, 254, 255
   GMR effect, 243–244
   magnetic recording system, 254
   NVE AA005–02 current sensor, 253
   OMR effect, 240–241
   in permalloy, 242
   TMR effect, 244–245
Magnetoresistive sensors, 25, 223; *see also* Hall sensors; Magnetic sensors
   AMR sensor, 246–250
   GMR sensor, 250–252
   MR sensor design, 252–253
   sensor applications, 253–256
Magnetosphere, 197, 207
Magnetostrictive
   fiber-optic magnetometers, 265, 266
   force sensor, 262, 263
   level transmitter, 260, 262
   position sensor, 260
   torque sensors, 263–265
Magnetostrictive delay lines (MDL), 260
Magnetostrictive effects, 256
   Joule effects, 256, 257
   Matteuci effect, 257–258
   principle, 256, 258
   Villari effects, 256, 257
   Wiedemann effect, 257–258
Magnetostrictive magnetoelastic sensors, 256
   design and applications, 259–266

magnetostrictive effects, 256–258
   materials and characteristics, 258–259
   principle, 256, 258
Malus's law, 278–279
Mass action law, 50
Matteuci effect, 257–258
Mattheisen's rule, 50
Maxwell bridge, 333
2-MBI, *see* 2-mercaptobenzimidazole (2-MBI)
MDL, *see* Magnetostrictive delay lines (MDL)
Measurement bandwidth, 285
MEG, *see* Magneto-encephalography (MEG)
Megaohm bridge, 326, 327, 328
Meissner effect, 287–288
Meissner–Ochsenfeld effect, 223
Membrane-type sensor, 71
MEMS, *see* Microelectromechanical systems (MEMS)
2-mercaptobenzimidazole (2-MBI), 135
Metal(s)
   alloy cores, 170
   foil strain gauges, 68
   piezoresistive effect in, 58–60
   thermoresistive effects, 33–34
   Wiedemann–Franz law for, 39–40
Microelectromechanical systems (MEMS), 2
microvolts (μV), 304
milliampere (mA), 304
millivolt (mV), 304
Mixed metal oxide semiconductor (MMOS), 79–80
MMF, *see* Magnetomotive force (MMF)
MMOS, *see* Mixed metal oxide semiconductor (MMOS)
MOKE, *see* Magneto-optical Kerr effect (MOKE)
Molybdenum (Mo), 40
Motor control, 237
MRAM, *see* Magnetic random-access memory (MRAM)
MR effect, *see* Magnetoresistance effect (MR effect)
MRI, *see* Magnetic resonance imaging (MRI)
Multi-point grounding, 314
Multiaxis configurations, 234
Multiplate capacitive sensor, 116
Multispeed synchros, 203
Multiturn potentiometer, 30
μV, *see* microvolts (μV)
mV, *see* millivolt (mV)

N

n-type semiconductor, 50
Naval Ordnance Laboratory (NOL), 259
NDD, *see* Nondestructive detection (NDD)
NDE, *see* Non-destructive evaluation (NDE)
Negative temperature coefficient (NTC), 37

NEP, *see* Noise equivalent power (NEP)
NF, *see* Notch filters (NF)
Nickel (Ni)
    characteristics, 40
    RTDs, 41
    temperature coefficients, 34
NMR sensor, *see* Nuclear magnetic resonance
        sensor (NMR sensor)
Noise, 230, 231
    factor, 10
    sources, 305, 306
    types, 309–313
Noise equivalent power (NEP), 55
Noise transmission mechanisms, 306
    capacitive coupling, 306–307
    conductive coupling, 306
    electromagnetic coupling, 308
    inductive coupling, 307
    pseudo impedance value and coupling
        type, 309
NOL, *see* Naval Ordnance Laboratory (NOL)
Non-destructive evaluation (NDE), 295
Nondestructive detection (NDD), 184
Nonlinearity, 9, 230
    compensation, 350
    of iron core, 174
Nonmagnetic core, 169
Nonshielded mounting, 177
Normalized noise, 312
Notch filters (NF), 351, 361–364
NTC, *see* Negative temperature coefficient (NTC)
Nuclear magnetic resonance sensor (NMR sensor),
        223, 266–267
Null offset, 229
NVE AA005–02 current sensor, 253
Nyquist noise, *see* Thermal—noise

O

Offset, 5, 169, 229
    error compensation, 350–351
    ohmic, 230
    shift, 176
Offset error, *see* Offset
Ohmic offset, 230
Ohm's law, 26
Oil analyzer, 142
$1/f$ noise, 230, 294, 311, 312–313
op-amp measurement circuit, 28
Ordinary magnetoresistance effect (OMR effect),
        240–241
Organic material sensors characteristics, 80–81
Outside diameter (OD), 187
Overshoot, 16
Ovulation predictor, 91
Owen bridge, 335
Oxidation process, 78

P

PA, *see* Polyamide (PA)
Parallel-plate capacitive sensors, 112
    area-variation-based sensors, 120–125
    dielectric-constant-variation-based sensors,
        125–132
    electrode-property-variation-based sensors,
        132–135
    spacing-variation-based sensors, 112–120
Parasitic capacitance, 307
Passive filters, 354–357; *see also* Active filters
Passive line, 310
Passive sensors, 99
PBT, *see* Polybutylene terephthalate (PBT)
PC, *see* Polycarbonate (PC)
PCB, *see* Printed circuit board (PCB)
PEEK, *see* Polyetheretherketone (PEEK)
Pellicular sensor, 129
Perfectly nonpolarized material, 88
Perfectly nonreversible material, *see* Perfectly
        polarized material
Perfectly polarized material, 88
Perfectly reversible material, *see* Perfectly
        nonpolarized material
Phase circuit, 214
Photocell, *see* Photoresistor
Photoconductor, *see* Photoresistor
Photoresistive sensors, 25, 49
    applications, 57–58
    design, 56–57
    examples, 51–54
    ISL2902 CdS photoresistor datasheet, 54
    light-controlled LED circuit, 58
    photoresistive effect, 49–51
    photoresistor characteristics, 54–56
Photoresistor, 57
    characteristics, 54–56
    circuit, 95
    materials in, 56
    unit step response, 22
Piezoelectric
    effect, 110–111
    pressure sensor, 134
    sensor, 111
Piezoresistive sensors, 25, 58
    accelerometer, 74–75
    in alloys, 58–60
    blood pressure sensor, 76
    characteristics, 63–67
    example, 72–74
    flow rate sensor, 75, 76
    force sensor, 76
    imaging sensor, 76
    LPM 560 micro force sensor, 76
    in metals, 58–60
    p-type piezoresistors, 62–63

pressure sensors, 75
in semiconductors, 60–62
strain gauges, 67–73
Pink noise, *see* 1/*f* noise
Planar chemoresistive sensor, 81
Planar core, 186
Platinum (Pt), 40
electrical resistivity, 167
gauge factor, 65
relative permeability, 156
RTD, 41
temperature coefficients, 167
ultimate elongation, 65
Polarimetric technique, 283
Polyamide (PA), 172
Polybutylene terephthalate (PBT), 179
Polycarbonate (PC), 172
Polyetheretherketone (PEEK), 172
Polymer chemoresistor, 82, 83
Polytetrafluoroethylene (PTFE), 172
Polyurethane (PUR), 173
Polyvinylchloride (PVC), 172
Polyvinylidene fluoride (PVDF), 110
Popcorn noise, 310
Position sensor, 124–125
Positive temperature coefficients (PTC), 33
Pot core, 186
Potentiometers, *see* Potentiometric sensors
Potentiometric sensors, 25, 77
airflow sensor, 31
biosensor, 32–33
circuitry, 26–29
configuration, 26–29
gas sensor, 31–32
linear potentiometer, 29–30
op-amp measurement circuit, 28
pressure sensors, 30, 31
rotary potentiometer, 30
sensing principle, 25–26
water level sensor, 28
Powdered metal cores, 170
Power spectral density (PSD), 312
Precision, 12–13
Pressure sensor, 122
capacitive, 2, 3, 118, 119, 122, 371
LVDT, 205
piezoelectric, 134
piezoresistive, 6, 7, 75, 369
piezoresistive blood, 76
potentiometric, 30, 31
resistive, 1, 351
strain-gauge-type blood, 64
Printed circuit board (PCB), 99, 179, 207,
228
Probe design, 186
inductive sensing probes, 189
internal structure, 187

tangential and pancake coil configurations, 188
two-coil, 187
Proximity effect, 166–168
PSD, *see* Power spectral density (PSD)
Pseudo impedance, 308, 309
PTC, *see* Positive temperature coefficients (PTC)
PTFE, *see* Polytetrafluoroethylene (PTFE)
Pulse width modulation signals (PWM signals),
214, 235
PUR, *see* Polyurethane (PUR)
PVC, *see* Polyvinylchloride (PVC)
PVDF, *see* Polyvinylidene fluoride (PVDF)
PWM signals, *see* Pulse width modulation
signals (PWM signals)

Q

Q factor, *see* Quality factor (*Q* factor)
Quad Hall sensors, 234
Quality factor (*Q* factor), 175

R

Radial displacement measurement, 138–139
Radiative coupling, *see* Electromagnetic coupling
Radio frequency (RF), 2, 44
Radio frequency interference (RFI), 306
RC, *see* Resistance and capacitance (RC);
Resistor–capacitor (RC)
RCIM sensors, *see* Resistive, capacitive,
inductive, and magnetic sensors
(RCIM sensors)
RCIM sensor circuitry, 303
capacitance values, 304
detectable magnetic fields, 305
grounding and shielding techniques, 313–316
inductive and magnetic sensor characteristics,
305
noise sources and forming mechanisms,
305–313
resistance values, 304
sensor compensation circuits, 349–351
sensor signal characteristics, 303–305
RCIM sensor output circuits, 337
amplifier circuits, 338
capacitance output circuits, 345–347
charge output circuits, 340–341
current output circuits, 339, 340
inductance output circuits, 347–349
resistance output circuits, 341–345
voltage output circuits, 337–339
Reduction process, 78
Reflection probe, 188–189
Relative accuracy, 7
Relative humidity (RH), 11, 126
Remanence, *see* Retentivity
Renal vascular resistive sensor (RVR sensor), 17

Repeatability error, 12–13
Reproducibility error, *see* Repeatability error
Residual magnetism, 171, 172
Resistance and capacitance (RC), 133
Resistance output circuits, 341
   resistance-to-current conversion, 342, 343
   resistance-to-frequency conversion, 344, 345
   resistance-to-time conversion, 343, 344
   resistance-to-voltage conversion, 341–342
Resistance temperature curves (R–T curves), 45
Resistance temperature device (RTD), 33, 40
   applications, 43–44
   Callendar–Van Dusen coefficients, 35
   characteristics, 40–41
   constructions, 42–43
   measurement, 41
   resistance–temperature curve, 34
   wiring configurations, 43
Resistance tolerance, 45
Resistive, capacitive, inductive, and magnetic
       sensors (RCIM sensors), 1; *see also*
       RCIM sensor circuitry
   accuracy, 7
   bandwidth, 16–18
   calibration, 14
   FS, 5–6
   FSO, 6–7
   hysteresis error, 8
   KP125 pressure sensor, 3
   nonlinearity, 9
   offset, 5
   one-point calibration, 14–15
   performance, 20–21
   precision, 12–13
   repeatability error, 12–13
   resolution, 11–12
   response time, 16–18
   sensitivity, 2–5
   sensor history, 1–2
   sensor lifespan, 18–20
   SNR, 9–11
   transfer function, 2
   two-point calibration, 15–16
Resistive sensors, 25, 75, 317, 341; *see also*
       Bioresistance/bioimpedance sensors;
       Chemoresistive sensors; Photoresistive
       sensors; Piezoresistive sensors;
       Potentiometric sensors
Resistive temperature sensors, 25, 33
   copper wire, 34–35
   metals thermoresistive effects, 33–34
   Pt100 sensor, 35–37
   RTDs, 40–44
   semiconductors thermoresistive effects,
      37–39
   thermistors, 44–49
   Wiedemann–Franz law, 39–40

Resistor–capacitor (RC), 118–119
Resolvers, 200
   design and application, 209–211
   features, 203–204
Resonant magnetic field sensors, 268–269
Response time, 16–18
Retentivity, 171
RF, *see* Radio frequency (RF)
RFI, *see* Radio frequency interference (RFI)
RF SQUID, 291–292
   for flux noise spectral density, 294
RH, *see* Relative humidity (RH)
Ring core, 186
rms, *see* Root mean square (rms)
Root mean square (rms), 9
Rotary potentiometer, 30
Rotary variable differential transformers
       (RVDT), 178, 193
   design and applications, 204
   features, 201
   sensing principles, 195, 197
   specifications, 202
R–T curves, *see* Resistance temperature curves
       (R–T curves)
RTD, *see* Resistance temperature device (RTD)
RVDT, *see* Rotary variable differential
       transformers (RVDT)
RVR sensor, *see* Renal vascular resistive sensor
       (RVR sensor)

S

Safety grounding, 314
SAL, *see* Soft adjacent layer (SAL)
Sallen–Key topology, 358
Schering bridge, 332–333
Schottky noise, *see* Shot noise
Screening current, 290
Self-contained sensor, 178, 179
Semiconductors
   Hall effect in, 227
   piezoresistive effect in, 60–62
   thermoresistive effects, 37–39
Sensitivity, 229, 230
   drift, 4
   error, 4
   factor, 104
Sensor compensation circuits, 349
   nonlinearity compensation, 350
   offset error compensation, 350–351
   temperature compensation, 349, 350
Sensor signal conditioning, 351
   active filters, 357–358
   amplification circuit, 359
   characteristics of filter, 352–354
   filtering, 351–352
   high-pass filter, 359–360

low-pass filter, 360–361
notch filter, 361–364
passive filters, 354–357
Q value of filter, 354
RLC filter, 353
Twin-T notch filter, 362
Shielded conductive enclosure, 316
Shielded mounting, 177
Shielding, 314
capacitive shielding, 314, 315
electromagnetic shielding, 316
inductive shielding, 315, 316
Shot noise, 313
Signal-to-noise ratio (SNR), 9–10, 285
Signal return, 314
Silver (Ag)
electrical resistivity, 167
gauge factor, 65
Lorenz numbers, 40
minimum thickness, 166
temperature coefficients, 167
temperature coefficients, 34
ultimate elongation, 65
Silver–silver chloride electrode (Ag–AgCl
electrode), 88
Single-axis PCB fluxgate sensor, 210
Single-crystal semiconductor strain gauges, 68, 69
Single-point grounding, 314
Single-turn potentiometer, 30
S–I–S junction, *see* Superconductor–insulator–
superconductor junction, (S–I–S
junction)
Skin effect, 165–166
Slug-tuned inductors, 155
SNR, *see* Signal-to-noise ratio (SNR)
S/N ratio, *see* Signal-to-noise ratio (SNR)
Soft adjacent layer (SAL), 252, 253
Spacing-variation-based sensors, 112; *see also*
Area-variation-based sensors
accelerometer, capacitive, 116–117, 119–120
characteristics, 112–113
defect detection, capacitive, 117
distance, capacitive, 117–118
presence sensing, 117–118
pressure sensor, capacitive, 118–119
sensing principle, 112–113
sensor design, 113–116
Span, 229
Spectral response curve, 55
Spherical capacitor-based sensors, 143–144
Spherically folded pressure sensor array, 145–146
SQUIDs, *see* Superconducting quantum
interference devices (SQUIDs)
Steinhart–Hart equation, 38
Steinmetz exponent, 174
STJs detectors, *see* Superconducting Tunnel
Junction Detectors (STJs detectors)

Strain gauges
gauge spring element, 70
GF range, 65
materials, 70–71
metal foil, 68
nominal resistance, 69
piezoresistive, 64
single-crystal semiconductor, 68–69
supporting structure and bonding methods,
71–74
temperature characteristics, 66
thin-film, 69
wire, 67
Superconducting quantum interference devices
(SQUIDs), 223, 287
battery monitors, 295
for biomagnetism, 295
DC SQUID, 290–291
design of, 294
materials and characteristics of, 292–293
noise, 293–294
RF SQUID, 291–292
superconductor quantum effects, 287–290
Superconducting Tunnel Junction Detectors (STJs
detectors), 295
Superconductor–insulator–superconductor
junction, (S–I–S junction), 289
Synchros, 198–200
design and application of, 209–211
features of, 203–204

T

TCR, *see* Temperature coefficient of resistance
(TCR)
TCS, *see* Temperature coefficient of sensitivity
(TCS)
Temperature coefficient
input and output resistance, 230
material-dependent, 33
metals, 33, 34, 167
Temperature coefficient of resistance (TCR), 66
Temperature coefficient of sensitivity (TCS), 66
Temperature compensation, 349, 350
Temperature sensor, 141–142
Terfenol-D stress sensor, 259, 263
TFD, *see* Time–frequency domain filter (TFD)
Thermal
noise, 230, 310, 311
sensitivity shift, 176
stability, 176
zero shift, 176
Thermistors, 44
applications, 48–49
characteristics, 44–46
design, 46–48
Thin-film RTDs, 42–43

Thin-film strain gauges, 69
Thomson bridge, *see* Kelvin bridge
Thread detector, 191
3-degree of freedom (3-DOF), 138
Time–frequency domain filter (TFD), 313
Tin (Sn), 293
    electrical resistivity, 167
    Lorenz numbers, 40
    minimum thickness, 166
    temperature coefficients, 167
Titanium, 211
    minimum thickness, 166
TMR effect, *see* Tunneling magnetoresistance
       effect (TMR effect)
TR, *see* Transformation ratio (TR)
Transfer function, 229
Transformation ratio (TR), 204
Transformer-type inductive sensors, 178, 192;
       see also Inductive sensors
    AD598 LVDT signal conditioner, 196
    fluxgate sensors, 197–198
    LVDTs/RVDTs, 195–197, 201
    resolvers, 200
    synchros, 198–200
Transimpedance amplifier, 340
Transition temperature, 46
Transmitter, 131
    quartz, 110
    synchro, 198, 199
Tubular sensors, 179
Tungsten
    electrical resistivity, 167
    temperature coefficients, 167
    temperature coefficients, 34
Tunneling magnetoresistance effect (TMR
       effect), 240, 244–245
2D Fourier transform technique (2DFT
       technique), 267

U

Ultra-precision spherical probe, 144
Ultraviolet (UV), 49
U shaped core, 185

V

Variable inductor, 155, 191
VEGA Technique, 122

Venous blood volume measurement, 91
Vertical configurations, 232, 233
Vertical Hall sensor (VHS), 232, 233
Victim line, *see* Passive line
Villari effects, 256, 257
Voigt effect, 277, 278
Voltage gradient, *see* Voltage sensitivity
Voltage output circuits, 337–339
Voltage sensitivity, 204

W

Water quality detector, 142–143
Waveguide, 260
Wavelet transform filter (WTF), 313
Wheatstone bridge-driven, 321
    by constant current, 322–324
    by constant voltage, 321, 322
Wheatstone bridge, 317
    advantage, 318
    resistive sensor, 319
    RTD wiring connections, 320
    sensitivity, 319–321
Wiedemann effect, 257–258
Wiegand sensors, 223, 273
    design and applications,
       274–276
    operating principle, 273–274
    Wiegand effect, 223, 273
Wien bridge, 335, 336
Winding loss, *see* Copper loss
Wire-wound RTD, 42–43
Wire strain gauge, 67
WTF, *see* Wavelet transform filter
    (WTF)

Y

Yttrium iron garnet, 284–285

Z

Zero or null drift, *see* Offset
Zero or null offset, *see* Offset
Zinc
    electrical resistivity, 167
    minimum thickness, 166
    temperature coefficients, 167

Printed in the United States
by Baker & Taylor Publisher Services